Spintronics-based Computing

Weisheng Zhao • Guillaume Prenat
Editors

Spintronics-based Computing

 Springer

Editors
Weisheng Zhao
Spintronics Interdisciplinary Center
Beihang University
Beijing, China

Guillaume Prenat
CEA
Grenoble, France

ISBN 978-3-319-34661-8   ISBN 978-3-319-15180-9   (eBook)
DOI 10.1007/978-3-319-15180-9

Foreword

A major issue today for the development of the Information and Communication Technologies is the challenge of the "beyond CMOS," how to go beyond the physical limits of the semiconductor-based conventional electronics, how to continue the progress of the electronic components in terms of size, speed, and price reduction, with the additional challenge of treating and storing efficiently a continually increasing quantity of information while reducing the total energy consumption. If we look at the 2013 International Technology Roadmap for Semiconductors (ITRS), we can see that a major perspective for the "beyond CMOS" comes from spintronic technologies exploiting not only the charge of the electrons but also their magnetism, their spin. The book "Spintronics Based Computing" edited by Prof. Weisheng Zhao and Dr. Guillaume Prenat is particularly timely as it describes some of these spintronic technologies: Spin Transfer Torque-Random Access Memory (STT-RAM) and several types of logic or communication devices based on Magnetic Tunnel Junctions, STT-RAM, or Spin Torque Nano-Oscillators.

The STT-RAM nonvolatile memories, owing mainly to their potential for the highly desirable reduction the energy consumption of computing, are included in the ITRS roadmap into a category of "Memory and Logic Technologies Highlighted for Accelerated Development." The recent advances in the technology of the STT-RAMs are described in the chapter by Dr. Naoharu Shimomura, the associated critical issue of the current-induced switching of magnetic tunnel junctions (MTJ) is discussed in the chapter written by Prof. Weisheng Zhao and coworkers, and we can learn about the perspective for electric control of magnetic devices in the chapter by Prof. Kang L. Wang.

The MTJ and STT-MRAM can be used to replace the DRAM of the today computers but can also be integrated more deeply into the logic circuits to obtain a new generation of processors of enhanced speed and higher density. These new architectures are discussed in the two chapters written by Prof. Lionel Torres and Prof. Takahiro Hanyu. Finally, as computing cannot be strictly separated from RF communication, I want to point out the particular interest of the last chapter by Prof. Mircea Stan on the Spin Torque Nano-Oscillators and their synchronization.

I can conclude that Spintronics Based Computing, by presenting several expected next steps in the development of computing technology together with the associated problems of fundamental research, can be extremely useful for the researchers in both the academic and industry communities.

Albert Fert

Emeritus Professor at Université Paris-Sud
2007 Nobel Laureate in Physics

Contents

About the Editors

Weisheng Zhao received the Ph.D. degree in physics from the University of Paris-Sud, France, in 2007. From 2004 to 2008, he investigated Spintronic devices-based logic circuits. Since 2009, he joined the CNRS as a tenured research scientist and his interest includes the hybrid integration of nanodevices with CMOS circuit and new nonvolatile memory (40 nm technology node and below) like MRAM circuit and architecture design. Weisheng has authored or co-authored more than 120 scientific papers (e.g., Advanced Material, Nature Communications, >40 I.E. Transactions); he is also the principal inventor of 4 international patents. Since 2014, he becomes "Youth 1000 plan" distinguished professor in Beihang University, Beijing, China where he leads the Spintronics Interdisciplinary Center. He is a senior member of IEEE.

Guillaume Prenat (36 years old) is a researcher in microelectronics now in charge of the design activity for Spintec Lab. He graduated from Grenoble Institute of Technology in France, in 2002. He obtained a Ph.D. degree in the field of analog and mixed signal testing in 2005. He joined the spintronics laboratory SPINTEC in November 2006 to take in charge the design activity. He holds 6 international patents and has authored or co-authored 45 scientific publications in this field.

Current-Induced Magnetic Switching for High-Performance Computing

Yue Zhang, Weisheng Zhao, Wang Kang, Eyra Deng, Jacques-Olivier Klein, and Dafiné Revelosona

1 Introduction

The shrinking of complementary metal oxide semiconductor (CMOS) fabrication node below 90 nm leads to high static power in memories and logic circuits due to the increasing leakage currents [1]. This power issue limits greatly the miniaturization and improvement of electronic devices. For example, the design of multicore microprocessors for CPU in computer is a proof for this point. In this background, novel technologies to replace the mainstream charge-based electronics are hot topics for both academics and industries. Beyond the electrical charge, the devices based on the spintronics attract a broad attention and show the performance advantages in many aspects [2].

Magnetic tunnel junction (MTJ), one of the most important spintronic devices, is the basic element of magnetoresistance random access memory (MRAM) which becomes a most promising candidate for the next generation of universal non-volatile memory. Among its various features, the magnetization switching in MTJ is a crucial point. Much of the academic and industrial research efforts are presently focused on developing efficient switching strategies. One promising method relies on using spin transfer torque (STT) [3, 4], which involves low threshold currents and well-understood mechanisms. Furthermore, only a bi-directional current is

Y. Zhang • W. Kang
Spintronics Interdisciplinary Center, Beihang University, Beijing 100191, China

IEF, Univ. Paris-Sud 11, UMR8622, CNRS, Orsay 91405, France

W. Zhao (✉)
Spintronics Interdisciplinary Center, Beihang University, Beijing 100191, China
e-mail: weisheng.zhao@buaa.edu.cn

E. Deng • J.-O. Klein • D. Revelosona
IEF, Univ. Paris-Sud 11, UMR8622, CNRS, Orsay 91405, France

© Springer International Publishing Switzerland 2015
W. Zhao, G. Prenat (eds.), *Spintronics-based Computing*,
DOI 10.1007/978-3-319-15180-9_1

1

needed in this approach, which simplifies greatly the CMOS switching circuits and thereby allows for higher density than the other approaches.

On the other hand, racetrack memory is an emerging spintronic concept based on current-induced domain wall (CIDW) motion in magnetic nanowires [5, 6]. Combining with MTJs as write and read heads, CMOS integrability and fast data access speed can be achieved. In this concept, the data are stored via the magnetizations of magnetic domains separated by domain walls (DWs). Due to STT mechanism, the DWs can be propagated consecutively in a direction by a spin-polarized current, which makes the racetrack memory possible to be widely applied for logic and memory designs.

However, some unexpected effects have been discovered using current-induced approach in the devices with small size (e.g., lateral size of 40 nm), such as erroneous state switching with reading currents and short retention times. These problems are mainly related to the in-plane magnetic anisotropy, which cannot provide a sufficiently high energy barrier to ensure thermal stability [7]. This issue limits greatly the potential for future miniaturization of spintronic devices. One compelling solution addressing this issue involves the perpendicular magnetic anisotropy (PMA) in certain materials (e.g., CoFeB/MgO), because it allows high energy barrier to be attained for small-size structures (<40 nm) while maintaining the possibility of fast-speed operations, high TMR ratios and low threshold currents [8, 9].

Thanks to the diverse advantages demonstrated by spintronics as well as various milestone breakthroughs of its related materials and techniques, hybrid spintronics/CMOS logic and memory circuits open a novel route to manipulate information more efficiently. Taking advantages of spintronic devices, the emerging circuits or systems can also realize low power, high density and high speed. For the past decade, many spintronics based logic and memory circuits and their prototypes have been designed and presented. From the relatively mature spin valve for HDDs [10–12] to recently commercialized STT-MRAM [13–17], from magnetic full adder (MFA) and Magnetic Flip-Flop (MFF) for magnetic processors [18–20] to magnetic content addressable memory (CAM) for internet router and search engines [21, 22], spintronics or concepts based on it has seeped into a majority of the advanced logic and memory systems.

One of the beneficial applications is the computing: the structural and technological limitations of conventional computing systems prevent them from reaching high frequency (~4 GHz) and limit power efficiency [23]. In this chapter, we describe an overview of the devices and circuits for high-performance computing, which are particularly based on current-induced magnetic switching. From the compact modeling to the circuit design and the optimization, the contributions of this chapter have been made at a series of levels.

Firstly we introduce the concepts of PMA STT MTJ and racetrack memory. Their fundamental physics, structures and performances, promising to achieve high performance computing, will be involved. In the following, we present spintronics based computing designs under intense R&D. MFA (1-bit one based on PMA STT MTJ and multi-bit one based on racetrack memory) and CAM are particularly investigated. They demonstrate the performance advantages in terms of area and/or speed and/or energy, compared with the CMOS based conventional ones.

2 Current-Induced Magnetic Switching Spintronic Devices

2.1 Perpendicular Magnetic Anisotropy Spin Transfer Torque Magnetic Tunnel Junction (PMA STT MTJ)

The MTJ nanopillar, as shown in Fig. 1a, is one of the important devices for current spintronics based integrated circuits. Particularly, it is the basic element of MRAM. According to the different switching mechanisms, MTJs can be categorized into certain generations. Field-induced magnetic switching (FIMS) [24, 25] and thermally-assisted switching (TAS) [26, 27] are two achievable and mainstream approaches. Some prototypes or even commercialized products are based on these mechanisms. However, the mandatory utilization of magnetic field in these approaches leads to drawbacks on speed, density and power consumption, which hinder the integration of MRAM for advanced computing or memory applications.

In this context, the current-induced magnetic switching comes into view. STT was proposed independently by Berger and Slonczewski in 1996 [3, 4]. They found that a spin-polarized current injected perpendicularly to the plane could equally influence the magnetizations. This interaction is attributed to angular momentum transferred from the polarized electrons to the local magnetization of the ferromagnetic (FM) layer. Once the amount of electrons exceeding the threshold value (often represented by critical current or critical current density), the STT exerted by the current will switch the magnetization of the free layer of MTJ [13, 28]. The STT switching approach was initially researched on giant magnetoresistance (GMR) effect based spin valve [29, 30], and then focused on the MTJ providing a significantly higher magnetoresistance [31, 32]. In MTJ, one FM layer acts as a polarizer for an electric current, which then transfers angular momentum by exerting a torque on the magnetization of the other FM layer. This current-only approach simplifies

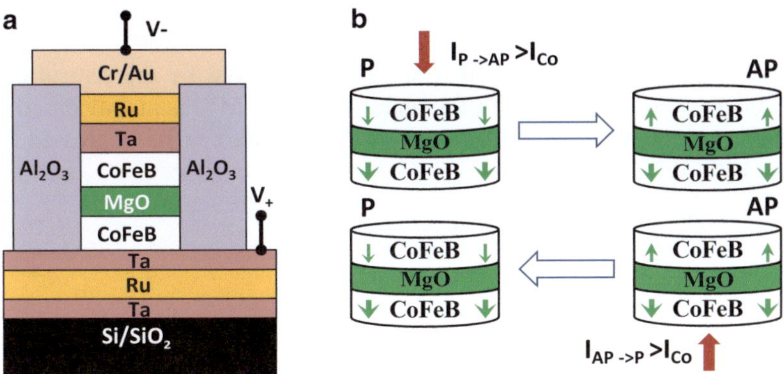

Fig. 1 (**a**) Vertical structure of an MTJ nanopillar composed of CoFeB/MgO/CoFeB thin films. (**b**) STT switching mechanism: the MTJ state changes from parallel (P) to anti-parallel (AP) as the positive direction current $I_{P \to AP} > I_{C0}$, on the contrast, its state will return to P state with the negative direction current $I_{AP \to P} > I_{C0}$

greatly the switching process as it only requires a bi-directional current (see Fig. 1b). Moreover, the magnitude of current for STT is normally less by an order than that for generating a large magnetic field. As a consequence, STT switching approach is widely considered the most promising one to be applied in the future MRAM applications [33].

The effect of STT on the free layer of MTJ can be described by the extra STT term in the Landau-Lifshitz-Gilbert (LLG) equation [34, 35] given by

$$\frac{d\vec{m}}{dt} = -\gamma \vec{m} \times \vec{H}_{eff} + \alpha \vec{m} \times \frac{d\vec{m}}{dt} - \beta J \left(\vec{m} \times \vec{m} \times \vec{M} \right) \tag{1}$$

where $\vec{m}$ and $\vec{M}$ are the unit vectors of the free and pinned layers' magnetizations, α is the damping constant, γ is the gyromagnetic ratio, β is the STT coefficient depending on both the spin polarization and the geometric configuration of the spin torque efficiency. $\vec{H}_{eff}$ is the effective field that includes the external field, the anisotropy field, the magnetostatic field, the Oersted field and the exchange coupling field.

In this equation, the first term on the right is to describe the precession of the field-induced magnetization. The second term describes the intrinsic damping process that results in a decrease of the precessional angle as a function of time. The last term on the right is the STT term whose vector direction is opposite to the damping direction. In the current-induced system, the magnetization switching on the free layer can be considered the competition between the damping term and the STT term (see Fig. 2). When the current density is small, the STT term is weaker than the damping term, then the magnetization dynamics maintain in an equilibrium state. In contrast, if the current density is high enough to make STT term stronger than damping term, the magnetization can be excited to larger precessional angles and further be switched. The critical current is defined as the threshold current to distinguish these two regimes, which is described by STT switching static model presented as follows.

The threshold for excitations driven by STT is given by the critical current. The static behavior to describe STT switching in PMA MTJ is mainly based on the calculation of threshold or critical current I_{C0}, which can be expressed by the Eqs. (2) and (3) [8].

$$E_p = \frac{\mu_0 M_S \times Vol \times H_K}{2} \tag{2}$$

$$I_{C0p} = \alpha \frac{\gamma e}{\mu_B g} (\mu_0 M_S) H_K Vol \tag{3}$$

where H_K is the perpendicular magnetic anisotropy field.

Note that the spin accumulation effects are neglected and the spin polarization efficiency factor g is firstly obtained with the following equation to describe the

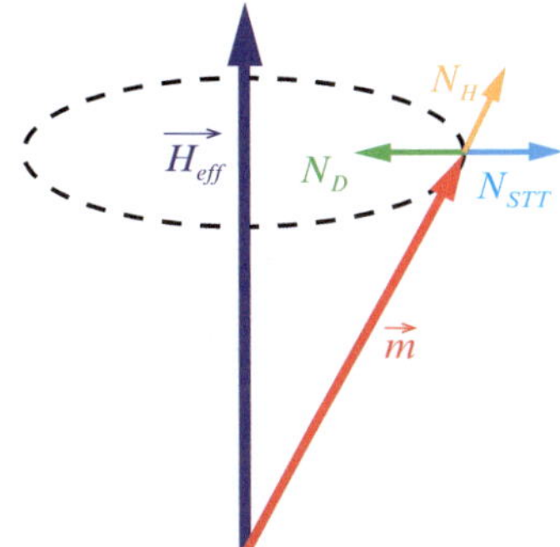

Fig. 2 Diagram of the LLG equation. N_D is the damping term, N_{STT} is the STT term and N_H is the field precession term

asymmetric current case [36]. It provides the best agreement with the experimental results illustrated in [8],

$$g = g_{SV} \pm g_{Tunnel} \tag{4}$$

where the sign depends on the free-layer alignment. g_{SV} and g_{Tunnel} are respectively the spin polarization efficiency in a spin valve and tunnel junction nanopillars. They are both predicted by Slonczewski,

$$g_{SV} = \left[-4 + \left(P^{-1/2} + P^{1/2} \right)^3 (3 + \cos\theta)/4 \right]^{-1} \tag{5}$$

$$g_{Tunnel} = (P/2)/\left(1 + P^2 \cos\theta\right) \tag{6}$$

where P is the spin polarization percentage of the tunnel current, θ is the angle between the magnetization of the free and the pinned layers [3, 37].

The good agreement between the physical model and experimental measurement has been verified. Figure 3 shows the verification of static model with the measured data reproduced by Ohno group [8]. The blue and red solid lines represent the STT switching static model for parallel to anti-parallel process and anti-parallel to parallel process, respectively. The blue squares and red points represent the experimental results. From Fig. 3, the overlaps between the lines and the squares (or points) show the good agreement and the feasibility of this physical model to describe the STT switching static behavior.

On the other hand, more recent experimental progress of IBM shows that an MTJ involving symmetric electrodes provides a single spin polarization efficiency factor g for both state change processes (anti-parallel state to parallel state process or parallel state to anti-parallel state process) of MTJ [38], which allows the same critical current for both parallel and anti-parallel states. In this mechanism, g is only related to TMR ratio and described as follows:

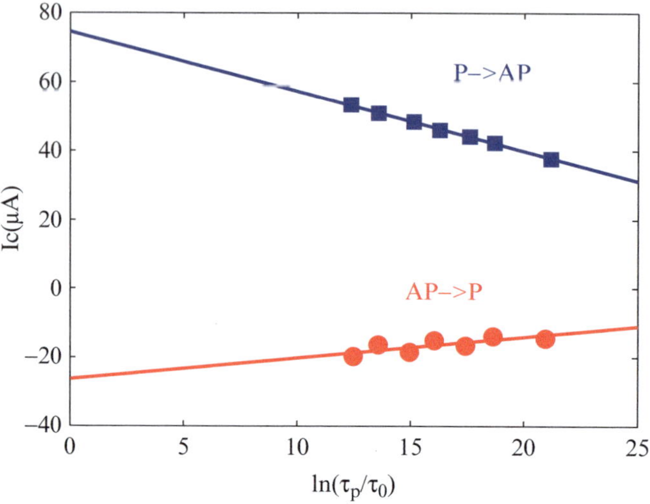

Fig. 3 Verification of the static model with measured data reproduced by [8]. Solid lines represent the STT static model, the red points and blue squares represent the experimental results

$$g = (TMR(TMR + 2))^{1/2}/2(TMR + 1) \tag{7}$$

The dynamic switching behavior of STT in PMA MTJ shows the dependence of switching current on switching duration. It is considered to be a complex process as it can be categorized into two regimes depending on the relative magnitude between switching current (I) and critical current (I_{C0}, calculated by Eq. (3) for static behavior): thermally assisted ($I < 0.8\ I_{C0}$) and precessional ($I > I_{C0}$) switching regimes. Thermally assisted regime can be described by Néel-Brown model and precessional regime can be described by Sun model [39–41]. Note that there are no clear experimental results and theories related to the range from $0.8I_{C0}$ to I_{C0}, we thus neglect this range and consider no effect occurs in this range.

For each model, the relationship between current and duration follows different laws. For practical applications, the two regimes have their own specific interest: the thermally assisted regime corresponds to low current density but slower switching, which is usually used for the sensing operation; the precessional regime corresponds to fast switching (sub 3 ns) but high current density, which is usually used for the writing operation.

In the sub-threshold condition where the current remains below the critical current ($I < 0.8I_{co}$), the switching can still occur thanks to thermal activation above the voltage/current-dependent barrier. In this case, the switching behavior can be described by Néel-Brown model [42]:

$$\frac{d\mathrm{Pr}(t)}{(1 - \mathrm{Pr}(t))dt} = \frac{1}{\tau_1} \tag{8}$$

$$\tau_1 = \tau_0 \exp\left(\frac{E}{k_B T}\left(1 - \frac{I}{I_{c0}}\right)\right) \tag{9}$$

where τ_0 is the attempt period, $\mathrm{Pr}(t)$ is the switching probability. Eq. (8) can be transformed to a simple formula:

$$t = -\tau_1 \ln(1 - \mathrm{Pr}(t)) \tag{10}$$

These equations demonstrate that the STT dynamic switching behavior is probabilistic or stochastic. However, from Eq. (10), it can convert this stochastic behavior to be deterministic by determining the switching probability. That means ones should apply a specifically long current pulse to get the determined switching probability. This assumption would greatly simplify the description and analyses of the thermally assisted regime. Meanwhile, the stochastic effect is still the key point for this regime, which will be described and integrated in the following part.

In the case that the switching current is near or exceeding the critical one, the STT excitation becomes more obvious and deterministic. The high current pulse drives the magnetization to process, then after reaching the switching time, a magnetization reversal occurs suddenly and quickly [9]. Considering a small thermal fluctuation in this regime with a relatively high thermal stability, the average switching time is given by

$$\frac{1}{\langle\tau\rangle} = \left[\frac{2}{C + \ln\left(\frac{\pi^2\xi}{4}\right)}\right]\frac{\mu_B P_{pin}}{e m_m\left(1 + P_{pin}P_{free}\right)}(I - I_{c0}) \tag{11}$$

where C is the Euler's constant, $\xi = E/k_{BT}$ is the activation energy in units of k_{BT}, P_{pin}, P_{free} are the tunneling spin polarizations of the pinned and free layers, we assume here that $P_{pin} = P_{free} = P$, m_m is the magnetic moment of free layer. Figure 4 shows the good agreement of this dynamic model with the experimental data extracted from [9]. From this figure, the increase of I and decrease of I_{C0} both contribute to scale down the switching latency. Considering the high currents are always ensured by the large-size transistors, this physical model also implies the alternatives to optimize the tradeoff between the overall area and the speed of hybrid spintronic/CMOS circuits.

Recently, a lot of experimental and theoretical results have shown that, although STT switching may allow sub-nanosecond switching duration, the switching process of STT is intrinsically stochastic, which results from the unavoidable thermal fluctuations of magnetization (see Fig. 5) [43–46]. They are responsible for large fluctuation in the switching duration, which can be proven by the Eqs. (8)–(11) describing the dynamic behavior. Moreover, the stochastic behavior can also be

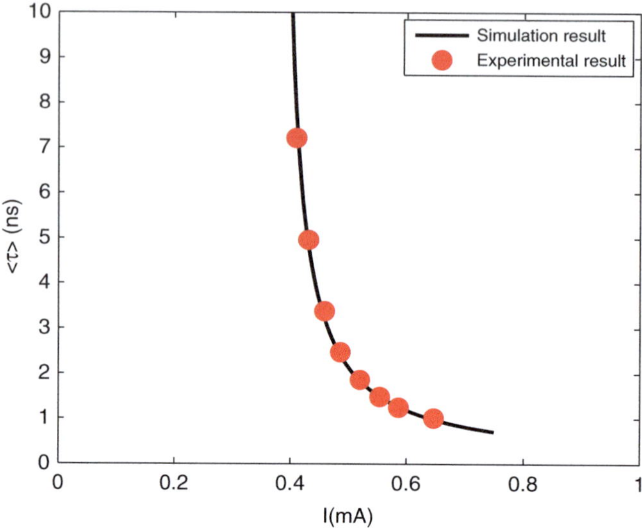

Fig. 4 Comparison of the dynamic model with measured data

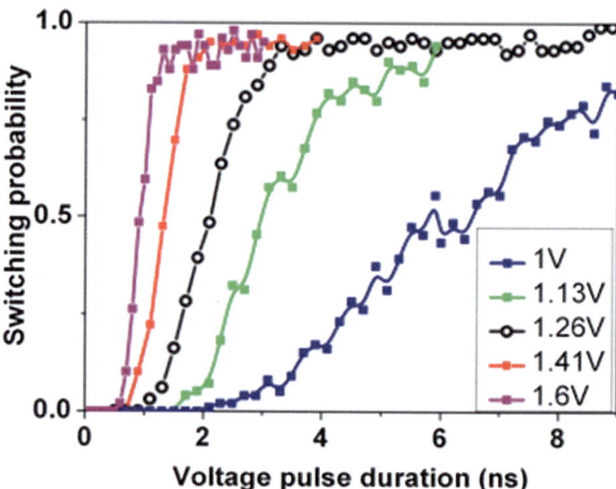

Fig. 5 Experimental measurements of STT stochastic switching behaviors, high writing current drives faster speed and higher switching probability [44]

divided into two regimes: thermally assisted ($I < 0.8\,I_{C0}$) and precessional ($I > I_{C0}$) switching regimes.

For the thermally assisted regime, we can transform Eq. (10) to another form:

$$\Pr(t) = 1 - \exp(-t/\tau_1) \tag{12}$$

It describes the probability density function (PDF) of the switching duration for this regime, which follows an exponential distribution with characteristic time τ_1 decreasing with the current density.

In the super-threshold region described as precessional regime, the stochastic switching is triggered by a thermal fluctuation which creates an initial angle between the current spin-polarization and the magnetization of magnetic layer. The switching duration then follows a specific exponential-like distribution centered on the average switching delay time calculated by Eq. (11) [47].

From the above expressions, it shows that, in both regions, increasing the switching probability requires to increase either the write current or the current pulse duration. It could also be of great benefit for tolerating the high mismatch and process variations [48, 49].

To address the requirement of high-performance MTJ for the future logic and memory applications, there are usually five criteria to evaluate: small area, high TMR ratio, low STT switching current, capacity to withstand the standard semiconductor processing and high thermal stability. With the shrinking of size, the conventional MTJ with in-plane magnetic anisotropy becomes more and more difficult to satisfy these criteria. Recent material progress showed that the MTJ with PMA could offer lower switching critical current, higher switching speed and higher thermal stability compared with that with in-plane magnetic anisotropy [7]. These can be explained by the following theories.

The barrier energy and critical current of STT switching in the materials with in-plane magnetic anisotropy can be expressed as:

$$E_i = \frac{\mu_0 M_S \times Vol \times H_C}{2} \tag{13}$$

$$I_{C0i} = \alpha \frac{\gamma e}{\mu_B g} (\mu_0 M_S) \left(H_{ext} \pm H_{ani} \pm \frac{H_d}{2} \right) Vol \tag{14}$$

where H_C is the coercive field, H_{ext} is the external field, H_{ani} is the in-plane uniaxial magnetic anisotropy field, H_d is the out-of-plane magnetic anisotropy induced by the demagnetization field, μ_0 is the permeability in the free space, M_s is the saturation magnetization, Vol is the volume of the free layer, μ_B is the Bohr magneton, γ is the gyromagnetic ratio, e is the electron charge, m is the electron mass.

By comparing Eqs. (2) and (13), as H_K is higher than H_C, PMA allows obtaining relatively high barrier energy with a small size. By comparing Eqs. (3) and (14), as H_K is much lower than H_d, the critical current for PMA materials can be significantly reduced.

From 2002, when the first MTJ with PMA was reported, this advantageous structure attracts a great deal of attentions from academics and industries [50]. A variety of material systems has been attempted, for example, rare-earth/transition metal alloys, multilayers and other alloy materials. However, they have not been able to truly realize low critical current and high thermal stability at the same time. This situation didn't change until the Ta/CoFeB/MgO structure was revealed in 2010 [8, 9]. Figure 6 demonstrates the excellent performances of this structure. It takes advantages of CoFeB-MgO interface anisotropy to provide a good tradeoff

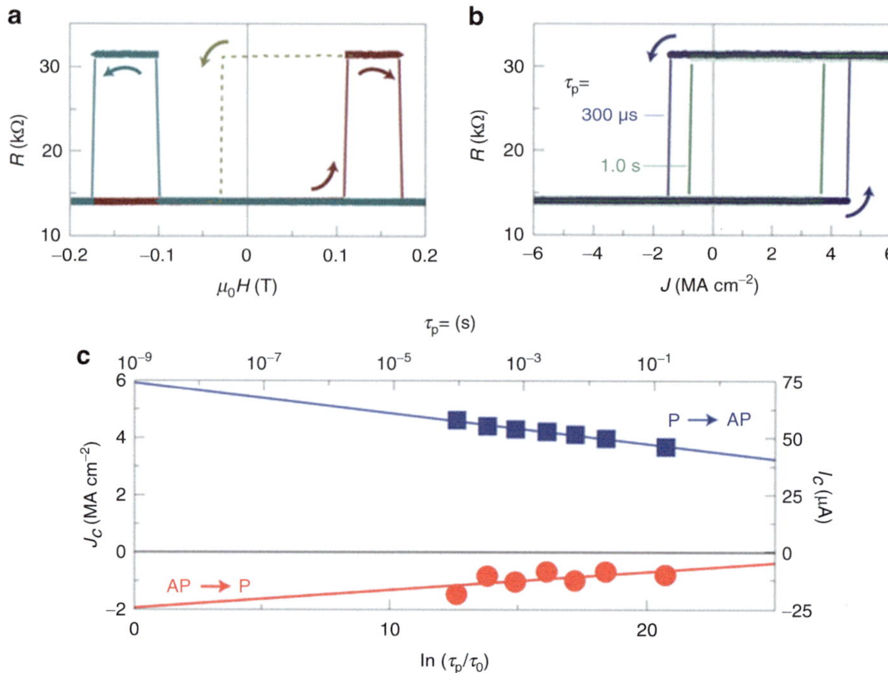

Fig. 6 TMR and current-induced magnetization switching for Ta/CoFeB/MgO structure MTJ with PMA. (**a**) Perpendicular R-H curve. (**b**) Typical results of current-induced magnetization switching at current pulse duration of 300 µs and 1.0 s. (**c**) Critical current density as a function of pulse duration [8]

among the area (40 nm), critical current ($\sim$50 µA), thermal stability (40 k_BT) and TMR ratio ($>$100 %). Thanks to the material and technical improvement of MTJ, especially MgO based PMA STT MTJ, a lot of persistent and intensive efforts have been made for the past years to develop the high-performance spintronic systems [51, 52].

2.2 Racetrack Memory

The observation of electrical CIDW motion in magnetic nanowires promises numerous perspectives [2, 53, 54] and the most interesting one is to build a novel ultra-dense non-volatile storage device, called "racetrack memory" (see Fig. 7).

The term "racetrack memory" was firstly proposed by Parkin in 2008 [5, 6]. In the concept that he proposed, write head nucleates a local domain in the magnetic nanowire and a current pulse drives the domain to move sequentially from write head to read head. Data or magnetization direction is stored between two artificial potentials or constrictions, which pin the DW as no current pulse is applied. The distance between two constrictions can be extremely small to some nanometers and

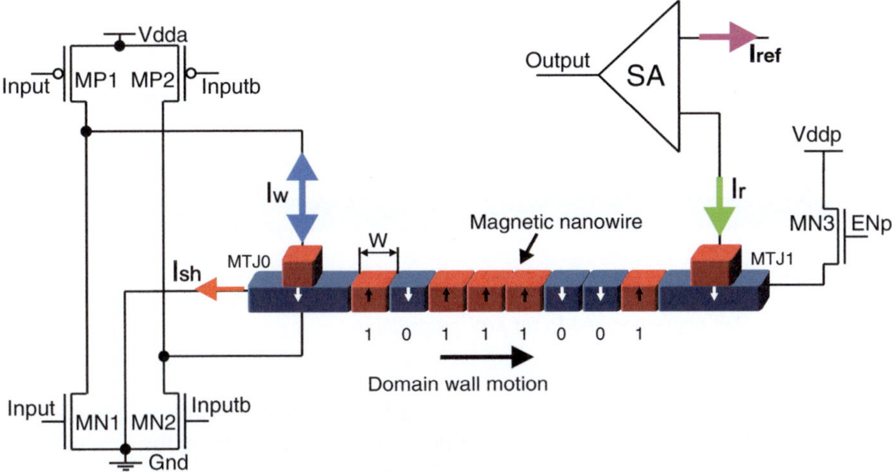

Fig. 7 Racetrack memory based on CIDW motion, which is composed of one write head (MTJ0), one read head (MTJ1) and one magnetic nano-stripe. Iw nucleates data or magnetic domain in the magnetic stripe through STT approach, Ish induces DW motion along the magnetic stripe and Ir detects the magnetization direction through TMR effect

this allows an enormous storage (>GB) in a small die area. Compared with other non-volatile memory candidates, the scalability potential of racetrack memory is evident. By using MTJ as write and read heads, its operations, such as DW motion, domain nucleation and detection, can be addressed directly by CMOS circuits [55]. This hybrid integration makes racetrack memory promise high performance like high speed (>100 MHz) and low power beyond classical STT-MRAM. The nanowire can be built in 3D or 2D, the latter one is easier to be fabricated and become the mainstream solution for the current research on this topic. Based on in-plane magnetic anisotropy, the first racetrack memory prototype was presented in 2011 by IBM despite of its small capacity 256 bits [56]. However the intrinsic low energy barrier separating the two in-plane magnetization directions of storage layer leads to short data retention in advanced technology node (e.g., 22 nm) [51]. This drawback limits its use for high-density racetrack memory. PMA in some structures (e.g., CoFeB/MgO) providing a high energy barrier [8, 57, 58] were demonstrated and PMA MTJ become one of the most promising candidates to realize a read head. Advantageous domain wall nucleation current and speed with PMA MTJ were also observed recently [9] and this makes it be a better write head than in-plane MTJ.

The Cross-section structure of racetrack memory is shown in Fig. 8, which includes mainly three parts: a magnetic stripe separated by constrictions to store data, two MTJs as write and read heads. The number of constrictions equals to the number of stored bits. It is noteworthy that the CMOS circuits dominate the whole area of this racetrack memory as the magnetic stripe is implemented at the back-end through 3D integration as MRAM.

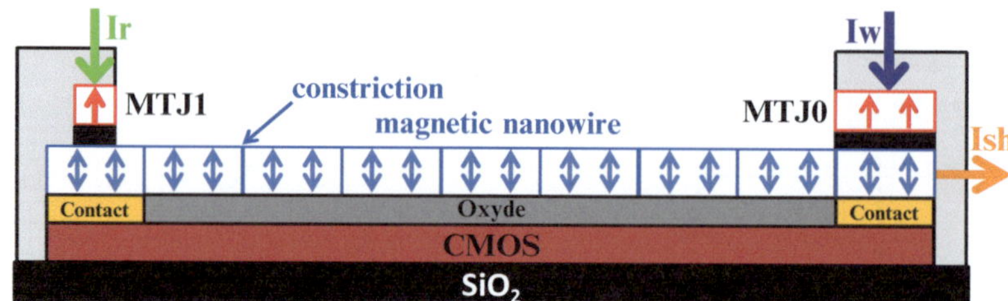

Fig. 8 Cross-section structure of racetrack memory. At the back-end process, the magnetic nanowire is implemented above the CMOS/MTJ interfacing circuits, which generate Ir for reading, Iw for DW nucleation and Ip for DW propagation

Figure 7 shows simultaneously one example of CMOS circuits to generate Iw and Ish, which are respectively bi-directional and uni-directional at the side of write head. Ir is driven by a sense amplifier [59] and it can convert the stored data from different magnetization directions to digital signal "0" or "1". In order to achieve the best write and read reliability, the width of write and read heads are different. For writing, a lower resistance of MTJ0 with larger width can reduce the rate of oxide barrier breakdown, which is one of the most significant constraints of the high-speed STT switching mechanism. On the contrary, high resistance of the MTJ1 with smaller width for reading can greatly improve the sensing performance.

For the racetrack memory, the speed performance is governed by the velocity of DW motion. The physical model to calculate DW velocity is indispensable for the compact modeling of racetrack memory. According to the previous literatures, the dependence of DW velocities on current and magnetic field can be described by the one dimensional (1D) model. This 1D model is deduced from the LLG equation in a 1D system [60], which can be described as:

$$\dot{\phi}_0 + \alpha\dot{X}/\lambda = \gamma H + \beta u/\lambda + f_{pin} \tag{15}$$

$$\dot{X} - \alpha\lambda\dot{\phi}_0 = v_\perp \sin 2\phi_0 + u \tag{16}$$

where X is the position of a DW, and ϕ_0 is the angle that the DW magnetization forms with the easy plane. λ is the width of DW, α is the Gilbert damping constant, β is the dissipative correction to the STT, H is the external field, γ is the gyromagnetic ratio, f_{pin} is the pinning force. The velocity constant $v_\perp$ comes from the hard-axis magnetic anisotropy $K_\perp (\sim K_\perp \lambda/h)$. u is spin current velocity. These two equations can describe a lot of qualitative features of DW motion driven by the field and the current. The field acts as a "force" to drive ϕ_0, the current acts as a "torque" to drive X. In addition, as the "torque" is also contributed from the hard-axis magnetic anisotropy, the state of ϕ_0 can determine whether there is intrinsic pinning or pure STT. Considering only the process after depinning, Eqs. (15) and (16) can be solved analytically and described in the forms of the influence of field and current on the velocity:

$$V = V_H + V_j \tag{17}$$

The velocity is the vector sum of field-induced (V_H) and current-induced velocities (V_j). Above the Walker breakdown field, the field-induced velocity contribution is given by

$$V_H = \alpha^2 \mu H \left\{ 1 - \frac{1}{1+\alpha^2} \sqrt{1 - \left(\frac{H_W}{H}\right)^2} \right\} \tag{18}$$

where the mobility $\mu = \gamma\lambda/\alpha$, H_w is the Walker breakdown field.

The general racetrack memory is based on CIDW motion, which means there is normally no magnetic field. Hence, the dependence of DW velocity on current is the key point. Regarding the relationship between α, the damping constant, and β, the nonadiabatic coefficient, the dependence can be categorized into three cases. Before introducing these three cases, we should indicate the definition of the spin current velocity [60], which is given by Eq. (19).

$$u = \frac{\mu_B P j_p}{e M_S} \tag{19}$$

where j_p is the propagation current density. Figure 9 shows the dependence of DW velocity on current according to different configurations of α and β, which depends on the material of the magnetic nanowire.

When $\beta > \alpha$,

$$u_{WB} = \frac{1}{2}\gamma H_K \Delta \frac{\alpha}{\beta - \alpha} \tag{20}$$

$$\langle v \rangle = \frac{\beta}{\alpha} u \qquad (u < u_{WB}) \tag{21}$$

$$\langle v \rangle = \frac{\beta}{\alpha} u - \frac{\sqrt{\left(1 - \frac{\beta}{\alpha}\right)^2 u^2 - \left(\frac{1}{2}\gamma\Delta H_K\right)^2}}{1+\alpha^2} \qquad (u > u_{WB}) \tag{22}$$

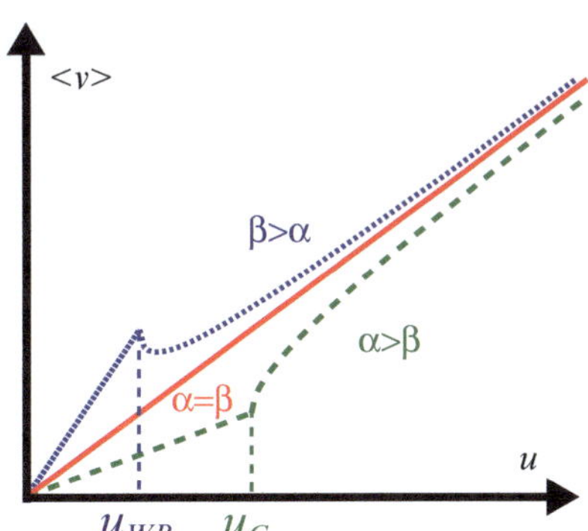

Fig. 9 Dependence of DW velocity on current described by 1D model

When $\alpha > \beta$,

$$u_C = \frac{1}{2}\gamma H_K \Delta \frac{\alpha}{\alpha - \beta} \tag{23}$$

$$\langle v \rangle = \frac{\beta}{\alpha} u \qquad (u < u_C) \tag{24}$$

$$\langle v \rangle = \frac{\beta}{\alpha} u + \frac{\sqrt{\left(1 - \frac{\beta}{\alpha}\right)^2 u^2 - \left(\frac{1}{2}\gamma \Delta H_K\right)^2}}{1 + \alpha^2} \qquad (u > u_C) \tag{25}$$

When $\alpha = \beta$,

$$\langle v \rangle = u \tag{26}$$

where u_{WB} is the Walker breakdown velocity, u_C is the critical velocity corresponding to the critical current density of DW motion. In order to achieve a high speed racetrack memory, the current density should be more or far more than the critical one. In both cases, when applying a much higher current, the DW velocity approaches to spin current velocity. Therefore, we take this assumption into account, which means we use the spin current velocity to directly represent DW velocity. Thus,

$$V_j = u = \frac{\mu_B P j_p}{e M_S} \tag{27}$$

We verified this physical model by comparing with the micromagnetic simulations done by Ohno group (see Fig. 10) [61]. In this case, we suppose that the DWs are definitely pinned when the current density is lower than the critical one, the velocity is thus kept to zero in this condition. From the figure, a current density of

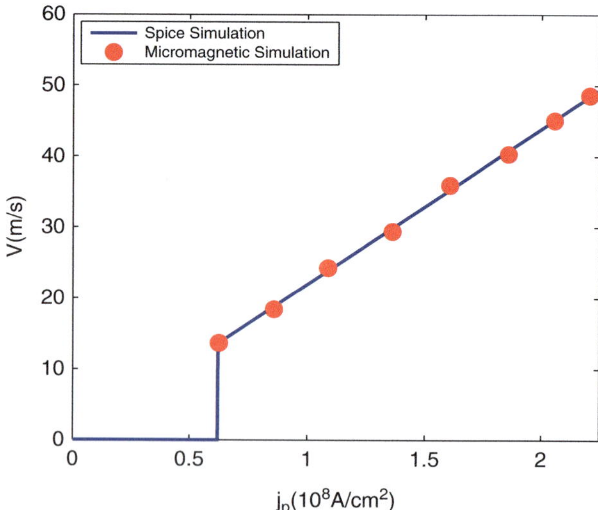

Fig. 10 Good agreements with micromagnetic simulation for DW motion velocity as a function of current density

~2×10^8 A/cm^2 can trigger a DW motion in 50 m/s, which is beneficial for the logic and embedded memory circuits.

By considering the distance W between two adjacent constrictions (see Fig. 7), we can calculate the necessary pulse duration for current to move one storage element by the Eq. (28). For example, when W is 40 nm, the DW velocity is 50 m/s, the pulse duration can be as small as 0.8 ns. If neglecting the nucleation process, the frequency of racetrack memory can thus be as high as 1 GHz. If considering the nucleation process time (e.g., 1–2 ns), the frequency can still be 500 MHz.

$$D = W/V_j \tag{28}$$

Caused by the thermal activation, stochastic nature has been found for DW motion in diverse structures and materials. With the reduction of the applied current or field by optimizing the techniques and the materials, the stochasticity of DW motion will be further enhanced [62]. DW velocity and displacement are susceptible to stochastic effect, which exerts a considerable influence to the feasibility and reliability of DW-based devices, not only racetrack memory. However, as there have not been some coherent experimental results or physical theories concerning the pure CIDW in PMA materials, we refer to the measurements of DW motions in spin valve induced simultaneously by current and field. Under this condition, after depinning, the DW motion velocity is found to follow a Gaussian-like specific distribution centered with the value calculated by Eq. (27) [63]. We analyze the dependence of cumulative probability of DW motion versus different current pulse durations and magnitudes in Fig. 11. It illustrates a coherent functionality of stochastic behavior where we can also find that higher and longer current pulse yields a more probable DW motion.

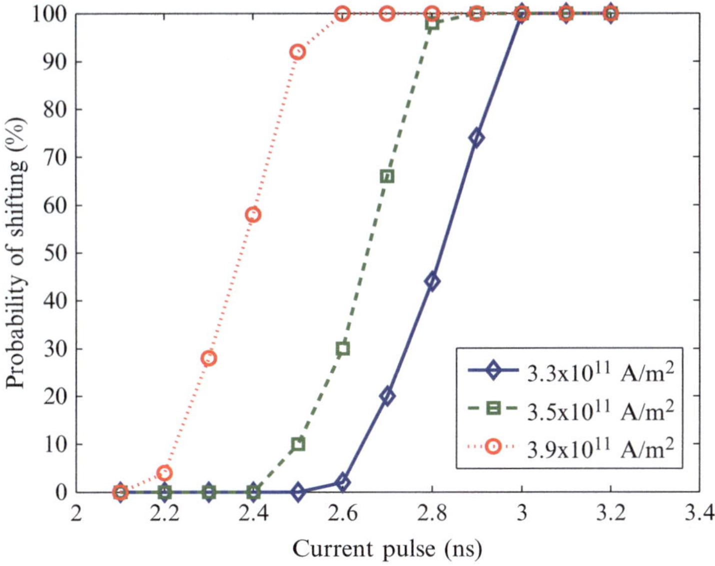

Fig. 11 Dependence of cumulative probability of DW motion versus shifting current pulse duration and magnitude

3 Current-Induced Magnetic Switching Based Hybrid Spintronics/CMOS Circuits for High-Performance Computing

3.1 Sensing Circuit

Due to the TMR effect, MTJ presents the property of resistance difference for different states. This resistance property allows MTJ to be compatible with CMOS sense amplifier circuit that detects the MTJ's configuration and amplifies them to logic level. Among various sense amplifiers [64–66], pre-charge sense amplifier (PCSA) is proposed to provide not only the best tradeoff between sensing reliability and power efficiency, but also high-speed performance [59]. Thereby we focus on PCSA and apply it for the hybrid logic circuits involved in this chapter.

The PCSA circuit (see Fig. 12) consists of a pre-charge sub-circuit (MP2-3), a discharge sub-circuit (MN2) and a pair of inverters (MN0-1 and MP0-1), which act as an amplifier. Its two branches are normally connected to a couple of MTJs with complementary states. It operates in 2 phases: "Pre-charge" and "Evaluation". During the first phase, "CLK" is set to "0" and the outputs ("Qm" and "/Qm") are pulled-up to "Vdd" or logic "1" through MP2-3 while MN2 remains off. During the second phase, "CLK" becomes "1", MP2-3 are turned off and MN2 on. Due to the resistance difference between the two branches, discharge currents are different. The lower resistance branch will be pulled-down to reach more quickly the threshold voltage of the transistor (MP0 or MP1), at that time, the other branch will be pulled up to "Vdd" or logic "1" and this low-resistance branch will continue to drop to "Gnd" or logic "0".

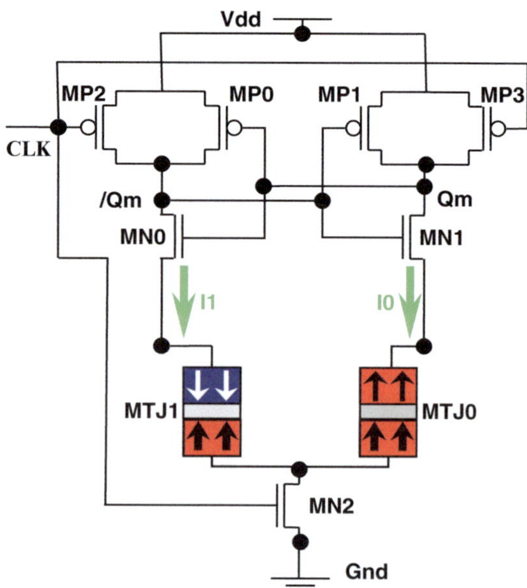

Fig. 12 Pre-charge sense amplifier (PCSA) for MTJ state detection and amplification to logic level

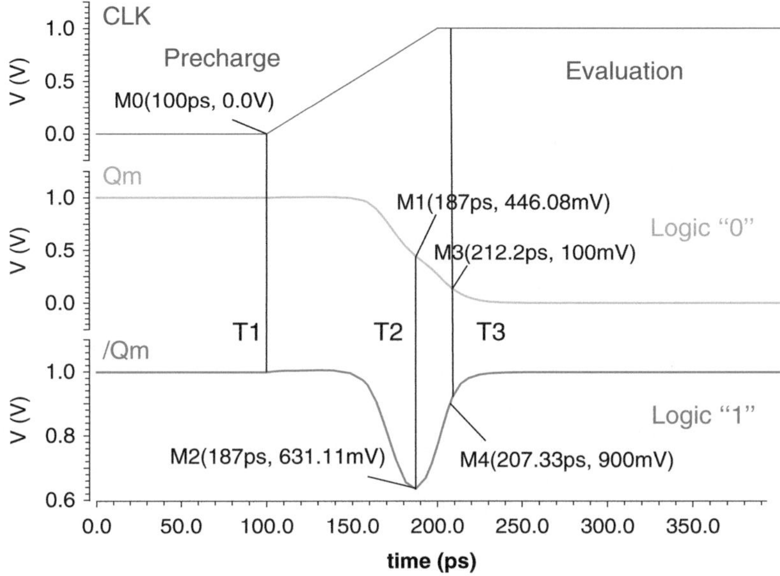

Fig. 13 PCSA sensing operation in the case of MTJ0 with "parallel" state and MTJ1 with "anti-parallel" state

Figure 13 shows a sensing operation of PCSA in the case of MTJ0 with "parallel" state and MTJ1 with "anti-parallel" state. Before the moment "T1", it is "Precharge" phase. Both outputs are pulled up to 1 V. Then the "Evaluation" phase starts from "T1", two branches begin to discharge after a small delay considering the rising time of "CLK" signal. At the time "T2", the branch "Qm" reaches the threshold firstly and this branch will continue to decrease to "0". At the same time, MP0 begin to work and recharge the complementary branch "/Qm" back to "1". This sensing operation is so speedy. From the figure, we can find the whole process costs a sensing delay less than 100 ps. From the point of view of consumption energy, a sensing operation can only cost as low as 10 fJ. This high-speed and low-power feature makes PCSA suitable for the logic applications.

We use PCSA circuit in the hybrid MTJ/CMOS design for the other reason: the read disturbance induced by sensing operations can be significantly decreased. It is important for embedded STT-MRAM as it is an intrinsic nature and difficult to correct in logic circuit where complex error correction circuit (ECC) is prevent to ensure fast computing speed (e.g., 1 GHz). The read disturbance can be regarded as the unexpected switches during the sensing operation. As the sensing current is usually much lower than the critical current, the switch probability can be described by Néel-Brown model. If there are N bits of MTJs in the chip, the chip failure rate F_{chip} can be calculated by Eq. (29).

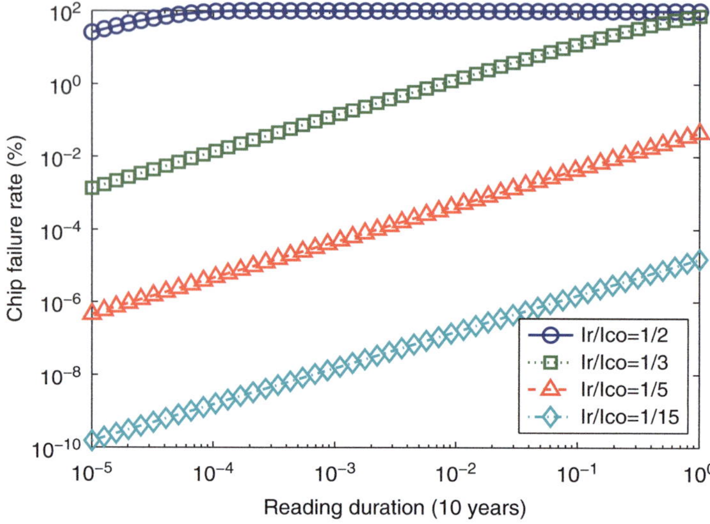

Fig. 14 Dependence of chip failure rate on reading duration for different reading current

$$F_{chip} = 1 - \exp\left[-N\frac{\tau}{\tau_0}\exp\left(-\Delta\left(1 - \frac{I_r}{I_{C0}}\right)\right)\right] \qquad (29)$$

where N is the number of bits per word, I_r is sensing current, I_{C0} is the critical current, τ is the read duration and τ_0 is the attempt period. As shown in Fig. 14, lower I_r and shorter τ can reduce greatly the chip failure rate for the STT-MRAM with the same thermal stability factor $\Delta = 40$.

In reality, numbers of words of memories (e.g., 1 k) normally share a sense amplifier. As shown in Fig. 15, a 16 k-bit (1 k words of 16 bits) PCSA sensing circuit has been studied. This enormous parallel structure leads to a huge capacitance, which drives the current pulse through the MTJ. As a result, an evaluation phase lasts almost 10 ns. By taking the effect of stochastic behavior into account, Monte-Carlo simulations after 1 μs of sensing duration (i.e., 1,000-time sensing operations) has been performed (see Fig. 16). We found that the 33 errors occurred among 100 simulations. They are caused by either mismatch and process variations of CMOS part or STT stochastic behaviors of MTJ, or sometimes by both of them.

To identify the impact from each of them, we also performed Monte-Carlo simulations for sensing circuit with only mismatch and process variations. We found that the read disturbance was ~11 %. Compared with the result presented in Fig. 16, we can conclude that the stochastic behavior of MTJ greatly increases the error probability for a long-pulse current, and that this PCSA is not suitable for very large memory systems.

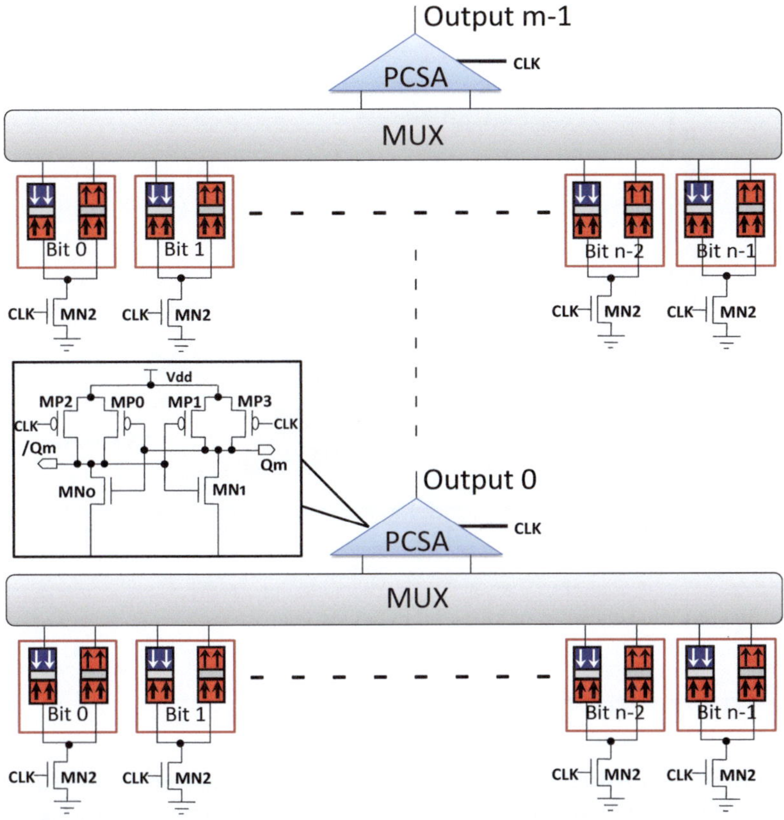

Fig. 15 Schematic for 16 k-bits PCSA sensing circuit (m = 16, n = 1 k)

To improve further the sensing reliability, a separated PCSA (SPCSA) was proposed, as shown in Fig. 17 [67]. The basic structure as well as the operation of SPCSA is similar to that of the PCSA. The main difference is that SPCSA separates the discharging and evaluation stages with two different paths, which alleviate greatly the voltage headroom problem, enabling it to operate at a relatively lower supply voltage. Meanwhile, thanks to the separated discharge and evaluation stages, we can amplify the input signals before entering the evaluation stage so as to tolerate the input-offset. In addition, two inverters (IV1 and IV2) and two NMOS transistors (MN2 and MN3) are added connecting between the discharging and evaluation stages, to amplify the limited current or voltage difference (due to the limited TMR ratio) between the two discharging paths flowing through MTJ0 and MTJ1, thus tolerating significantly the process variations and increasing greatly the sensing margin. Figure 18 shows the statistical sensing error rate of SPCSA compared to PCSA. As can be seen, SPCSA provides a much higher sensing reliability with the same hardware.

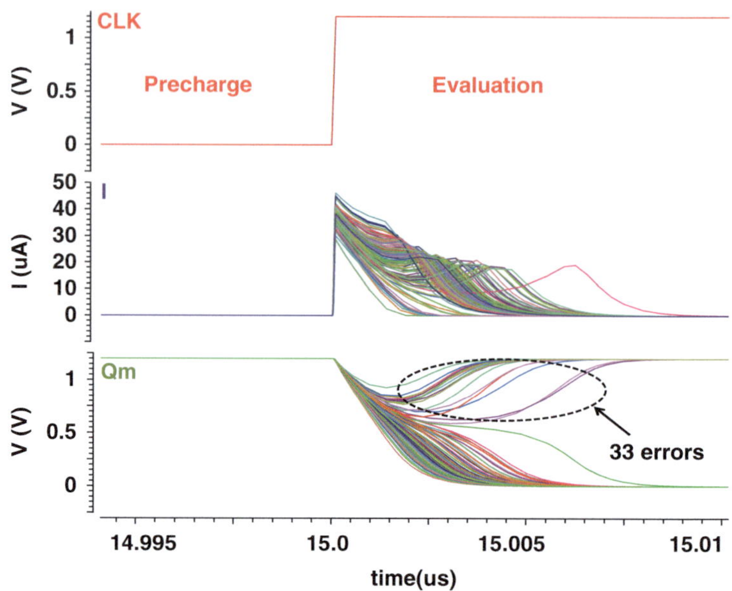

Fig. 16 Monte-Carlo simulation of a 16 k-bit PCSA circuit

3.2 Writing Circuit

According to STT switching mechanism, a bi-directional current is required to switch the magnetization in free layer of MTJ. In order to achieve high-speed logic design, high current is required to ensure the speed. In contrast to the low power and low area of the sensing circuit, the writing circuit for PMA STT MTJ occupies the main area and power of the whole circuit. As a result, the study on writing circuit is of importance to hybrid MTJ/CMOS circuit design.

In concert with the sensing circuit for a couple of MTJs with complementary states, a writing circuit to generate the bi-directional current for switching a couple of MTJs is designed as Fig. 19. Two NMOS (MN0-1) and two PMOS (MP0-1) transistors construct the main circuit. Each time one NMOS and one PMOS are always left open and the others closed, which creates a path to make the current pass from "Vdda" to "Gnd". Through two NOR and three NOT logic gates, the signals "Input" and "EN" control respectively the current direction and activation. Normally, it requires a "Vdda" higher than "Vdd" for logic operations to avoid the area overhead in the write circuit.

In order to generate the maximum current flowing through the couple of MTJs, both the transistors (one PMOS and one NMOS) should operate in their linear region above the threshold voltage V_{TH} to obtain the relatively lower resistances. In this case, they should satisfy the conditions: $V_{DS} \ll 2(V_{GS}-V_{TH})$ for NMOS and $V_{DS} \gg 2(V_{GS}-V_{TH})$ for PMOS. Their resistances, R_{on} and R_{op}, can be

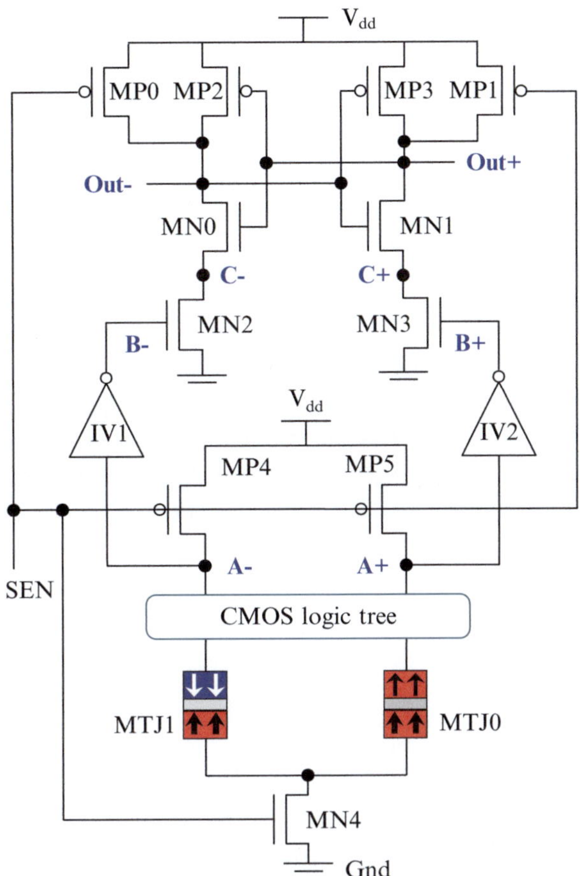

Fig. 17 The schematic of the separated pre-charge sense amplifier (SPCSA)

approximately expressed by Eqs. (30) and (31), and the generated current can be obtained through the Eq. (32),

$$R_{on} = \frac{1}{\mu_n C_{ox} \frac{W}{L} (V_{GS} - V_{TH})} \tag{30}$$

$$R_{op} = \frac{1}{\mu_p C_{ox} \frac{W}{L} (V_{SG} - |V_{TH}|)} \tag{31}$$

$$I_{write} = \frac{Vdda}{R_p + R_{ap} + R_{on} + R_{op}} \tag{32}$$

where μ_n is the electron mobility, μ_p is the hole mobility, C_{ox} is the gate oxide capacitance per unit area, W is the channel width, L is the channel length, V_{GS} is the gate-source voltage.

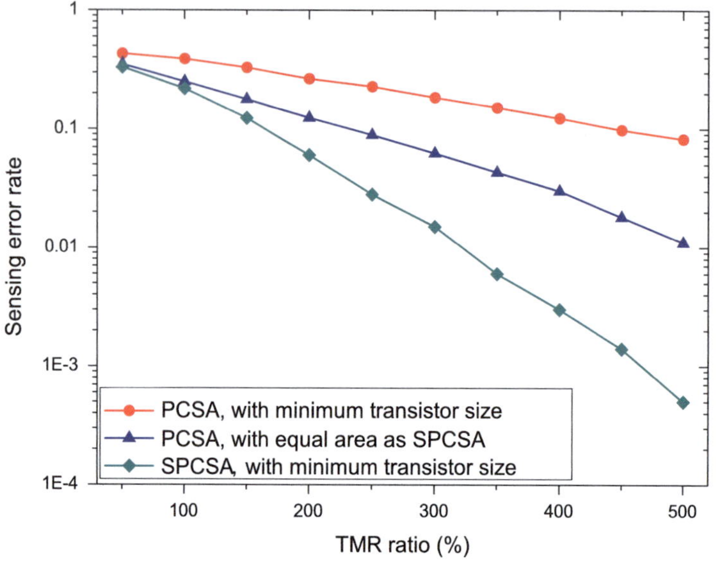

Fig. 18 Sensing error rate of the SPCSA circuit with Monte-Carlo simulations

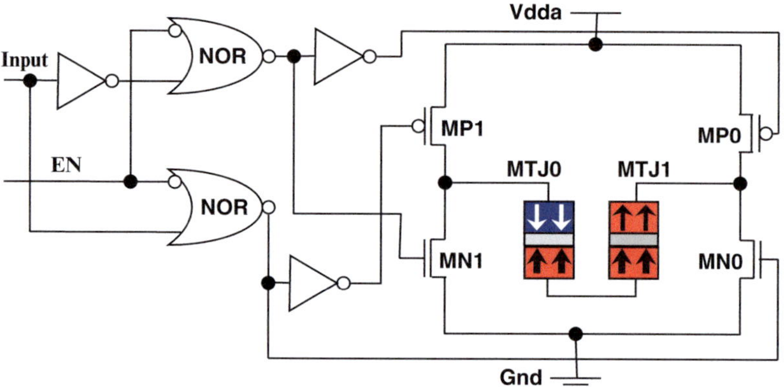

Fig. 19 Full writing schematic for STT writing approach, which is composed of two modified inverters and logic control circuits

By simulating a writing operation including anti-parallel to parallel switching and parallel to anti-parallel switching (see Fig. 20). We can find that the writing operation is not activated until the signal "EN" is set to "1". The states of the couple of MTJs remain always opposite and the switching direction follows the signal "Input".

From Eqs. (30)–(32), we find that the most efficient method to improve the current value is by increasing W, but this leads to significant area overhead. Figure 21 shows a study of area, speed and energy performance for this circuit. Here, only the area of four transistors (MN0-1, MP0-1) has been taken into account as the area of logic control circuit is the same for different simulation and is often in

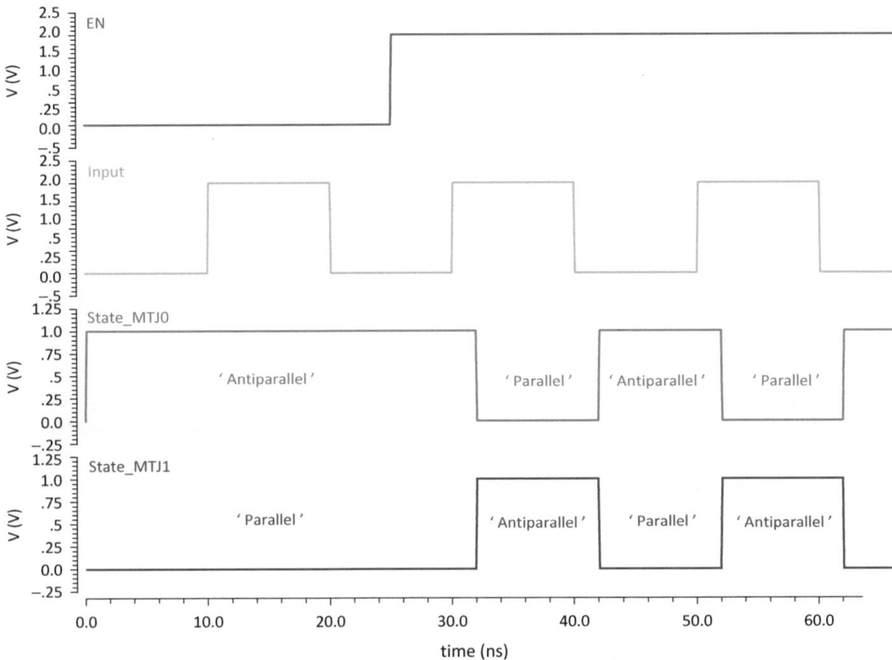

Fig. 20 Writing operation for a couple of MTJs with complementary states. The signals "Input" and "EN" control respectively the current direction and activation

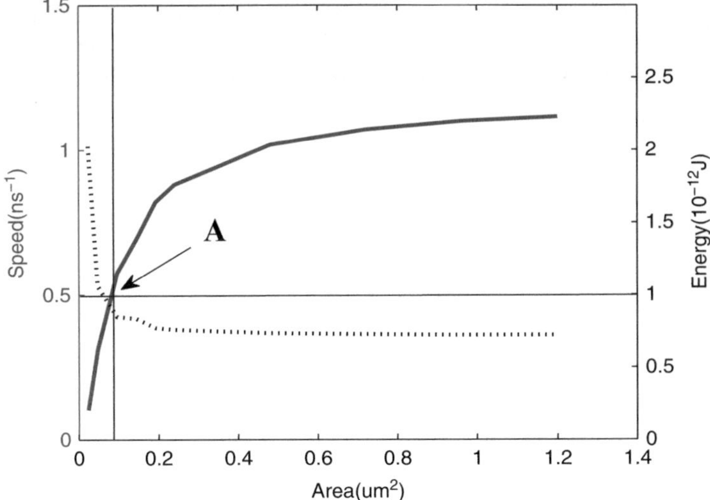

Fig. 21 Dependence of circuit switching speed (*solid line*) and energy dissipation (*dotted line*) versus die area with four transistors (MN0-1, MP0-1)

the minimal size. A strong dependence between area and speed can be found, especially when the area is smaller than 0.2 μm^2. The speed improvement becomes less significant for larger areas and saturates at ~1.1 GHz, which is different from the 2 GHz obtained with a single cell. There are two reasons for explaining this: first, "Vdda" is set to 2 V as 2.2 V is the breakdown limit for CMOS technology node [68]; second, there are a couple of MTJs, so the bias-voltage for each one cannot be larger than 1 V as there is also bias on the transistors in the circuit.

The energy of each switching operation has been calculated with Eq. (33). We also find a turning point, ~0.1 μm^2, below which the energy will be increased rapidly with a smaller area due to the extremely long switching duration as the current I_{write} approaching to the threshold I_{C0}. Contrarily, the energy is nearly the same for whatever the size larger than ~0.1 μm^2. This is firstly because that the writing current and speed approach to be saturated. Even if the writing current can increase continuously, from Eq. (11), the current is inversely proportional to the switching duration when the current is much higher than the critical one. Therefore the energy will inevitably be saturated for a high writing current.

$$E_{operation} = Vdda \times I_{write} \times Duration \tag{33}$$

The region around the crossing point of the two curves (point "A" in Fig. 21) can be localized. It can be considered as a good tradeoff among the area (~0.096 μm^2 or 30 F^2), power (1 pJ) and speed (~500 MHz) performance of this switching circuit, and be suitable to build up both logic chip and memory. This analysis can also help to investigate the circuits with special requirements like 800 MHz operating frequency.

For the advanced node below 90 nm, high reliability is becoming more and more crucial for the IC design [69–72]. Thanks to the integration of STT stochastic behavior into this model, an overall reliability investigation becomes possible. Figure 22 shows the statistical Monte-Carlo simulations of 100 complete writing

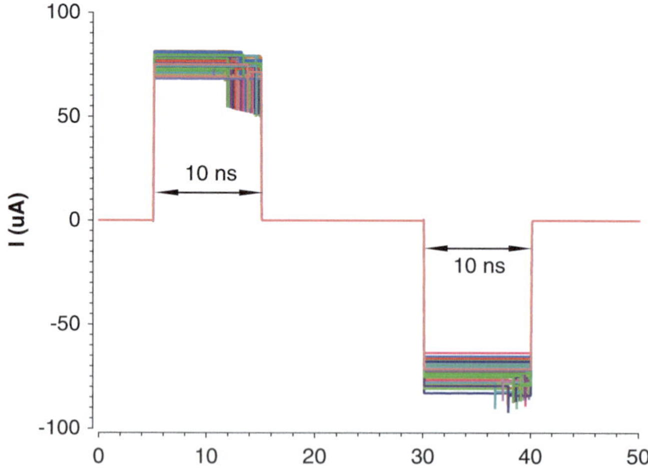

Fig. 22 Monte-Carlo simulation of a whole writing operation implemented by the writing circuit

operations using the writing circuit shown in Fig. 19 (however, instead of a couple of MTJs, there is only one MTJ connected in the circuit for this part of study). The complete writing operation includes the switching from parallel to anti-parallel and from anti-parallel back to parallel. Similarly to the case of the sensing circuit, the writing current at each write event is different because of the mismatch and process variation of CMOS part. The switching delay times vary randomly due to the stochastic behavior of MTJ cell. Since writing current is normally larger than sensing current, the stochastic effect in writing operation is relatively weaker than that in sensing operation. This can be proven by Fig. 22, the variation of every event is not so enormous.

Writing current magnitude and pulse duration are two key factors for the writing operation. As mentioned above, the writing current magnitude is dependent on the die area of writing circuit. We then perform the Monte-Carlo simulations for different writing pulse durations (5, 10 and 20 ns) to observe the dependence of writing Bit Error Rate (BER_W) versus die area of writing circuit (four main transistors: MP0-1 and MN0-1). The simulation results shown by Fig. 23 demonstrate their tradeoff relation: the increase of area can improve the BER performance. The reason is that a larger circuit allows larger write current, which in average reduces the time required to switch. For a given pulse duration, this increases the switching probability. Correlatively, it is observed that a longer pulse can also increase the reliability, which confirms the explanation mentioned above.

On the other hand, in order to overcome both power and reliability issues of conventional switching circuits due to the STT stochastic behavior, a self-enabled "error-free" strategy was proposed [73]. The corresponding circuit schematic is shown in Fig. 24. A sense amplifier (S.A) associated to the MTJ detects its state and outputs the data in logic level. The "self-enable" signal depends on the comparison

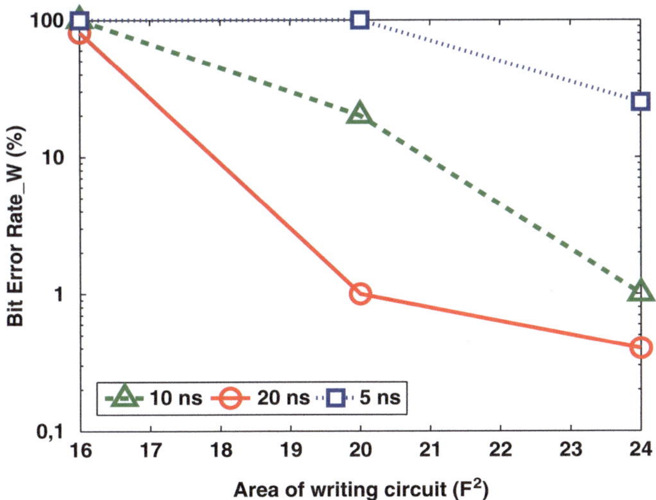

Fig. 23 Dependence of writing Bit Error Rate (BER_W) versus die area of writing circuit

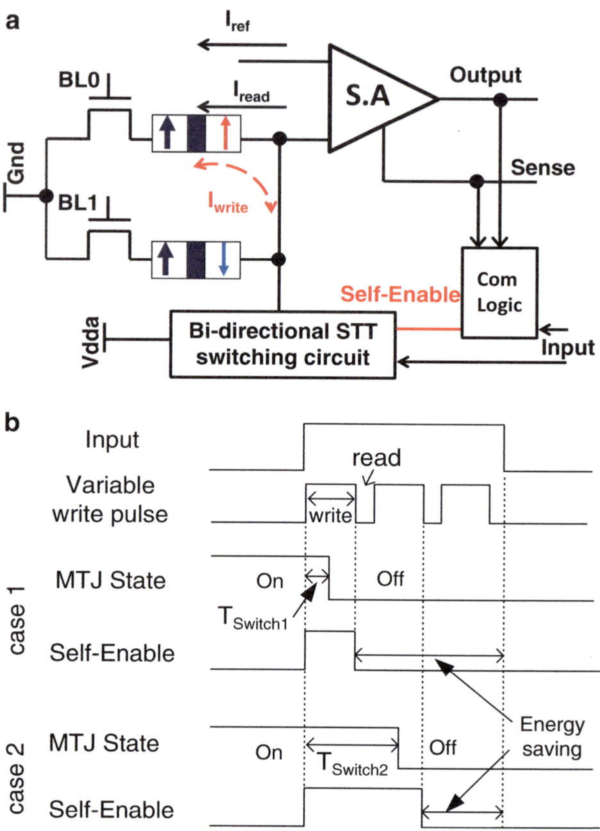

Fig. 24 Self-enabled "error-free" switching circuits and strategy with adaptive driver pulse duration (**a**) Scheme of proposed switching circuit (**b**) Varied duration for "self-enable" signal

result between output and "Input" data. For instance, it becomes "ON" as output is different from "Input" data. The fixed long writing pulse is replaced by a sequence of short duration T_{SS} including both switching and sensing operations. Thanks to the stochastic behaviors of STT magnetic switching, the state of MTJ can be changed just after one short write pulse, as shown in Fig. 24b. After that, "self-enable" is set to "OFF" and no current flows through the MTJ. Different from a self-adaptive write circuit designed for memristor, the proposed circuit takes benefits from the stochastic behaviors of STT switching. Moreover, periodic sensing is used to obtain the STT-MRAM storage in logic level for the comparison with "Input" data. This is due to the relatively low TMR or R_{Off}/R_{on} ratio of MTJ (e.g., 150–250 %). The frequency of read operations equals normally to the global clock (e.g., 500 MHz).

This switching circuit with self-enable mechanism presents a number of advantages. Firstly, it allows "error-free" as the switching operation becomes fully deterministic instead of stochastic behaviors caused by the intrinsic STT and PVT variations. As the write pulse duration is shortened and the number of switching operation is also reduced, the lifetime of oxide barrier can be greatly improved.

As mentioned above, the state of MTJ may be erroneously changed by a read current, "self-enable" becomes automatically "ON" to correct this error. Thereby, this proposed circuit provides evident high reliability.

Secondly, high power efficiency can be achieved by eliminating completely the additional power to tolerate the process voltage temperature (PVT) variations and stochastic behaviors. Another power saving comes from the reduced switching numbers as the "self-enable" signal is activated only while the stored data is different from "Input" data. On average, half of the switching operations can be economized, but exact power saving depends greatly on applications. Note that, for asynchronous applications, in addition to power saving, better operating speed could also be expected.

3.3 Magnetic Full Adder

Aiming to overcome the issue of rising standby and dynamic power, magnetic processor based on spintronic devices is thus expected. Since addition is the basic operation of the arithmetic/logic unit of any processors, MFAs attract a lot of attention and several designs based on diverse technologies are proposed in the last years [18, 19]. Here, we present a 1-bit MFA based on PMA STT MTJ (STT-MFA) [74] and a multi-bit MFA based on PMA racetrack memory [75].

3.3.1 1-Bit MFA Based on PMA STT MTJ

Figure 25 shows a 1-bit STT-MFA circuit, which is based on the generic logic-in-memory structure [74]. To evaluate the logic function, PCSA circuit is used. The inputs are "A", "C_i" and "B", and the outputs are "SUM" and "C_o". Among them, the input "B" relates to non-volatile storage PMA STT MTJ. The MOS tree is designed according to Eqs. (34)–(37) and the truth table shown in Table 1.

$$SUM = A \oplus B \oplus C_i = ABC_i + A\overline{B}\,\overline{C_i} + \overline{A}B\overline{C_i} + \overline{A}\,\overline{B}C_i \tag{34}$$

$$\overline{SUM} = AB\overline{C_i} + \overline{A}\,\overline{B}\,\overline{C_i} + \overline{A}BC_i + A\overline{B}C_i \tag{35}$$

$$C_o = AB + AC_i + BC_i \tag{36}$$

$$\overline{C_o} = \overline{AB} + \overline{AC_i} + \overline{BC_i} \tag{37}$$

For "SUM" logic, the MOS tree corresponds directly to the logic relationship among the inputs "A", "C_i" and "B", we can simply adapt it to the general structure with a couple of complementary PMA STT MTJ. However, it is a little difficult for "C_o" logic as there is the term AC_i in the logic function Eq. (36) and we cannot adapt the schematic to the general "logic-in-memory" structure. It can be inferred that the impact of the term AC_i on the resistance is equivalent to a sub-branch

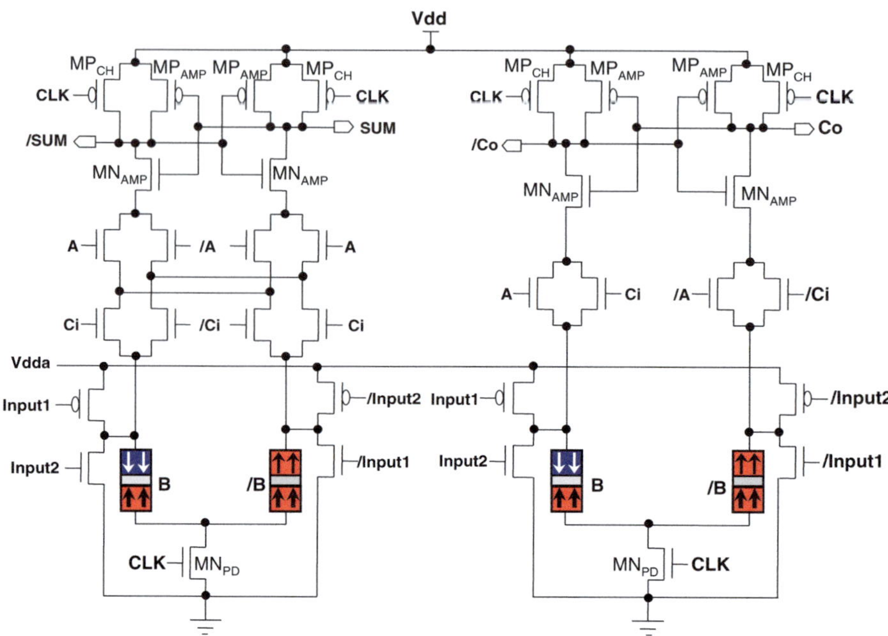

Fig. 25 STT-MFA architecture with "*SUM*" (*left*) and output carry "*C_o*" (*right*) sub-circuits, "A" is volatile data for computing, "B" is non-volatile data using as quasi-constant [74]

Table 1 Truth table of "*SUM*" and "*C_o*" logic gate for MFA

A	B	C	SUM	C_o
0	0	0	0	0
0	0	1	1	0
0	1	0	1	0
0	1	1	0	0
1	0	0	1	0
1	0	1	0	1
1	1	0	0	1
1	1	1	1	1

connecting PCSA and the discharging transistor (*MN2* in Fig. 12). Table 2 exhibits the true table and the resistance configuration of "C_o" logic. R_{OFF} and R_{ON} are respectively the close and open resistances of MOS transistor. R_L and R_R are respectively the whole resistance of the left and right branch of PCSA. We can find that whatever the value of "A" and "C_i", the sub-branches AC_i and $\overline{AC_i}$ have no impact on the output. If "A" and "C_i" are different, the resistances of the two sub-branches are the same. If they are the same, their comparison corresponds to that of R_L and R_R in the condition of $R_{ON} > R_{AP}$, which is always true for PMA STT MTJ under present technology condition. This allows the term AC_i to be deleted from Eq. (36) and we can obtain the "C_o" logic circuit shown in Fig. 25.

Table 2 Truth table and resistance configuration of "C_o" for MFA

A	B	C	Resistance comparison	C_o	Sub-branch AC_i	Sub-branch $\overline{AC_i}$
0	0	0	$R_L > R_R$	0	$2R_{OFF}$	$2R_{ON}$
0	0	1	$R_L > R_R$	0	$R_{OFF} + R_{ON}$	$R_{OFF} + R_{ON}$
0	1	0	$R_L > R_R$	0	$2R_{OFF}$	$2R_{ON}$
0	1	1	$R_L < R_R$	0	$R_{OFF} + R_{ON}$	$R_{OFF} + R_{ON}$
1	0	0	$R_L > R_R$	0	$R_{OFF} + R_{ON}$	$R_{OFF} + R_{ON}$
1	0	1	$R_L < R_R$	1	$2R_{ON}$	$2R_{OFF}$
1	1	0	$R_L < R_R$	1	$R_{OFF} + R_{ON}$	$R_{OFF} + R_{ON}$
1	1	1	$R_L < R_R$	1	$2R_{ON}$	$2R_{OFF}$

The PMA STT MTJs connect serially with a common central point. In order to program MTJs, we use a writing circuit composed of pass transistors, which are connected respectively to the bottom and top electrodes of the serial branch. In such a manner, as a control signal ("*Input1*" or "*Input2*") is activated, the first PMA STT MTJ noted "*B*" is put in high resistance state (R_{AP}) or low resistance state (R_P) while the second PMA STT MTJ noted "*/B*" is put in the complementary state R_P or R_{AP}.

It is noteworthy that there is neither capacitance for the data sensing and nor magnetic field for data programming in this new structure beyond the previous structures [18, 19]. Therefore, this design allows efficient area minimization and is suitable for advanced fabrication nodes below 65 nm.

Figure 26 illustrates the transient simulation of 1-bit STT-MFA shown in Fig. 25. It is performed by using PMA STT MTJ compact models introduced above and CMOS 40 nm design kit. The time-dependent behaviors of outputs ("*SUM*" and "C_o") confirm the logic functionality of full addition. For instance, for the operation "*A*" = "1", "*B*" = "0", "C_i" = "0", the result is "1" and no carry yields; for the operation "*A*" = "1", "*B*" = "0", "C_i" = "1", the result is "0" and the carry is "1".

Figure 27 emphasizes one sensing operation of this STT-MFA and shows the analog behaviors. It confirms the pre-charge, evaluation and amplification process described previously. Moreover, we find that the sensing delay of "*Output_C_o*" (~127 ps) is shorter than that of "*Output_SUM*" (~147 ps). This is due to the higher resistance of the branch associated with "*Output_SUM*", leading to lower current and slower amplification.

The delay time and dynamic energy are generally two crucial parameters to evaluate the performance of computation system. We have studied the effects of three possible factors: the size of discharge transistor (MN_{PD} in Fig. 25), PMA STT MTJ resistance-area product (RA) and TMR ratio. Figure 28 demonstrates the performance dependence of this STT-MFA in terms of delay time and dynamic power on the size of discharge transistor. We can find a tradeoff between the speed and power performance by varying the die area. A larger discharge transistor can drive a higher sensing current and faster amplification of PCSA circuit, but cost more energy.

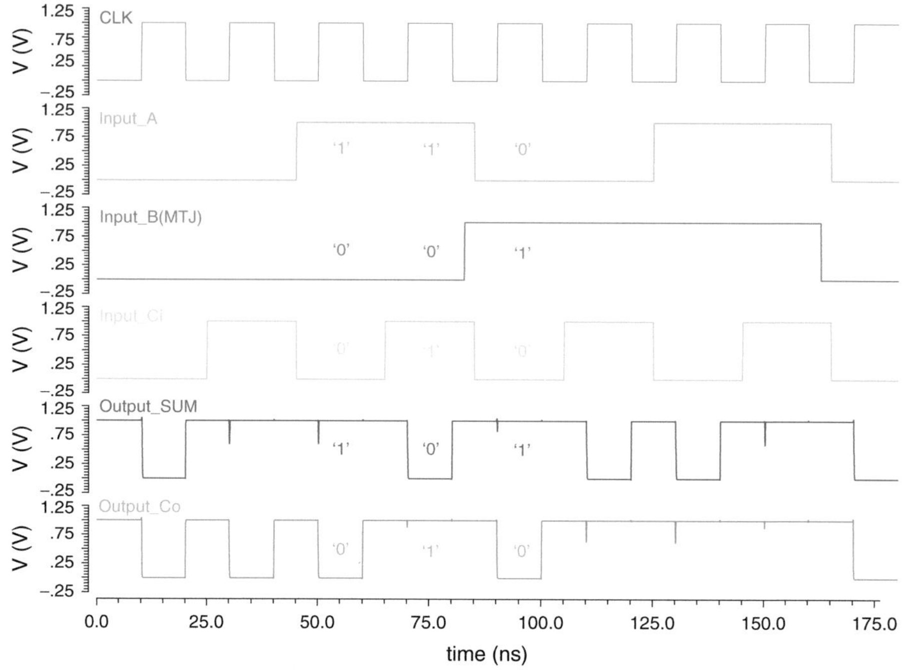

Fig. 26 Transient simulations of 1-bit STT-MFA in 40 nm node

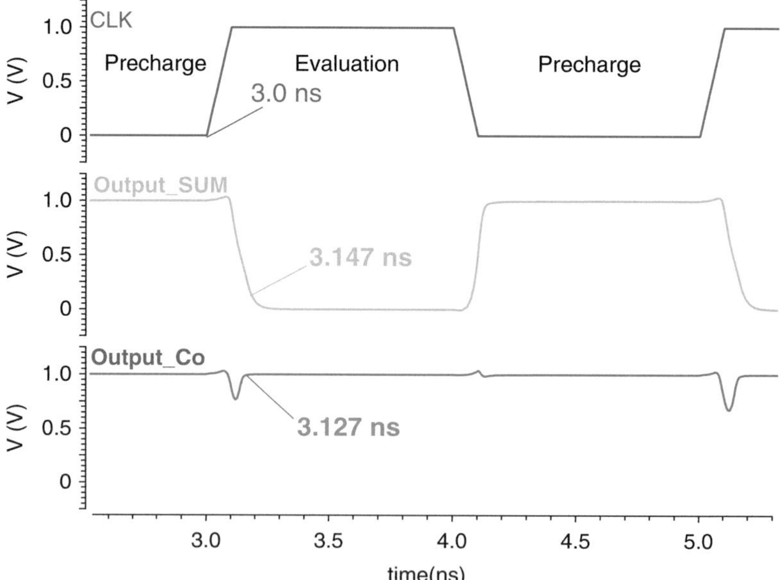

Fig. 27 One sensing operation of the PCSA based STT-MFA: outputs are pre-charged as CLK is set to "0" and are evaluated as CLK is set to "1"

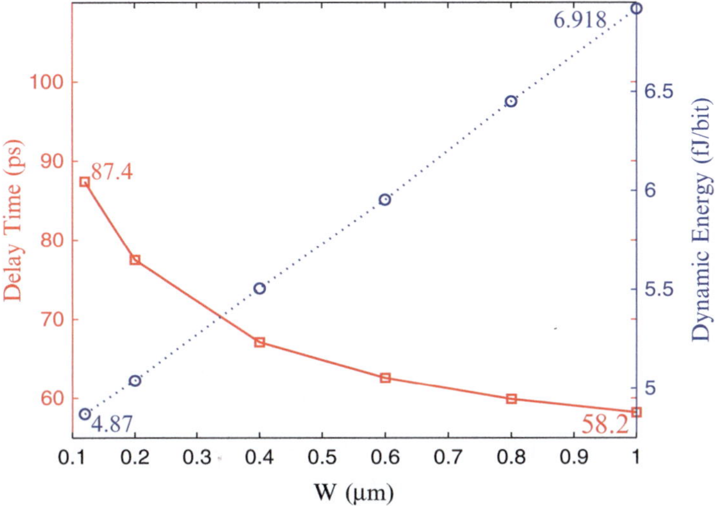

Fig. 28 Dependence of delay time (*red solid line*) and dynamic energy (*blue dotted line*) on the width of discharge transistor for STT-MFA

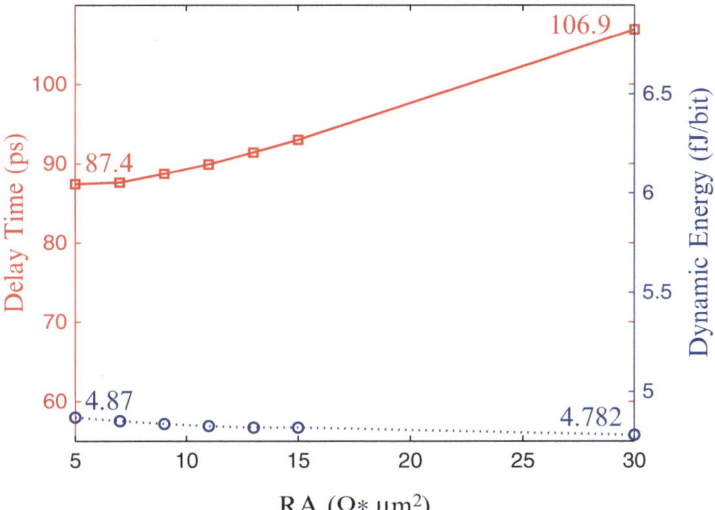

Fig. 29 Dependence of delay time (*red solid line*) and dynamic energy (*blue dotted line*) on the resistance-area product (*RA*) of PMA STT MTJ

Figure 29 shows the *RA* dependence for this STT-MFA. By decreasing *RA*, the delay time becomes shorter while keeping a relatively steady dynamic power performance. This confirms that the speed advantage of using low *RA*.

We also investigate the dependence between TMR ratio of PMA STT MTJ and STT-MFA performance. Figure 30 shows that faster speed is possible by increasing the TMR ratio while the dynamic energy changes slightly.

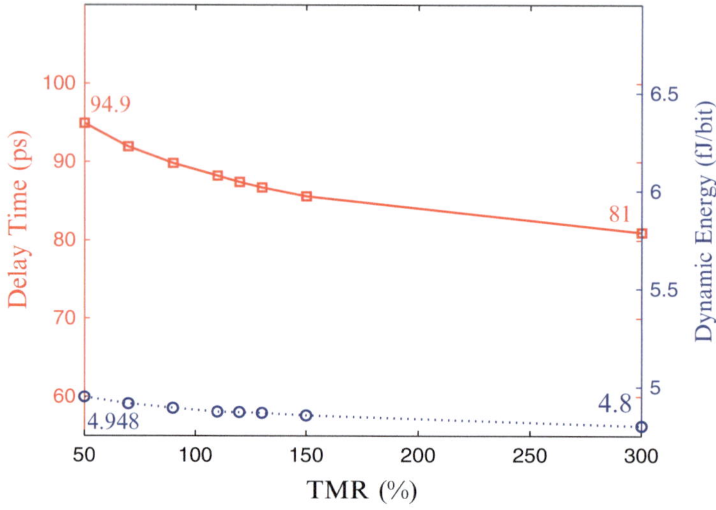

Fig. 30 Dependence of delay time (*red solid line*) and dynamic energy (*blue dotted line*) on PMA STT MTJ TMR ratio for 1-bit STT-MFA

Table 3 Comparison of 1-bit STT-MFA with CMOS only full adder

Performance	CMOS full adder (40 nm)	STT-MFA
Delay time	75 ps	87.4 ps
Dynamic power @500 MHz	2.17 μW	1.98 μW
Standby power	71 nW	<1 nW [31]
Data transfer energy	>1 pJ/bit	<1 fJ/bit
Die area	46 MOS	38 MOS + 4 MTJs

According to the above analyses, a PMA STT MTJ with lower RA and higher TMR ratio is expected to perform fast computation while keeping nearly the same dynamic energy. In the recent experimental demonstration of the MTJ, a low RA (e.g., 5 $\Omega\mu m^2$) and high TMR ratio (e.g., 200 %) can be achieved in PMA STT MTJ [51, 76, 77].

We compare the STT-MFA with conventional CMOS only full adder in terms of delay time, dynamic power, standby power, data transfer energy and die area (see Table 3). The CMOS-only full adder is taken from the standard cell library of STMicroelectronics 40 nm design kit. Two full latches are added to synchronize the outputs with clock signal.

In conventional computing architectures, logic and memory are completely separated [78]. In order to perform a logic operation, both the instruction and data need to be read from memory units (i.e., cache and main memories), and then moved to logic unit. The results are transferred back to the memory units after the computing. In the STT-MFA circuit based on "logic-in-memory" architecture, logic operations are processed directly with the magnetic data stored in MTJs and the addition result is written to other MTJs for the next operations. Long latency and high dynamic power due to data moving can be significantly economized. For example, the data transfer energy (~1 pJ/mm/bit @22 nm [1]) becomes much lower thanks to the shorter distance between memory and computing unit, which is about some μm or below in STT-MFA instead of some mm for CMOS only logic circuit.

Furthermore, thanks to the 3D integration of STT-MRAM, the die area of this design (38 MOS + 4 MTJs) is advantageous compared to those of the CMOS full-adder (46 MOS). However, its energy-delay product (EDP) exceeds that of a CMOS full-adder by approximately 10 % since it takes more time for PCSA amplification process. Due to the non-volatility of PMA STT MTJ, the new chip can be powered off completely and this allows the standby power to be reduced significantly down to 0.75 nW [77]. Thereby, the STT-MFA can greatly reduce the consumption in a full computing system, especially for those normally in OFF state.

Another critical idea of this design is to use a programming frequency (e.g., 1 kHz) of STT-MRAM much lower than the computing frequency. Thereby, the switching power for non-volatile storage becomes insignificant to other power consumption in a full system. We can continue to reduce it by shortening the non-volatile data retention (e.g., 1 day). Moreover, the programming energy for the non-volatile data (bit "B" in Fig. 25) can be reduced, following the area minimization [79] and new material development for MTJs (e.g., ~0.1 pJ/bit).

3.4 Multi-Bit MFA Based on Racetrack Memory

PMA racetrack memory is distinguished as it can store and shift multiple bits of data through CIDW motion along a magnetic nanowire. This advantageous feature makes it possible to design a high speed and compact multi-bit serial MFA.

Figure 31 shows the detailed schematic of CARRY circuit of the multi-bit MFA based on PMA racetrack memory including MTJ writing circuit [75]. "A" and "B" are multi-bit input data stored in different nanowires. Each data is designed to be stored in dual magnetic nanowires with exactly opposite configuration to minimize the variation between two complementary data (e.g., "A" and "$\bar{A}$") as the same I_{shift} is used in the dual nanowires to move the DWs [3]. At each rising edge of CLK, "C_o" and "$\bar{C}_o$" are evaluated through the PCSA circuit and become inputs of a writing circuit, which generates writing current I_{write} to reverse or just conserve the state of nucleation MTJs ("$C_{nucleation}$" and "$\overline{C_{nucleation}}$"). At each falling edge of CLK, propagating current I_{shift} induces the DW motion of all magnetic nanowires ("A", "$\bar{A}$", "B", "$\bar{B}$", "C", "$\bar{C}$") simultaneously, moves next magnetic domains under the read MTJ for next adding operation.

The operation of SUM circuit is similar to that of CARRY: the SUM output and its complement are evaluated through the SUM PCSA circuit and become inputs of SUM writing circuit, which generates the writing current to write these values into

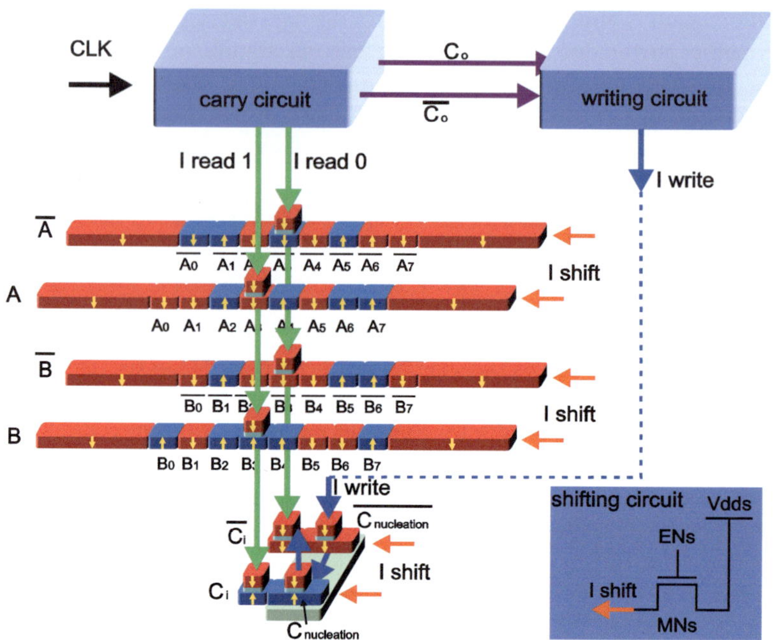

Fig. 31 CARRY circuit of multi-bit MFA based on PMA racetrack memory including MTJ writing circuit

the most significant bits of SUM nanowire. These bits are then shifted in the way that at the end of all additions, the sooner the bit is calculated, the greater its weight is.

The transient simulation of this multi-bit MFA shows the addition operation of two random 8-bit words: "A" = "01110011" (Fig. 32c) and "B" = "01011010" (Fig. 32d). "CLK" (Fig. 32a) drives PCSA circuit and "I_{shift}" (Fig. 32b) induces DW motion in the magnetic nanowire. The outputs "SUM" (Fig. 32e) and "C_o" (Fig. 32f) are firstly pre-charged to logic "1" when "CLK" = "0" and are evaluated when "CLK" is set to "1". The DW motions are implemented in the pre-charge phase in order to avoid the disturbance to the output evaluation. The serial addition is performed from the least significant bit and the simulation result "SUM" = "11001101" and "C_o" = "01110010" confirms the correct operation of MFA.

Between two addition evaluations, there is a data transition process to achieve multiple bits operation. Figure 33 demonstrates the CARRY transition including DW nucleation and motion. Carry-out "C_o" (Fig. 33b) is firstly pre-charged to "Vdd" before the time "M0", after the rising edge of "CLK" (Fig. 33a), "C_o" is evaluated by the PCSA and becomes the input signal of writing circuit (Fig. 33b). I_{write} is generated to nucleate DW in the magnetic nanowire (Fig. 33c, d). I_{shift} is in the following activated to propagate the DW and replace the value of carry-in "C_i" with "C_o" at the time "M3" for next cycle of addition (Fig. 33e).

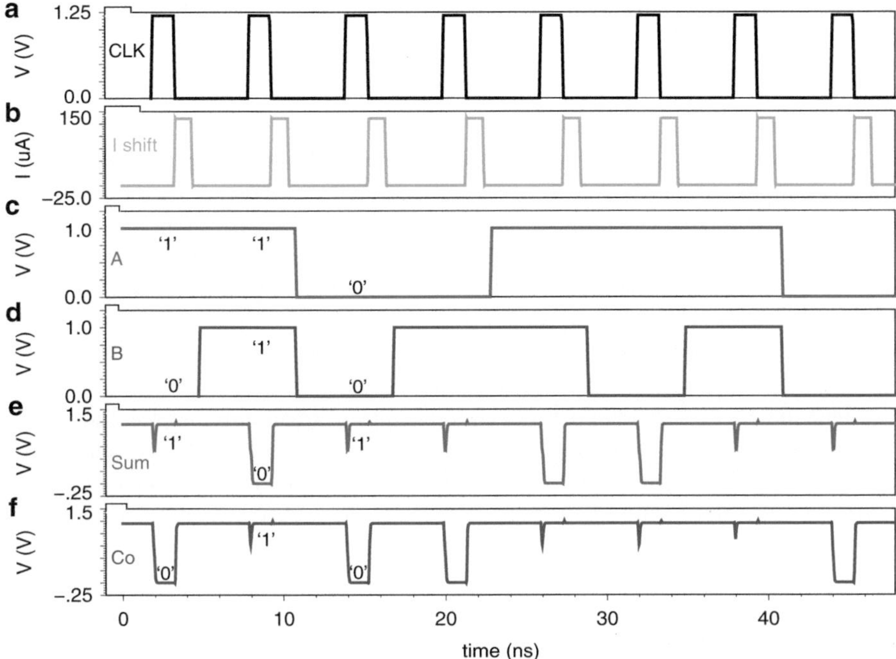

Fig. 32 Transient simulation of the multi-bit MFA. (**a**) CLK signal (**b**) Data shifting current pulse I_{shift} (**c**) Input data "A" (**d**) Input data "B" (**e**) "SUM" (**f**) "C_o"

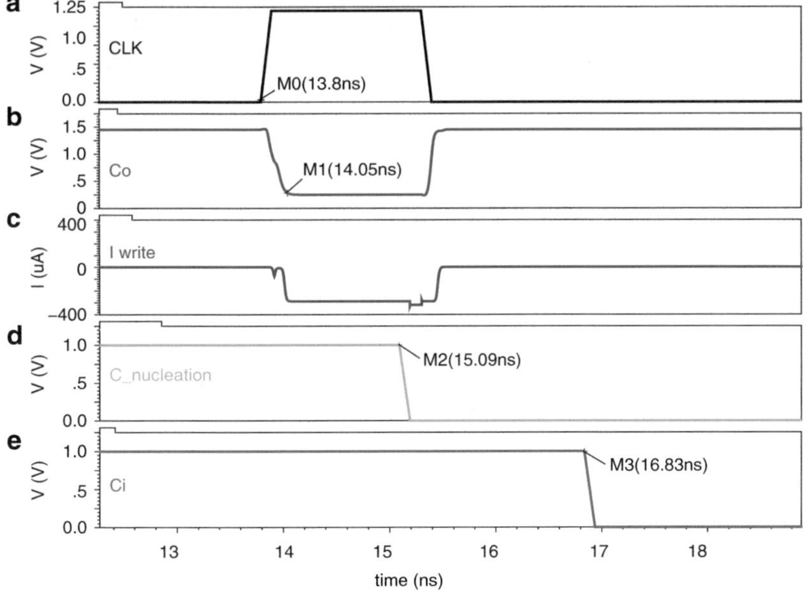

Fig. 33 Transient simulation of CARRY storage (**a**) CLK signal (**b**) "C_o" (**c**) DW nucleation current I_{write} (**d**) State of DW nucleation MTJ (**e**) Carry in for next adding operation

We then analyze the performance of this multi-bit MFA in terms of delay and power dissipation. Indeed, several parameters, such as the size of transistors and "Vdd", can affect greatly on them. A first look at the repartition of consumed energy in this MFA shows that the energy for nucleation and propagation is of the same order of magnitude, and higher than that of data sensing or logic computing.

In order to propagate the magnetic domains to their next positions (shift 1 bit), we must supply a current I_{shift} in a period t_{pulse}. The period t_{pulse} is the necessary time for all magnetic domains to move from their current positions to their next positions. It corresponds to the propagation delay, which is inversely proportional to I_{shift}. Consequently, the propagation energy does not vary much (seeing that this energy is the integral of the product $V_{pulse} \times I_{shift} \times t_{pulse}$ and the power supply voltage V_{pulse} is kept invariable). Simulations show that energy needed for shifting all racetrack memories 8 bits is about 29 pJ.

Since the energy needed for propagation is almost invariable, we can reduce the propagation delay by increasing I_{shift}. Normally, one transistor based current source is used to generate the DW propagation current, thus the size of transistor determines the generated current in propagation circuit. Figure 34 shows the tradeoff dependence of propagation delay on the width of transistor MN2: the reduction of propagation delay at the cost of satisfying the area.

The writing circuit nucleates domain walls under the MTJ write head (e.g., MTJ0 in Fig. 7) by passing through a bi-directional current I_{write}. I_{write} is proportional to both supply voltage V_{write} and transistors' size. V_{write} will be set as high as possible in order to minimize the size of transistors (MN0-1 and MP0-1) while keeping the switching current at fixed value. In this setup, V_{write} is set to 2 V to avoid the breakdown of oxide barrier at 65 nm technology node. A higher I_{write} can reduce the switching delay, but increase the power consumption.

A study of the tradeoff among the width of transistors, switching speed and power dissipation have been made to find out optimal operation point (see Fig. 35). In this analysis, the width of transistors W is started at 0.35 μm because I_{write} is not

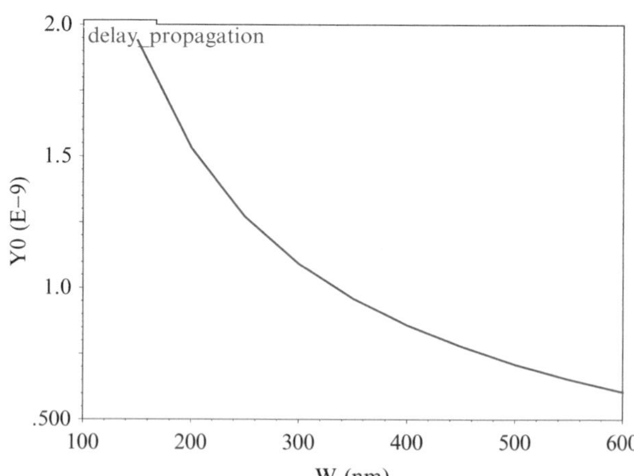

Fig. 34 Dependence of domain wall propagation delay on the transistor width of propagation circuit

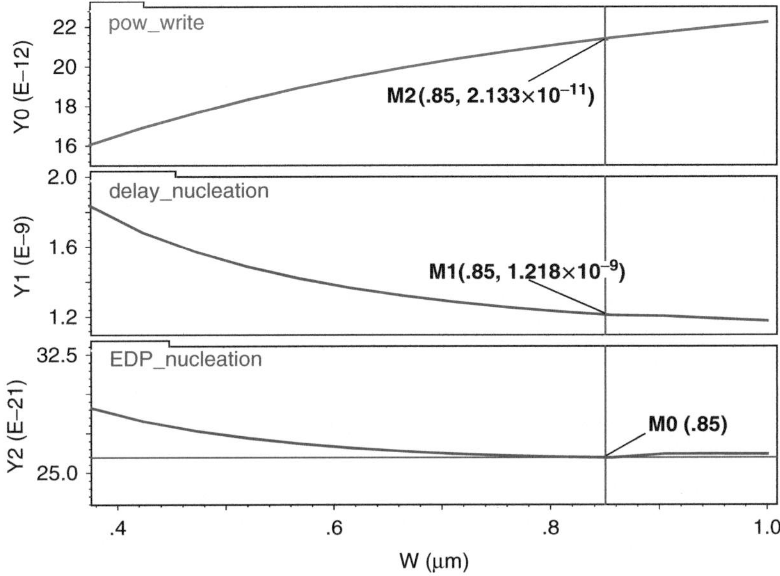

Fig. 35 Dependence of energy consumption (pow_write), switching speed (delay_nucleation) and the energy-delay product (EDP_nucleation) versus the width of four transistors using in writing circuit

Table 4 Comparison of 8-bit MFA based on racetrack memory with CMOS only full adder included transferring and writing data

Performance	CMOS full adder (65 nm)	MFA
Write time	200 ps	2 ns
Write energy	16 fJ/8 bits	$(21.39 + 29)$ pJ/8 bits
Transfer time	~ns	0
Transfer energy	8 pJ/mm (for 8 bits)	~0
Die area	310 MOS	23 MOS + 18 MTJs

high enough to switch the state of MTJ below this value. This curve shows that one can increase W to reduce the factor of merit EDP until the point "M0" $(W = 0.85 \ \mu m)$ and then it slightly goes up. Thereby the operating points should be chosen around the optimum (e.g., $W = 0.75–1.0 \ \mu m$) to address different applications. For instance, the two markers "M1" and "M2" show that when $W = 0.85 \ \mu m$, the switching power and latency are 21.33 pJ and 1.22 ns, respectively. The switching current I_{write} equals to 291 μA in this case.

In order to understand the advantages and disadvantages of this multi-bit MFA based on PMA racetrack memory, we compare its performance with that of a CMOS-only series adder (see Table 4), which uses a full adder taken from the library of STMicroelectronics 65 nm design kit.

For the comparison with CMOS only multiple bits full adder regarding writing and transferring data, we see that the chip area of the MFA based on PMA racetrack memory is significantly reduced. The 8-bit MFA uses only 23 MOS transistors, 18 MTJs and 8 magnetic nanowires instead of 22 MOS plus 8×3 Flip-Flops (310 MOS transistors totally) for an 8-bit series CMOS full adder. Although the number of transistors decrease 13 times, the area reduction is about 4.5 times since the writing circuit and propagating circuit requires the transistors with 6.3 times and 3.3 times minimum width (0.135 µm@65 nm technology node), respectively. The total delay of one operation of the new MFA is ~2.1 ns, composed of DW nucleation (~1.2 ns), motion (~0.7 ns) and detection (~180 ps). It can be thus driven by a CLK frequency up to 470 MHz, which can be further increased with the feature size shrinking. This latency is of the same order with that of CMOS circuit (read time + transfer time + operate time + transfer time + write time). On the contrary, the MFA consume six times dynamic energy more than the CMOS only full adder since energy needed for nucleation and propagation is still too large with current technology. However, we have not yet addressed the static energy in this comparison. Regarding that power must be supplied in order to maintain stored data in CMOS-only storage circuit, the MFA does not require energy to conserve information thanks to its total non-volatility. This allows the circuit to be turned off safely in "idle" mode without data backup. All the operations can be retrieved instantly after power-on. This instant on/off capability promises to overcome completely the rising standby power issue due to leakage currents and could be very useful for normally-off systems [77].

It is important to note that for this non-volatile MFA, operations are performed directly with the data ("A", "$\overline{A}$", "B", "$\overline{B}$") stored in magnetic nanowires, which plays the role of shift registers. We do not take into account the writing circuits of "A" and "B" to keep the same comparison condition as the writing circuits of data are considered in the CMOS shift register part, not in the adder. The number of writing circuit is then reduced to two for respectively SUM and CARRY circuits, which are shared by the eight bits.

3.5 Content Addressable Memory (CAM) Based on Racetrack Memory

CAM is a computer memory that can output the address of search data. It compares search data with stored data and returns the match location with its high-speed fully-parallel manner. Therefore it is widely used in mobile, internet routers and processors to provide fast data access and ultra-high density [21]. The mainstream CAMs are composed of large-capacity volatile SRAM blocks (see Fig. 36a), which lead to high static power and large die area [80]. These become the key challenges for the future R&D of CAM. Replacing volatile memories by non-volatile memories or applying hybrid non-volatile logic-in-memory circuits is a promising

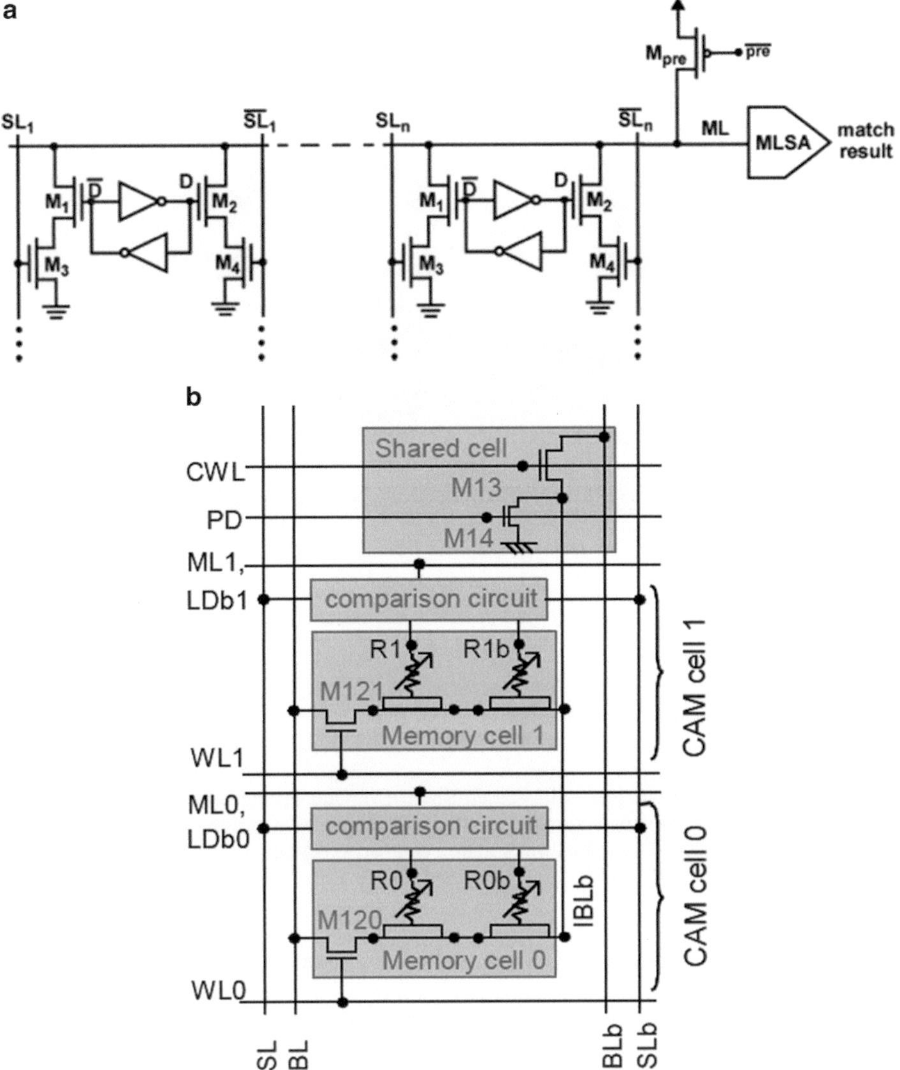

Fig. 36 Conventional CAM. (**a**) SRAM based CAM. (**b**) DW-CAM [21, 22]

solution to build non-volatile CAM and overcome both these drawbacks. This topic is currently under intense investigation. For instance, a DW motion MRAM based CAM (DW-CAM) was prototyped recently (see Fig. 36b), which demonstrated important progress in terms of power and density [22]. However, this DW-CAM used a three-terminal MTJ as storage element and every memory cell had one comparison circuit and one selected transistor, which lead to a high bit-cell cost and cannot allow the expected ultra-high density.

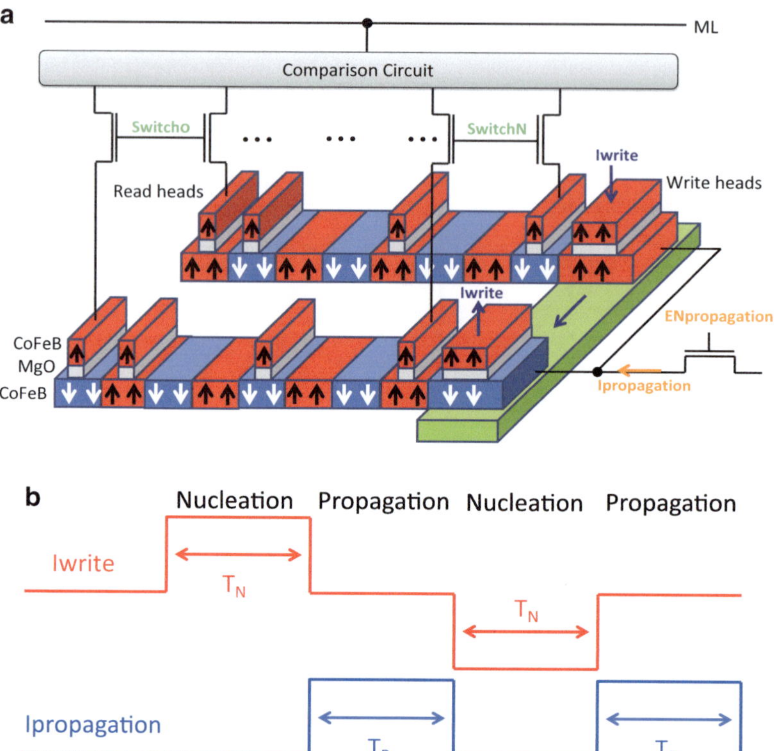

Fig. 37 (**a**) Structure of dual nanowires based RM-CAM. One writing current pulse nucleates a couple of MTJs with complementary configurations. A propagation current pulse drives the dual nanowires synchronously. Every dual wires share a comparison circuit. (**b**) One example of current pulse configuration for I_{write} and $I_{propagation}$. T_N and T_P are respectively their pulse durations

In this situation, a design of CAM based on complementary dual PMA racetrack memories (RM-CAM) was proposed [81]. Its non-volatile feature can reduce the static power due to leakage currents. The CMOS based DW nucleation and sensing circuits are globally shared to scale down the cell area. The complementary dual nanowires structure allows the local sensing and fast data search operation.

The RM-CAM is composed of comparison circuits, PMA racetrack memories and DW nucleation/propagation circuits. A couple of complementary magnetic nanowires are used to present one word (see Fig. 37a) in order to obtain the most reliable and fast access operation for CAM applications as this solution benefits the maximum TMR value instead of TMR/2 for conventional single nanowire structures. We design the comparison circuit based on PCSA, which allows minimum power and sensing errors. This RM-CAM includes a couple of PMA MTJs connected together as the write heads. Due to the different directions of the writing current pulse I_{write} through these two MTJs, they can nucleate the complementary configurations through STT switching mechanism under the same I_{write} pulse.

One of the critical challenges for complementary magnetic nanowires is to synchronize precisely the domain wall positions. Here, the same current pulse $I_{propagation}$ propagates domains in the dual nanowires and we implement the DW pinning constrictions with the same distance in the magnetic nanowires [82]. To avoid the interference between the DW nucleation and the previous data, write heads do not hold the data storage and there is always a $I_{propagation}$ pulse following each DW nucleation (see Fig. 37b). There are also a couple of PMA MTJs at each bit of storage elements as read heads. Since lower resistance can reduce the rate of breakdown and higher resistance can improve the sensing performance, the size of the read heads should therefore be smaller than that of the write heads to obtain the best switching and sensing reliability.

The comparison circuit (see Fig. 38) consists of two parts: a PCSA detects the complementary magnetizations of the read heads by two reading current pulses (I_{read} and I_{readb}) and outputs a logic value; the transistors MN3-MN6 build a classical NOR-type CAM. The signal "MLpre" is used to pre-charge the match line (ML). In case that the search line "SL" ("SLb" is its complementary signal) matches the stored data, there is no path to discharge and ML will thus be asserted. In contrast, ML will be discharged.

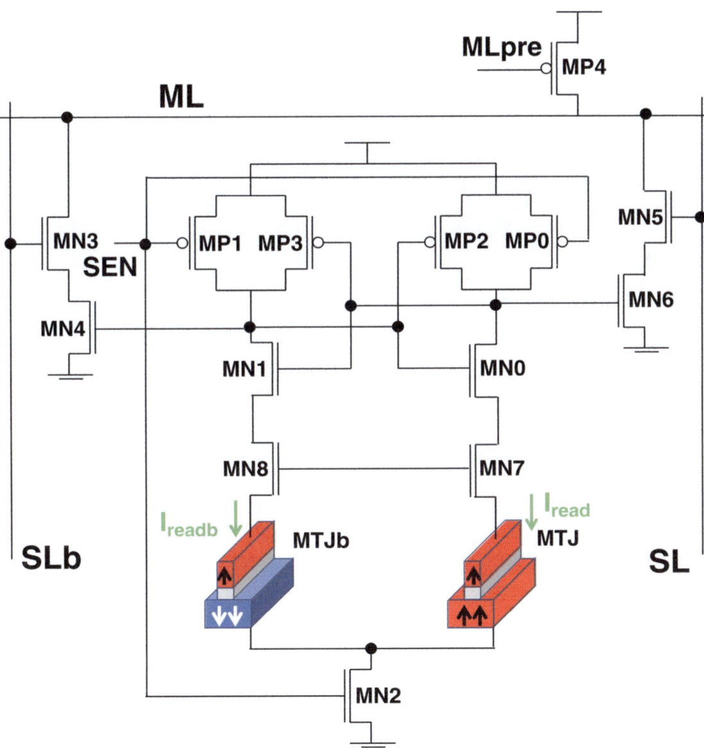

Fig. 38 Schematic of the comparison circuit. It outputs the logic value "1" or "0" according to the configuration of complementary MTJs. MN3-6 transistors build up a NOR-type CAM

The fast search operation as shown in [22] can be also expected in the RM-CAM. At first, we program the magnetic nanowires, and the switch signals then select each bit of magnetic nanowires to be loaded in the comparison circuit. By sequentially triggering the switch signals, all the words can be explored. If there is no match case, DW nucleation and propagation will be carried out to enter new words for the next search. The programming speed of magnetic nanowires depends on T_N and T_P, which are respectively the pulse durations of I_{write} and $I_{propagation}$. They can be both sped up to ~1 ns. According to the current pulse configuration shown in Fig. 37b, the worst case of programming duration is $N \times (T_N + T_P)$, where N is the number of pinning potentials in the magnetic nanowire. We can benefit a higher speed for the repeated bits such as "111" and "000" when only one DW nucleation is required for three bits.

In order to improve the area efficiency, every couple of dual nanowires shares the comparison circuit in this RM-CAM (see Fig. 37a). Unlike the DW-CAM where there is a large transistor for nucleation for every storage cell, the same write head is shared for one magnetic nanowire in RM-CAM, and the CMOS area dedicated for each storage cell becomes ignored for a long track with numerous pinning constrictions. This structure thus allows an ultra-high density.

An 8-bits-width-8-words-depth PMA RM-CAM shown in Fig. 39 has been designed. Firstly, we implement the transient simulation for the search operation without DW propagation (see Fig. 40a). The clock signal "CLK" involves the

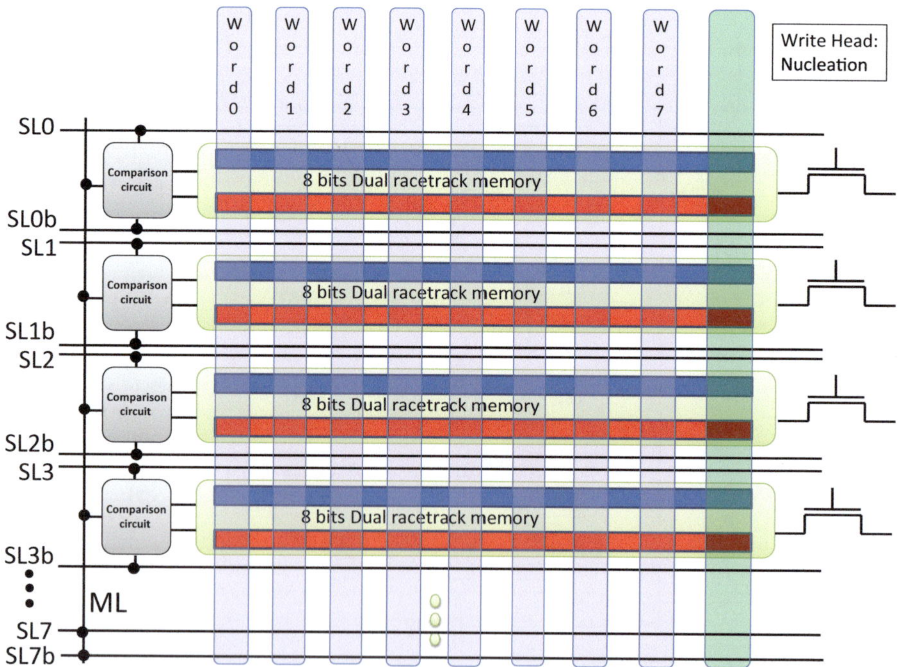

Fig. 39 Schematic of an 8×8 bits RM-CAM. Each word is composed of the bits at the same positions in 8 different dual nanowires; they can be driven to move simultaneously by the propagation currents

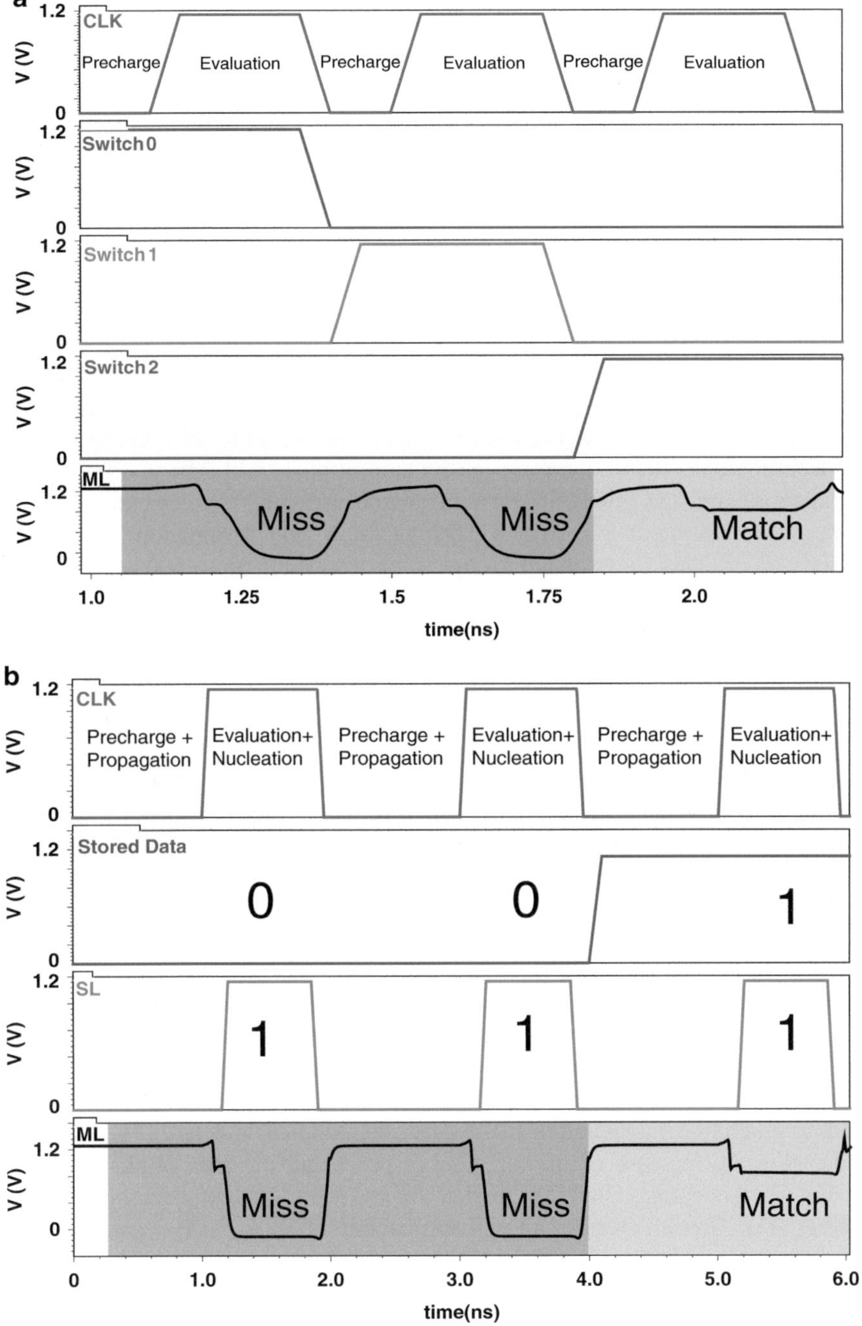

Fig. 40 Transient simulations of the RM-CAM: (**a**) Without DW nucleation and propagation. (**b**) With DW nucleation and propagation

"Pre-charge" phase and the "Evaluation" phase. During the "Pre-charge" phase, both of the signals "SEN" and "MLpre" (see Fig. 38) are set low to pre-charge the PCSA circuit and the match line "ML". The first word "Word0" has been loaded by enabling the signal "Switch0". With the response of the signal "Miss", "Switch1" will be then activated and so on. This process doesn't stop until the appearance of the match case. We find that this search operation needs only ~0.45 ns, which is faster than that of conventional SRAM-based CAM and DW-CAM. In addition, the energy consumption of searching is as low as ~12 fJ/bit/search, which can be further reduced by the decrease of activity rate thanks to the segmentation of the match line [83].

In case that no storage data can match the search word, a new word will be nucleated and propagated into the magnetic nanowire for the next round of search. Figure 40b shows the transient simulation result of the worst case: 1-bit miss process. It means that the rest 7 bits of the search word match the stored data, only one bit is different from the stored data. As shown in Fig. 40b, the search bit is "1", if no match is found, the propagation current pulse will start to drive the DW propagation, until "SL" and "Stored data" match each other. We can find the whole operation, consisting of "Pre-charge", "Propagation" and "Evaluation" phases, only requires ~2 ns. This suggests a high operating frequency up to 500 MHz, comparable to that of traditional CAM [84].

We estimate the cell area for RM-CAM with Eq. (38):

$$A_C = \frac{A_{CO} + A_{NU} + A_{PR} + N \times MAX(A_{BT} + A_{LS})}{N} \tag{38}$$

where A_{CO} denotes the area of a comparison circuit, which is ~50 F^2, A_{NU} denotes the area of a DW nucleation circuit, which is ~48 F^2, A_{PR} denotes the area of a propagation current generating circuit, which is ~7 F^2, A_{BT} is the area of every bit in racetrack memory, A_{LS} is the area of two load selecting transistors for every bit and N is the number of bits per word.

Due to the 3D integration of MTJs above CMOS circuit, only the larger one between the MTJs' area and the selecting transistors' area will be involved for calculating the full area. For this design, A_{BT} is ~6 F^2 considering 2 F between two adjacent constrictions. Coincidentally, A_{LS} is also ~6 F^2 with the minimum size. If the distance between two adjacent constrictions exceeds 2 F, only A_{BT} would be taken into account in Eq. (38). As $N = 8$, the cell area per bit is therefore ~19 F^2, which is much lower than that of SRAM-based CAM or DW-CAM [21, 22]. Meanwhile, with the increase of the bit number per word, the area of shared CMOS circuits for data comparison, DW nucleation and motion would become negligible (see Fig. 41). The cell area per bit will approach to $MAX(A_{BT} + A_{LS})$ (e.g., ~6 F^2 for this design).

With the performance analyses above, the comparison of CAMs based on different technologies is summarized. From Table 5, we can find that non-volatility of racetrack memory allows RM-CAM to eliminate the static power. DW propagation in the racetrack memory benefits for improving the search speed. Most importantly,

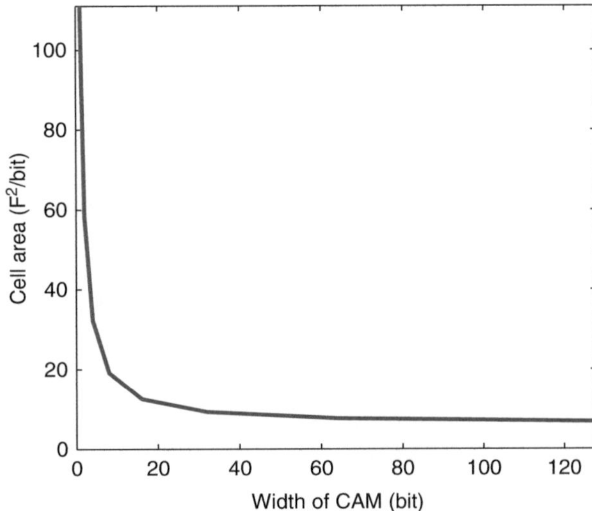

Fig. 41 Dependence of full area versus number of bits per word

Table 5 Comparison of CAMs based on different technologies

Type	SRAM based CAM [21]	DW-CAM [22]	RM-CAM
Cell area (F^2/bit)	540	~815	~19
Cycle time (ns)	2	5	~0.45
Energy (fJ/bit/search)	9.5	~30	~12
Static power	Yes	No	No

RM-CAM shows a great advantage in terms of density thanks to 3D integration and sharing of CMOS circuits (comparison circuit, DW propagation circuit and DW nucleation circuit). However, we have to mention that the cell area of DW-CAM shown in Table 5 does not consider the 3D integration. If it applies 3D integration, its cell area could reach N times that of RM-CAM. For example, if $N = 8$ for RM-CAM, the cell area of DW-CAM would be about 160 F^2/bit.

4 Conclusions and Perspectives

This chapter focused on two current-induced magnetic switching technologies for high-performance computing: the PMA STT MTJ and the racetrack memory. The work covers from theoretical study to hybrid circuit design and performance analyses. Through this work, the integration functionality of the current-induced spintronic devices based on PMA materials has been proven. The performance analyses of related hybrid logic circuits demonstrated that spintronic devices could provide various advantages compared with conventional systems, such as scalability, low switching current and high operation speed. Thanks to these, the application potential of PMA spintronic devices to achieve future low-power high-density high-speed electronic systems can be confirmed.

In details, non-volatility allows the hybrid systems to be powered off while saving the data, and then to eliminate the static power consumption. This feature can reduce greatly the overall power consumption, especially for normally-off systems. 3D integration technology can improve the system's density efficiency. Moreover, it can shorten the distance between logic and memory, which helps to save considerably the transfer energy and time. Although the switching speed of MTJ doesn't show an evident advantage compared to conventional CMOS, it is still sufficient for logic and memory application. To overcome this challenge, using of the CIDW motion is an alternative solution. Considering that the distance between two adjacent DW is 40 nm and the propagation speed can be as high as 100 m/s, switching a state by propagating DW can be as fast as 400 ps. That is why we believe the CIDW motion based racetrack memory design has a great potential for the future high-speed low-power systems.

The emergence of spintronics is to achieve more efficient and reliable applications, which could overcome the issues of mainstream charge-based electronics. The term "efficient" here concerns many factors, which involve power, density and frequency, etc. This aim is the "beacon", which indicates the direction of the progress of spintronics. On this route, the innovative technologies are appearing ceaselessly, and an emerging mechanism would be replaced by a more emerging alternative.

Along with the downscaling pace of MTJs beyond sub-volume limit (~40 nm), MTJ displays a relatively high thermal stability factor and low STT critical current. This so-called "high spin torque efficiency" is a strong stimulus for high density MTJ application. Spin torque efficiency is defined here as [85]:

$$\kappa = \frac{E}{I_{c0}} \tag{39}$$

where E is barrier height (or thermal stability factor), and I_{c0} is the average critical current. Spin torque efficiency reflects the capability of spin polarized current to reverse the barrier height. Practically, when the lateral size of MTJ scaling down to the sub-volume limit, sub-volume activation effects make the leading term guiding the magnetization switch in devices, which is negligible when lateral diameter is larger than the limit. Thanks to this effect, the scaling gain (faster operation, higher density and improved spin torque efficiency) can be further continued, which benefits greatly for the miniaturization of MTJ. As a result, the high spin torque efficiency should be considered in the future work involving the small-size MTJs and hybrid circuits.

Beyond STT, spin orbit torque (SOT) is demonstrated to be able to switch magnetization and nucleate DWs. Two main effects referred to SOT have been observed: spin hall effect (SHE) [86, 87] and Rashba effect [88] (see Fig. 42). Compared with the STT switching mechanism, these effects are exhibited with assets in terms of power, speed and reliability. For example, three-terminal devices

Fig. 42 (**a**) Three-terminal device based on the giant spin Hall effect in β-Ta/CoFeB. (**b**) Kerr micrographs showing the DW motion induced by Rashba effect [87, 88]

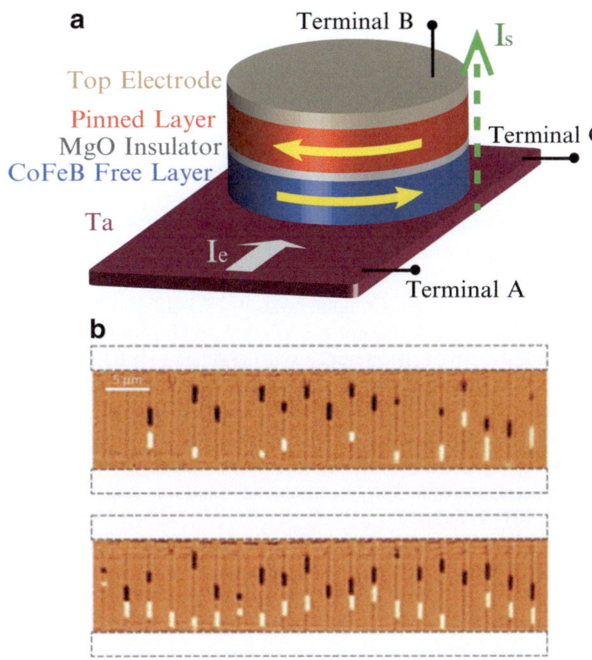

based on SHE can yield a more efficient spin torque which means to require a lower current [89, 90]. As a consequence, the power consumption can be further reduced. In addition, the current is not applied through the tunnel junction, which avoids the damage caused by the high current density. Furthermore, the separation of writing path and reading path can improve greatly the reliability performance.

References

1. International roadmap for semiconductor (ITRS) (2010)
2. C. Chappert, A. Fert, F. Nguyen Van Dau, The emergence of spin electronics in data storage. Nat. Mater. **6**, 813–823 (2007)
3. J.C. Slonczewski, Current-driven excitation of magnetic multilayers. J. Magn. Magn. Mater. **159**, L1–L7 (1996)
4. L. Berger, Emission of spin waves by a magnetic multilayer traversed by a current. Phys. Rev. B **54**, 9353–9358 (1996)
5. M. Hayashi, L. Thomas, R. Moriya, C. Rettner, S.S.P. Parkin, Current-controlled magnetic domain-wall nanowire shift register. Science **320**, 209–211 (2008)
6. S.S.P. Parkin, M. Hayashi, L. Thomas, Magnetic domain-wall racetrack memory. Science **320**, 190–194 (2008)
7. S. Ikeda et al., Recent progress of perpendicular anisotropy magnetic tunnel junctions for nonvolatile VLSI. SPIN **02**, 1240003 (2012)
8. S. Ikeda et al., A perpendicular-anisotropy CoFeB–MgO magnetic tunnel junction. Nat. Mater. **9**, 721–724 (2010)

9. D.C. Worledge et al., Spin torque switching of perpendicular Ta/CoFeB/MgO-based magnetic tunnel junctions. Appl. Phys. Lett. **98**, 022501 (2011)
10. B. Dieny et al., Giant magnetoresistive in soft ferromagnetic multilayers. Phys. Rev. B **43**, 1297–1300 (1991)
11. S. Parkin et al., Magnetically engineered spintronic sensors and memory. Proc. IEEE **91**, 661–680 (2003)
12. A. Barthélémy et al., Magnetoresistance and spin electronics. J. Magn. Magn. Mater. **242–245** (Part 1), 68–76 (2002)
13. Z. Diao et al., Spin-transfer torque switching in magnetic tunnel junctions and spin-transfer torque random access memory. J. Phys. Condens. Matter **19**, 165209 (2007)
14. D. Apalkov et al., Spin-transfer torque magnetic random access memory (STT-MRAM). J. Emerg. Technol. Comput. Syst. **9**, 13:1–13:35 (2013)
15. Y. Chen et al., A 130 nm 1.2 V/3.3 V 16 kb spin-transfer torque random access memory with nondestructive self-reference sensing scheme. IEEE J. Solid-State Circuits **47**, 560–573 (2012)
16. Y. Chen, X. Wang, H. Li, H. Xi, Y. Yan, W. Zhu, Design margin exploration of spin-transfer torque RAM (STT-RAM) in scaled technologies. IEEE Trans. Very Large Scale Integr. (VLSI) Syst. **18**, 1724–1734 (2010)
17. W. Zhao, S. Chaudhuri, C. Accoto, J. Klein, C. Chappert, P. Mazoyer, Cross-point architecture for spin-transfer torque magnetic random access memory. IEEE Trans. Nanotechnol. **11**, 907–917 (2012)
18. S. Matsunaga et al., Fabrication of a nonvolatile full adder based on logic-in-memory architecture using magnetic tunnel junctions. Appl. Phys. Exp. **1**, 091301 (2008)
19. Y. Gang, W. Zhao, J.-O. Klein, C. Chappert, P. Mazoyer, A high-reliability, low-power magnetic full adder. IEEE Trans. Magn. **47**, 4611–4616 (2011)
20. Y. Lakys, W. Zhao, J. Klein, C. Chappert, Low power, high reliability magnetic flip-flop. Electron. Lett. **46**, 1493–1494 (2010)
21. K. Pagiamtzis, A. Sheikholeslami, Content-addressable memory (CAM) circuits and architectures: a tutorial and survey. IEEE J. Solid-State Circuits **41**, 712–727 (2006)
22. R. Nebashi et al., A content addressable memory using magnetic domain wall motion cells. in *2011 Symposium on VLSI Circuits (VLSIC)* (2011), pp. 300–301
23. W. Zhao et al., High performance SOC design using magnetic logic and memory. in *VLSI-SOC: Advanced Research for Systems on Chip* (2012), pp. 10–33.
24. W.J. Gallagher, S.S.P. Parkin, Development of the magnetic tunnel junction MRAM at IBM: from first junctions to a 16-Mb MRAM demonstrator chip. IBM J. Res. Develop. **50**, 5–23 (2006)
25. S.A. Wolf et al., Spintronics: a spin-based electronics vision for the future. Science **294**, 1488–1495 (2001)
26. I.L. Prejbeanu et al., Thermally assisted MRAM. J. Phys. Condens. Matter **19**, 165218 (2007)
27. I.L. Prejbeanu et al., Thermally assisted MRAMs: ultimate scalability and logic functionalities. J. Phys. D Appl. Phys. **46**, 074002 (2013)
28. D.C. Ralph, M.D. Stiles, Spin transfer torques. J. Magn. Magn. Mater **320**, 1190–1216 (2008)
29. E.B. Myers, D.C. Ralph, J.A. Katine, R.N. Louie, R.A. Buhrman, Current-induced switching of domains in magnetic multilayer devices. Science **285**, 867–870 (1999)
30. J.A. Katine, F.J. Albert, R.A. Buhrman, E.B. Myers, D.C. Ralph, Current-driven magnetization reversal and spin-wave excitations in Co/Cu/Co pillars. Phys. Rev. Lett. **84**, 3149–3152 (2000)
31. J.Z. Sun, D.C. Ralph, Magnetoresistance and spin-transfer torque in magnetic tunnel junctions. J. Magn. Magn. Mater. **320**, 1227–1237 (2008)
32. A. Kalitsov, M. Chshiev, I. Theodonis, N. Kioussis, W.H. Butler, Spin-transfer torque in magnetic tunnel junctions. Phys. Rev. B **79**, 174416 (2009)
33. T. Kawahara, K. Ito, R. Takemura, H. Ohno, Spin-transfer torque RAM technology: review and prospect. Microelectron. Reliab. **52**, 613–627 (2012)
34. J.Z. Sun, Spin-current interaction with a monodomain magnetic body: a model study. Phys. Rev. B **62**, 570–578 (2000)

35. J. Xiao, A. Zangwill, M.D. Stiles, Macrospin models of spin transfer dynamics. Phys. Rev. B **72**, 014446 (2005)
36. G.D. Fuchs et al., Adjustable spin torque in magnetic tunnel junctions with two fixed layers. Appl. Phys. Lett. **86**, 152509 (2005)
37. J.C. Slonczewski, Currents, torques, and polarization factors in magnetic tunnel junctions. Phys. Rev. B **71**, 024411 (2005)
38. J.Z. Sun et al., Effect of subvolume excitation and spin-torque efficiency on magnetic switching. Phys. Rev. B **84**, 064413 (2011)
39. Y. Zhang et al., Compact modeling of perpendicular-anisotropy CoFeB/MgO magnetic tunnel junctions. IEEE Trans. Electron Devices **59**, 819–826 (2012)
40. R.H. Koch, J.A. Katine, J.Z. Sun, Time-resolved reversal of spin-transfer switching in a nanomagnet. Phys. Rev. Lett. **92**, 088302 (2004)
41. R. Heindl, W.H. Rippard, S.E. Russek, M.R. Pufall, A.B. Kos, Validity of the thermal activation model for spin-transfer torque switching in magnetic tunnel junctions. J. Appl. Phys. **109**, 073910 (2011)
42. L.-B. Faber, W. Zhao, J.-O. Klein, T. Devolder, C. Chappert, Dynamic compact model of spin-transfer torque based magnetic tunnel junction (MTJ). in *4th International Conference on Design Technology of Integrated Systems in Nanoscal Era*, 2009. DTIS'09 130–135 (2009)
43. Y. Zhang et al., Electrical modeling of stochastic spin transfer torque writing in magnetic tunnel junctions for memory and logic applications. IEEE Trans. Magn. **49**, 4375–4378 (2013)
44. T. Devolder et al., Single-shot time-resolved measurements of nanosecond-scale spin-transfer induced switching: stochastic versus deterministic aspects. Phys. Rev. Lett. **100**, 057206 (2008)
45. J.J. Nowak et al., Demonstration of ultralow bit error rates for spin-torque magnetic random-access memory with perpendicular magnetic anisotropy. IEEE Magn. Lett. **2**, 3000204 (2011)
46. Z. Wang, Y. Zhou, J. Zhang, Y. Huai, Bit error rate investigation of spin-transfer-switched magnetic tunnel junctions. Appl. Phys. Lett. **101**, 142406 (2012)
47. H. Tomita et al., Unified understanding of both thermally assisted and precessional spin-transfer switching in perpendicularly magnetized giant magnetoresistive nanopillars. Appl. Phys. Lett. **102**, 042409 (2013)
48. K. Lee, S.H. Kang, Development of embedded STT-MRAM for mobile system-on-chips. IEEE Trans. Magn. **47**, 131–136 (2011)
49. Y. Kim et al., Integration of 28nm MJT for 8~16Gb level MRAM with full investigation of thermal stability. in *2011 Symposium on VLSI Technology (VLSIT)* (2011), pp. 210–211
50. N. Nishimura et al., Magnetic tunnel junction device with perpendicular magnetization films for high-density magnetic random access memory. J. Appl. Phys. **91**, 5246–5249 (2002)
51. M. Gajek et al., Spin torque switching of 20 nm magnetic tunnel junctions with perpendicular anisotropy. Appl. Phys. Lett. **100**, 132408 (2012)
52. D.C. Worledge et al., Recent advances in spin torque MRAM. in *Memory Workshop (IMW)*, 2012 4th IEEE International (2012), pp. 1–3
53. D. Ravelosona et al., Domain wall creation in nanostructures driven by a spin-polarized current. Phys. Rev. Lett. **96**, 186604 (2006)
54. S. Mangin et al., Current-induced magnetization reversal in nanopillars with perpendicular anisotropy. Nat. Mater. **5**, 210–215 (2006)
55. Y. Zhang et al., Perpendicular-magnetic-anisotropy CoFeB racetrack memory. J. Appl. Phys. **111**, 093925 (2012)
56. A.J. Annunziata et al., Racetrack memory cell array with integrated magnetic tunnel junction readout. in *Electron Devices Meeting (IEDM)*, 2011 I.E. International (2011), pp. 24.3.1–24.3.4
57. Y. Zhang et al., Current induced perpendicular-magnetic-anisotropy racetrack memory with magnetic field assistance. Appl. Phys. Lett. **104**, 032409 (2014)
58. Y. Zhang et al., Implementation of magnetic field assistance to current-induced perpendicular-magnetic-anisotropy racetrack memory. J. Appl. Phys. **115**, 17D509 (2014)

59. W. Zhao, C. Chappert, V. Javerliac, J.-P. Noziere, High speed, high stability and low power sensing amplifier for MTJ/CMOS hybrid logic circuits. IEEE Trans. Magn. **45**, 3784–3787 (2009)
60. A. Thiaville, Y. Nakatani, J. Miltat, Y. Suzuki, Micromagnetic understanding of current-driven domain wall motion in patterned nanowires. EPL **69**, 990 (2005)
61. S. Fukami et al., Current-induced domain wall motion in perpendicularly magnetized CoFeB nanowire. Appl. Phys. Lett. **98**, 082504 (2011)
62. H. Tanigawa et al., Domain wall motion induced by electric current in a perpendicularly magnetized Co/Ni nano-wire. Appl. Phys. Exp. **2**, 053002 (2009)
63. X. Jiang et al., Enhanced stochasticity of domain wall motion in magnetic racetracks due to dynamic pinning. Nat. Commun. **1**, 25 (2010)
64. W. Kang et al., High reliability sensing circuit for deep submicron spin transfer torque magnetic random access memory. Electron. Lett. **49**, 1283–1285 (2013)
65. W. Kang et al., Variation-tolerant and disturbance-free sensing circuit for deep nanometer STT-MRAM. IEEE Trans. Nanotechnol. **13**, 1088–1092 (2014)
66. W. Kang et al., Variation-tolerant high-reliability sensing scheme for deep submicrometer STT-MRAM. IEEE Trans. Magn. **50**, 1–4 (2014)
67. W. Kang et al., Separated pre-charge sensing amplifier for deep submicron MTJ/CMOS hybrid logic circuits. IEEE Trans. Magn. **6**, 3400305–5 (2014)
68. CMOS065 Design Rule Manual, STMicroelectronics, Geneva, Switzerland (2010)
69. W. Zhao et al., Failure and reliability analysis of STT-MRAM. Microelectron. Reliab. **52**, 1848–1852 (2012)
70. W. Kang et al., A low-cost built-in error correction circuit design for STT-MRAM reliability improvement. Microelectron. Reliab. **53**, 1224–1229 (2013)
71. W. Kang et al., DFSTT-MRAM: dual functional STT-MRAM cell structure for reliability enhancement and 3D MLC functionality. IEEE Trans. Magn. **6**, 3400207 (2014)
72. W. Kang et al., A radiation hardened hybrid spintronic/CMOS nonvolatile unit using magnetic tunnel junctions. J. Phys. D Appl. Phys. **47**, 405003 (2014)
73. Y. Lakys, W. Zhao, T. Devolder, Y. Zhang, J. Klein, D. Ravelosona, C. Chappert, Self-enabled "error-free" switching circuit for spin transfer torque MRAM and logic. IEEE Trans. Magn. **48**, 2403–2406 (2012)
74. E. Deng et al., Low power magnetic full-adder based on spin transfer torque MRAM. IEEE Trans. Magn. **49**, 4982–4987 (2013)
75. H.-P. Trinh et al., Magnetic adder based on racetrack memory. IEEE Trans. Circuits Syst. I: Regular Papers **60**, 1469–1477 (2013)
76. W. Kim et al., Extended scalability of perpendicular STT-MRAM towards sub-20 nm MTJ node. in *Electron Devices Meeting (IEDM)*, 2011 I.E. International (2011), 24.1.1–24.1.4
77. H. Yoda et al., Progress of STT-MRAM technology and the effect on normally-off computing systems. in *Electron Devices Meeting (IEDM)*, 2012 I.E. International (2012), pp. 11.3.1–11.3.4
78. D.A. Patterson, J.L. Hennessy, *Computer Organization and Design: The Hardware/Software Interface* (Elsevier, Amsterdam, 2012)
79. F. Ren, D. Markovic, True energy-performance analysis of the MTJ-based logic-in-memory architecture (1-bit full adder). IEEE Trans. Electron Devices **57**, 1023–1028 (2010)
80. N.S. Kim et al., Leakage current: Moore's law meets static power. Computer **36**, 68–75 (2003)
81. Y. Zhang, W. Zhao, J.-O. Klein, D. Ravelsona, C. Chappert, Ultra-high density content addressable memory based on current induced domain wall motion in magnetic track. IEEE Trans. Magn. **48**, 3219–3222 (2012)
82. W. Zhao, D. Ravelosona, J. Klein, C. Chappert, Domain wall shift register-based reconfigurable logic. IEEE Trans. Magn. **47**, 2966–2969 (2011)
83. S. Matsunaga et al., Fully parallel 6T-2MTJ nonvolatile TCAM with single-transistor-based self match-line discharge control. in *2011 Symposium on VLSI Circuits (VLSIC)* (2011), pp. 298–299.

84. H. Kadota, J. Miyake, Y. Nishimichi, H. Kudoh, K. Kagawa, An 8-kbit content-addressable and reentrant memory. IEEE J. Solid-State Circuits **20**, 951–957 (1985)
85. J.Z. Sun et al., Spin-torque switching efficiency in CoFeB-MgO based tunnel junctions. Phys. Rev. B **88**, 104426 (2013)
86. J.E. Hirsch, Spin hall effect. Phys. Rev. Lett. **83**, 1834–1837 (1999)
87. L. Liu et al., Spin-torque switching with the giant spin hall effect of tantalum. Science **336**, 555–558 (2012)
88. I.M. Miron et al., Fast current-induced domain-wall motion controlled by the Rashba effect. Nat. Mater. **10**, 419–423 (2011)
89. W. Kang et al., An overview of spin-based integrated circuits. in *Asia and South Pacific Design Automation Conference (ASP-DAC)* (2014), pp. 676–683
90. Z.H. Wang et al., Perpendicular-anisotropy magnetic tunnel junction switched by spin-hall-assisted spin-transfer torque. J. Phys. D Appl. Phys. **48**, 065001 (2015)

Electric Control of Magnetic Devices for Spintronic Computing

Jianshi Tang, Qiming Shao, Pramey Upadhyaya, Pedram Khalili Amiri, and Kang L. Wang

1 Introduction

The scaling of CMOS technology has followed Moore's Law for more than five decades [1], enabling the continuous growth of the semiconductor industry. This trend, however, currently faces a major challenge due to increasing energy dissipation per unit area [2]. The increase in power dissipation results from the increase of static (standby) leakage power, as well as the continued increase of density as transistors are scaled down [3]. The former is a result of the fact that power needs to be continuously applied to CMOS elements in order for them to retain their information. In addition, the dynamic switching energy per unit area has also been increasing due to the increase of device density. The search for novel low-dissipation solutions at the device-, circuit- and system-levels is thus critical to the future of the electronics industry.

Spintronic devices, i.e., those utilizing the interactions of charge and spin at the nanoscale may offer a solution to this challenge [4]. This is in particular a result of their inherent nonvolatility, which allows for power to be removed without loss of information, hence potentially eliminating the standby power issue. Spintronics has been proposed both as a complement to CMOS to realize hybrid CMOS-magnetic logic, as well as a means to realize various proposals for fully magnetic, beyond CMOS computing [5–14]. For any of these proposed approaches to magnetic nonvolatile logic (NVL) to be applicable in practice, however, low dynamic power dissipation is the key requirement.

In most state-of-the-art device concepts, magnetic nano-elements are currently manipulated via spin transfer torque (STT), where passing of direct currents creates magnetization switching or oscillations. The lowest energy of STT-induced

J. Tang • Q. Shao • P. Upadhyaya • P.K. Amiri • K.L. Wang (✉)
Department of Electrical Engineering, Device Research Laboratory, University of California, Los Angeles, CA 90095, USA
e-mail: wang@ee.ucla.edu

© Springer International Publishing Switzerland 2015
W. Zhao, G. Prenat (eds.), *Spintronics-based Computing*,
DOI 10.1007/978-3-319-15180-9_2

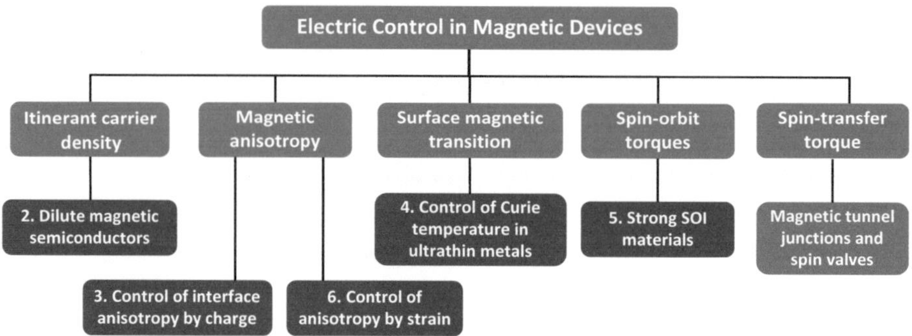

Fig. 1 Various approaches of electric control in magnetic devices. The topics in *violet boxes* will be covered in this chapter

switching, however, is typically on the order of ~100 fJ, which is still more than 100 times larger than that of a scaled CMOS transistor (at technology nodes below 32 nm). Hence, it is clear that while standby power dissipation may be reduced or even eliminated by spintronics, dynamic power dissipation is not, making most STT-based logic concepts unattractive from an overall energy perspective. As a result, there is a fast-growing interest in beyond-STT mechanisms to control magnetization, which may allow for magnetization control with energy budgets of less than 1 fJ per operation.

Figure 1 summarizes several of the main emerging contenders for electric control of magnetization in spintronic devices, which will be discussed in the following sections. Dilute magnetic semiconductors are one example, where magnetization can be controlled by electrically manipulating the itinerant carrier density. Since this effect can be achieved by electric fields, it is in principle much lower in power dissipation than current-controlled mechanisms, and may be used to create devices such as spin-based field-effect transistors (spin FETs), as will be discussed in Sect. 2. Alternatively, voltage control to eliminate Ohmic energy dissipation is also possible in ultrathin ferromagnetic metals. In particular, magnetic anisotropy can be controlled in this manner by modifying the occupancy of atomic orbitals at interfaces of ferromagnetic metal thin films, as will be described in Sect. 3, or using mechanical strain, which will be discussed in Sect. 6. These effects have been recently used to demonstrate ultralow-power switching of magnetoelectric memory elements by voltage, and may provide a pathway beyond the current STT memory and logic to improve energy efficiency and density. It has also been demonstrated that ferromagnetic phase transition can be controlled in such thin metal films by a gate voltage. This effect, although still fairly small to be of practical use in present material systems, will be discussed in Sect. 4. Finally, new types of current-induced torques—generally referred to as spin-orbit torques (SOTs)—can be created using non-magnetic materials with large spin-orbit interaction, including heavy metals and topological insulators. These effects have generated much attention due to their potential to dramatically lower the current densities required for device operation compared to existing STT devices. These types of SOT effects and their applications to the development of new spintronic devices will be the focus of Sect. 5. Finally, Sect. 6 will discuss the perspectives of emerging applications enabled by the spintronic elements discussed in this chapter, and will go over a few of these applications in further detail.

2 Electric Field Control of Ferromagnetism in Dilute Magnetic Semiconductors for Novel SpinFETs

2.1 Dilute Magnetic Semiconductors with Electric Field Controlled Ferromagnetism

In the past two decades, the search for high-temperature dilute magnetic semiconductors (DMS) has attracted tremendous efforts in both theoretical calculations and experimental implementations [15, 16]. The idea of making nonmagnetic semiconductors ferromagnetic, mainly through introducing magnetic dopants into the semiconductor lattice, provides great opportunities to make use of both charge and spin of electrons in semiconductors for novel spintronic devices [17]. For example, a variety of low-temperature functional devices have been demonstrated using DMS materials to realize electrical spin injection into nonmagnetic semiconductor [18], control of magnetization [19, 20], tunneling anisotropic magnetoresistance [21], current-induced domain wall switching [22], etc. The early study of DMS was mainly focused on the $Mn_xIn_{1-x}As$ and $Mn_xGa_{1-x}As$ thin films [23, 24], and then the DMS family quickly extends to a broad range of transition mental doped III–V [17], II–VI [25], and group IV semiconductors [26], as shown in Fig. 2a. Despite the fact that the origin of ferromagnetism in DMS is still under debate, one widely accepted picture is the Ruderman-Kittel-Kasuya-Yosida (RKKY) interaction between the itinerant carriers (mostly holes) and the magnetic impurities (such as Mn) helps align their magnetic moments along one direction and leads to a long-range ferromagnetic ordering [15, 16]. This implies the unique feature of DMS materials: their ferromagnetism is mediated by itinerant carriers, and hence their ferromagnetic property can be controlled through a gate electrode to modulate the carrier density [27, 28]. Experimentally, as shown in Fig. 2b, c, the electric field control of ferromagnetism has been demonstrated in $Mn_xIn_{1-x}As$ DMS thin film at low temperature [28], which is limited by the low Curie temperature of $Mn_xIn_{1-x}As$ itself.

For room-temperature spintronics applications, DMS materials with high Curie temperature T_c above 300 K are desired. The mean-field p-d Zener model formulated by Dietl et al. has predicted various Mn-doped DMS semiconductors with high T_c [29], as shown in Fig. 2a. However, it should be pointed out that there is apparent discrepancy between theoretical predications and experimental results of DMS materials, especially in the Curie temperature and the magnetic moment [15]. This could be mainly attributed to the non-ideal DMS material quality with

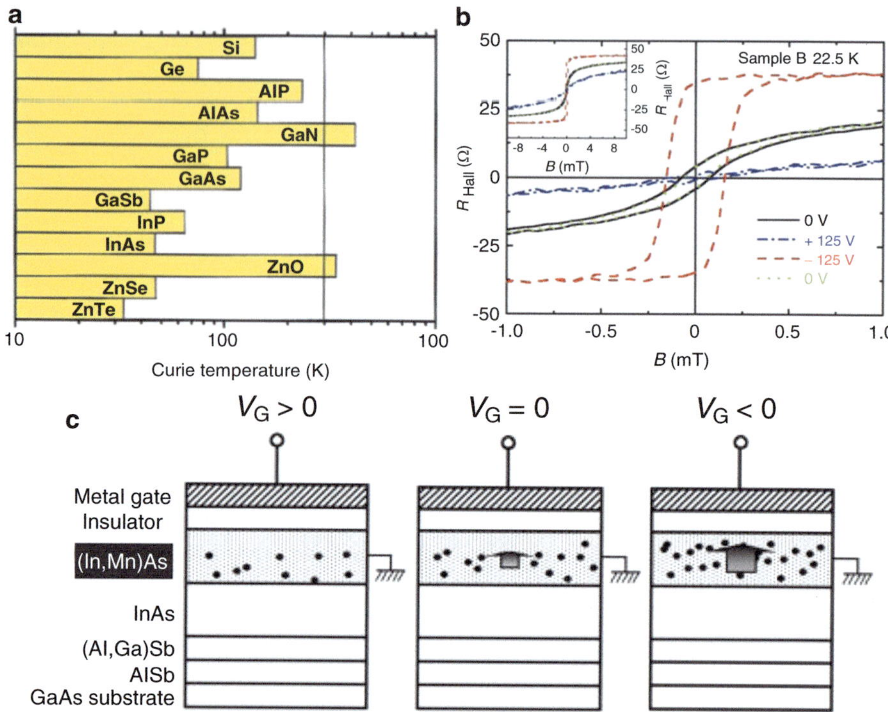

Fig. 2 Dilute magnetic semiconductors. (**a**) Curie temperature of various Mn-doped DMS semiconductors predicted by the mean-field *p-d* Zener model, assuming a Mn doping concentration of 5 % and hole density of 3.5×10^{20} cm^{-3}. (**b**) Anomalous Hall resistance of the Mn$_x$In$_{1-x}$As DMS thin film at different gate voltages at $T = 22.5$ K (**c**) Schematic illustration of the gate control of the hole-induced ferromagnetism in the Mn$_x$In$_{1-x}$As. (**a**) From Ref. [29]. Reprinted with permission from AAAS. (**b–c**) Reprinted by permission from Macmillan Publishers Ltd: Nature Ref. [28], copyright 2000

substantial crystal defects. According to the Zener model, one effective approach to improve the T_c of DMS is to increase the substitutional magnetic doping density, which provides both localized magnetic moments and itinerant holes that mediate the ferromagnetic ordering. However, such a strategy could easily lead to the formation of ferromagnetic precipitates or secondary phases [30], because of the low solubility of magnetic impurities in semiconductors.

To circumvent the defects problem, various growth techniques and device structures were employed. One effective approach is the growth of DMS nanostructures (quantum dots, nanowires, etc.) using molecular beam epitaxy (MBE). The non-equilibrium technique of MBE growth allows us to incorporate high density of magnetic impurities in the semiconductor matrix beyond the solubility limit. Moreover, the use of nanostructures helps minimize the defect-induced formation of secondary phases, and furthermore, their substantial quantum confinement of carriers enhances the carrier-mediated ferromagnetism through RKKY interaction, facilitating electric field controlled devices. Therefore, using

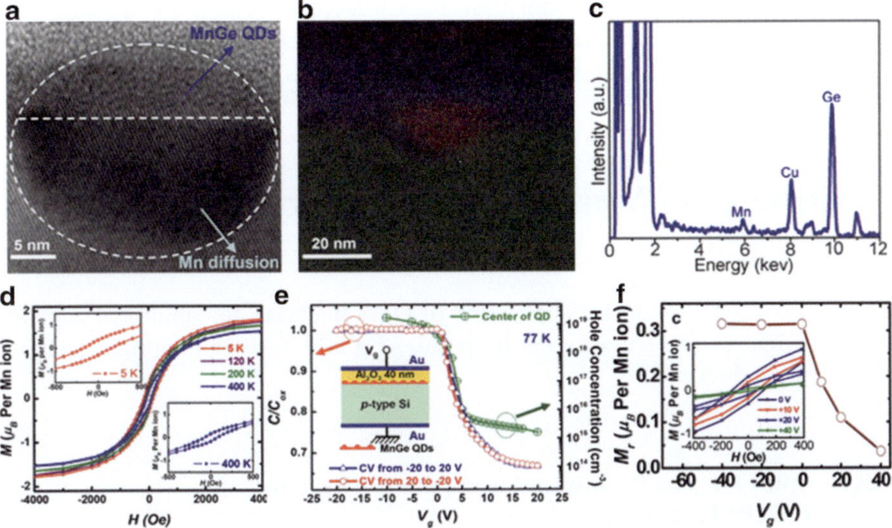

Fig. 3 Electric field control of ferromagnetism in $Mn_{0.05}Ge_{0.95}$ DMS quantum dots. (**a**) High-resolution transmission electron microscopy (HRTEM) image of a $Mn_{0.05}Ge_{0.95}$ quantum dot showing no apparent lattice defects. (**b**) EELS composition mapping of the Mn distribution. Mn is found to distribute uniformly in the $Mn_{0.05}Ge_{0.95}$ quantum dot and the Mn-diffused Si area. (**c**) EDAX composition spectrum showing that both Mn and Ge are present in $Mn_{0.05}Ge_{0.95}$ quantum dot. (**d**) Magnetic hysteresis loops measured at different temperatures from 5 K to 400 K. (**e**) C-V curves measured at 77 K with a frequency of 100 kHz and the simulated gate-dependent hole concentration. It clearly shows a transition between hole accumulation at negative bias and hole depletion at positive bias. The inset is a schematic drawing of the MOS device. (**f**) A representation of remnant moments with respect to the gate bias. Inset is a zoom-in view to clearly show the change of remnant moment. Reprinted by permission from Macmillan Publishers Ltd: Nature Materials Ref. [27], copyright 2010

this strategy, it is likely to produce high-quality room-temperature DMS nanostructures with T_c beyond theoretical prediction. Indeed, it has been experimentally demonstrated the MBE growth of single-crystalline Mn-doped Ge quantum dots and also nanowires with T_c above 400 K [27, 31]. At the same time, the electric field control of ferromagnetism in Mn-doped Ge quantum dots has been demonstrated up to 300 K, as shown in Fig. 3. Similar behavior have also recently been demonstrated in Mn-doped Ge nanowires grown by MBE [31]. These results point out a unique advantage of using DMS nanostructures over thin films and bulk materials in fabricating practical spintronic devices that can operate at room temperature.

As discussed above, one of the appealing features of DMS materials is that their ferromagnetism is mediated by itinerant carriers through exchange interaction with localized spins of magnetic impurities [29], and hence can be modulated through a gate electrode [27, 28]. This is of particular interest for low-power logic and memory applications, because it provides the possibility of building energy-efficient magnetic devices with the control of ferromagnetism using an electric field rather than an electric charge current. Indeed, it has been recently proposed to build a nonvolatile spin transistor (transpinor) device based on a DMS nanowire that can minimize the charge current flow and hence reduce the power dissipation [32, 33], which will be elaborated in details in the following.

2.2 *DMS-Based Nonvolatile Transpinor for Spintronic Computing*

Before discussing the device application of DMS materials for spintronic computing, let's first revisit several earlier proposals of spinFETs as an important category of spintronic devices.

2.2.1 SpinFET as a Promising Candidate for Beyond CMOS Technology

As discussed in the Introduction section, the aggressive scaling in modern Si technology is approaching the ultimate physical limit soon. Several critical challenges are highlighted in the *International Technology Roadmap of Semiconductors* [34], including the need to decrease power dissipation and manufacturing variability, as well as the increase in functionality. First of all, there is troublesome energy dissipation from static leakage due to the fact that the CMOS state is volatile and power supply must be on all the time. Besides, the voltage cannot be scaled down further due to the finite threshold voltage V_{th} (limited by finite subthreshold swing). Novel physics, materials and devices are in urgent need to resolve those critical challenges in the near future. Spintronics have emerged as a promising solution by utilizing the spin of electrons as another degree of freedom in devices for information processing. It could enable advanced electronic devices with nonvolatility and low threshold voltage to achieve low power dissipation along with increased functionalities [4, 35]. In particular, a variety of spin-based transistors have been proposed and extensively studied as an appealing substitute for Si CMOS devices, such as bipolar spin switch [36], spin-valve transistor [37, 38], unipolar spin transistor [39], magnetic bipolar transistor [40, 41], spin gain transistor [42], and other spinFETs [14, 32, 33, 43, 44]. SpinFETs are of particular interest for low-power logic applications because of the nonvolatility nature and additional control of current from the ferromagnetic state other than the gate electrode. A typical spinFET is composed of two ferromagnetic contacts on a semiconductor channel, in which the transistor's current drivability is controlled by the magnetization orientation of the two ferromagnetic contacts. The operation of spinFETs typically involves the injection, manipulation and detection of electron spins in the spin/charge transport process.

In 1985, Johnson and Silsbee were the first to successfully demonstrate electrical spin injection and detection from a ferromagnetic metal (permalloy) into a

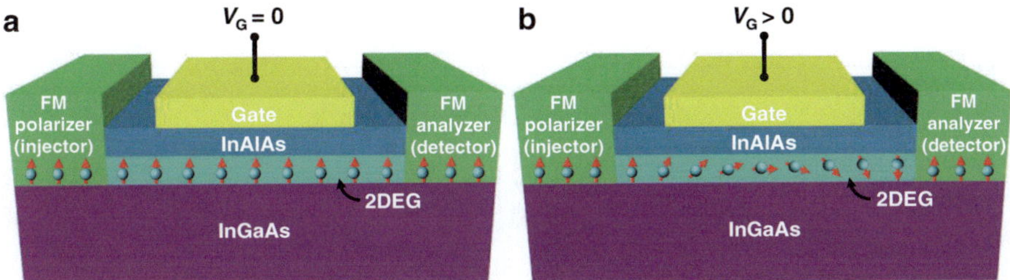

Fig. 4 Datta-Das type spin-polarized FET. A prototype of the spin-polarized FET can be built on a high-mobility InAlAs/InGaAs heterostructure transistor with a two-dimensional electron gas (2DEG) channel and ferromagnetic source/drain contacts. (**a**) ON state under zero gate voltage; (**b**) OFF state under a positive gate voltage that induces the electron spin precession through the Rashba effect, causing the electron spin is rotated by 180°. This device uses both charge and spin currents and the V_{th} is similar to those of CMOS and other 2DEG devices. Reprinted with permission from Ref. [43]. Copyright 1990, AIP Publishing LLC

nonmagnetic metal (Al) [45]. Johnson then extended the spin injection structure to develop an all-metal three-terminal bipolar spin switch [36], consisting of two ferromagnetic metals (F1 and F2) sandwiching one paramagnetic metal (P). The spin state of the injected carriers in P depends on the magnetization of F1 and the detected voltage is determined by the magnetization orientation of F1 and F2, which is controlled by an external magnetic field. This type of all-metal spintronic devices can take advantage of the nonvolatility of magnetism to reduce the static power dissipation; and at the same time to improve variability. However, the issue of this bipolar spin switch is that the device operation yields only a small change of the output voltage with no signal gain due to its all-metal construction.

Other than the all-metal construction, several proposals of spinFETs have been attempted in semiconductor structures to exploit the spin-dependent transport of charge carriers with a high spin-current gain [14, 42–44]. The manipulation of single spins suffers from having both charge and spin currents and the associated power dissipation. For example, Nikonov and Bourianoff devised a spin gain transistor based on ferromagnetic semiconductor, whose spontaneous magnetization is initiated by a small spin-polarized current, generating a spin gain of more than 1,000 at the output [42]. In 1990, Datta and Das proposed a spin-polarized FET [43], an electronic analog to the electro-optic modulator, as depicted in Fig. 4. In this Datta-Das type spinFET, spin-polarized electrons are injected into a semiconductor channel from one ferromagnetic contact (spin injector or polarizer), then the electron spin precession is modulated by the gate voltage through the Rashba effect [46], and is finally probed by the other ferromagnetic contact (spin detector or analyzer). Here the electron transmission coefficient through the spin detector is determined by the relative orientation between the electron spin after precession and the magnetization of the spin detector. The gate controlled electron spin precession has been experimentally observed in an InAs high-electron mobility transistor (HEMT) [47]; however, the full functionality of spin-polarized FET has not been experimentally verified [48]. It should be noted that, while semiconductors with

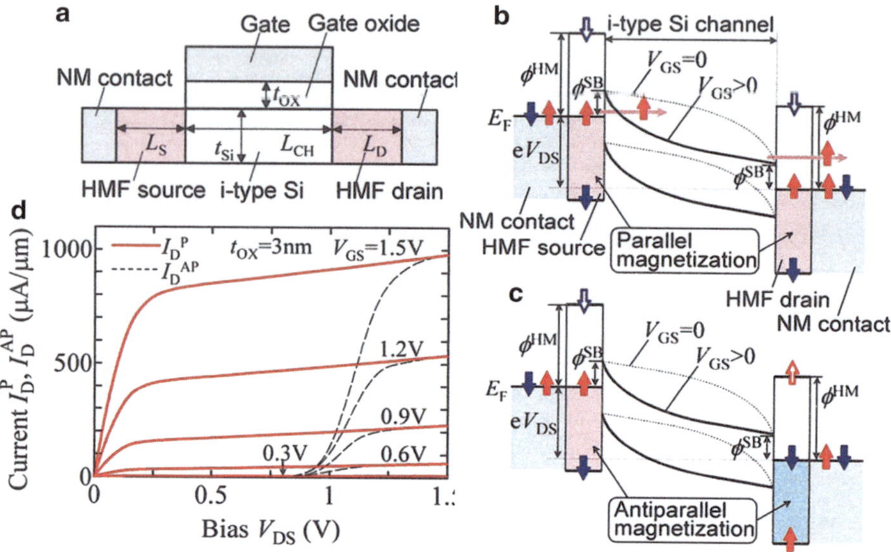

Fig. 5 Sugahara-Tanaka type spin-MOSFET. (**a**) Schematic device structure of the spin-MOSFET, composed of an ordinary MOSFET with half-metal source/drain contacts. (**b, c**) Energy band diagrams of the spin-MOSFET in the parallel and anti-parallel magnetic configurations, respectively. (**d**) Simulated output characteristics of the spin-MOSFET in the parallel and anti-parallel configurations. Reprinted with permission from Ref. [44]. Copyright 2004, AIP Publishing LLC

strong SOI are usually favored for making spin-polarized FET in order to have an enhanced Rashba effect, the strong SOI could also lead to dramatic degradation in the spin lifetime and spin diffusion length. Therefore, in practice, the semiconductor channel material should be carefully selected for a trade-off between the two.

Another variant of using single spins plus metallic ferromagnets shown in Fig. 5, proposed by Sugahara and Tanaka [44], is a spin-MOSFET comprised of an ordinary MOSFET with half metal as ferromagnetic source/drain contacts. The gate electrode modulates the Schottky barrier shape at the source/drain junctions, and hence the current. In the parallel magnetic configuration, the spin-MOSFET operates like an ordinary Schottky barrier MOSFET; while in the anti-parallel magnetic configuration, the half-metal contacts present another barrier for the carrier transport from the source to drain. The barrier filters carriers with an opposite spin direction to the ferromagnetic drain contact, and it can be overcome by a high drain bias. The resulting output characteristic in the anti-parallel magnetic configuration shows a threshold voltage in the drain bias V_{DS}, compared with that in the parallel configuration. Despite of progressive simulation work based on the spin-MOSFET, the full functionality of spin-MOSFET has not been experimentally demonstrated so far [49–52]. The growth of stable half metals with high Curie temperatures and the fabrication of a high-quality half metal/semiconductor interface to achieve effective spin injection remains a big technological challenge.

The power issues of these devices are similar to those of scaled CMOS with V_{th}, and thus limited by finite V_{dd}. Likewise, the variability is the same as CMOS.

As we can see, both the spin-polarized FET and the spin-MOSFET modulate single spin of individual electron, and hence they have the similar V_{th} limitation and are less energy efficient compared with the manipulation of collective spins [53], as to be discussed later. Also, the electron spin is mainly used as another handle to modulate the current drivability in the above proposed devices, which are expected to be governed by the same scaling limit of CMOS [34]. All these spinFETs still use electronic charge current along with spin transport that inevitably consumes power through both transport mechanisms. As a result, the potential advantage in reducing the active power dissipation is very limited compared with conventional charge-based CMOS devices [54].

2.2.2 DMS-Based SpinFET: Nonvolatile Transpinor

In order to make a spinFET that can truly meet the challenges and outperform scaled Si CMOS, one practical approach is to reduce or eliminate the charge current flow. Another important issue is to operate spin at room temperature, by which magnetic devices (or collective behavior of spins) must be invoked. A novel transpinor has been recently proposed using DMS nanostructures whose ferromagnetism can be controlled by an external electric field [32, 33]. The use of DMS may provide a promising solution to minimize the charge current flow. It allows us to manipulate the magnetic state of the DMS channel through a voltage signal without current flow in the channel (see Fig. 6). In this transpinor device, the magnetic moments are transferred from the source to the channel, and then to the drain through spin injection and dipole or exchange interaction, respectively. As the gate electrode manipulates the ferromagnetism of the DMS channel, it controls the communication between the source and drain. Based on the recent progress in the Ge-based DMS material growth and spin transport studies [32, 33], two possible implementations are schematically illustrated in Fig. 6a, b with Mn_5Ge_3 and Fe/MgO as source/drain contacts, respectively.

Unlike the Datta-Das and Sugahara-Tanaka types of spinFET relying on the control of the spin of individual electron, the proposed transpinor rather manipulates a collection of spins via the carrier-mediated paramagnetic-to-ferromagnetic transition as a single identity and thus is more energy-efficient and robust [42]. The use of N collective spins can be treated as a single identity and will enable the ultimate power dissipation of a single switching element (nanomagnet) of $k_B T \ln 2$, instead of $N k_B T \ln 2$ per switch for conventional equilibrium computing [53]. Moreover, the input and output information can be stored in the ferromagnetic contacts (nanomagnets); therefore, it inherently provides nonvolatility in the transpinor and hence eliminates the standby power dissipation—one major issue of scaled CMOS.

Compared with conventional CMOS devices, the transpinor could provide several important advantages, including a low V_{th} and nonvolatility (the information is stored as the magnetization of nanomagnets), to achieve lower power

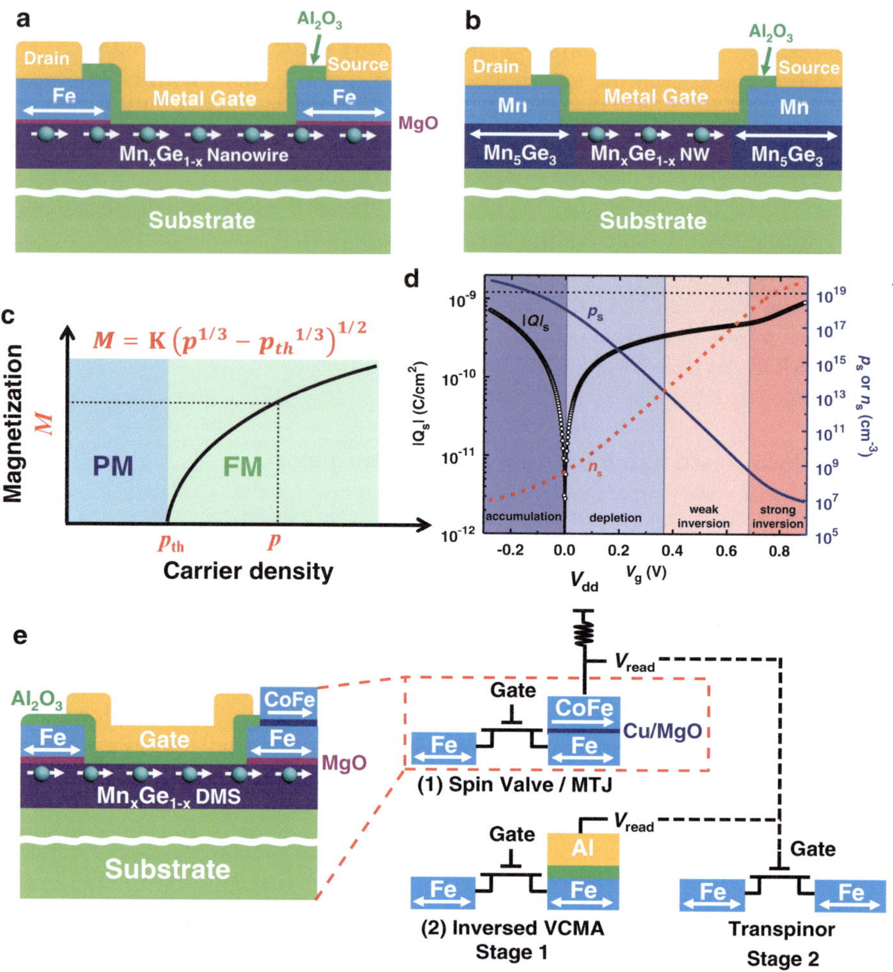

Fig. 6 DMS nanowire-based nonvolatile transpinor. Two possible implementations of the proposed diluted magnetic Ge nanowire-based nonvolatile transpinor with (**a**) Fe/MgO tunnel junctions, and (**b**) ferromagnetic Mn$_5$Ge$_3$ Schottky junctions, respectively. The magnetic moments are transferred from the source to the channel, and then to the drain with a spin gain. The Mn$_x$Ge$_{1-x}$ DMS nanowire channel, whose paramagnetism-to-ferromagnetism phase transition is electrically controlled by the gate voltage, communicates with the input and the output nanomagnets through spin injection/extraction and dipole/exchange interaction, respectively. (**c**) Schematic of the paramagnetism-to-ferromagnetism phase transition. (**d**) Simulated carrier densities (Q_s is the sheet charge density, p_s is the surface hole density, and n_s is the surface electron density) in a Ge MOS structure with $N_A = 10^{18}$ cm^{-3}, $V_{FB} = 0$ V, and $t_{ox} = 2$ nm. (**e**) Two possible schemes to convert the magnetization to a voltage signal in read out of a transpinor circuit. The devices can be cascaded, very much like CMOS, but in the magnetic state. The first scheme adopts either a metallic CoFe/Cu/Fe(Mn$_5$Ge$_3$) spin valve or a CoFe/MgO/Fe(Mn$_5$Ge$_3$) MTJ connected to the power supply V_{dd} through a resistor. The other scheme uses the inverse VCMA effect, in which the switching in the output magnet will induce a voltage pulse in the read out. Reproduced with permission from ECS Trans., **64**, 613 (2014). Copyright 2014, The Electrochemical Society

dissipation (minimize charge current flow), along with added functionalities (spin freedom and gate-controlled ferromagnetism), and faster switching (critical behavior in ferromagnetic phase transition). To see this more clearly, let's first take a quantitative analysis of the gate modulation in the DMS-based transpinor device. For simplicity, according to the Zener model, the Curie temperature of the Mn_xGe_{1-x} DMS nanowire can be approximately described as [29]:

$$T_c = A\,p^{1/3} \tag{1}$$

Here p is the itinerant hole density. On the other hand, near the critical temperature T_c of the ferromagnetic phase transition, the critical behavior of the magnetization is governed by a power-law relation [55]:

$$M(T) = B(T_c - T)^{1/2} \tag{2}$$

Here A and B are material-related constants. Equation (2) can be re-written as:

$$M(p) = B\left(A\,p^{1/3} - T\right)^{1/2} = K\left(p^{1/3} - p_{th}^{1/3}\right)^{1/2} \tag{3}$$

where $p_{th} = (T/A)^3$, and $K = BA^{1/2}$, as plotted in Fig. 6c. In the fabrication of transpinor, one can design the background doping density in the vicinity of the phase transition point p_{th} (the minimum carrier density to mediate long-range ferromagnetism, $p_{th} \approx 10^{18}$–10^{19} cm^{-3} for Mn_xGe_{1-x} DMS [27]). As the gate voltage is swept to increase p to reach p_{th}, the paramagnetism-to-ferromagnetism phase transition occurs abruptly, and the resulted ferromagnetic DMS channel leads to a simultaneous switch of the output nanomagnet. In this process, the change in the gate voltage could be very small (less than 0.15 V as shown below, and there is no theoretical minimum); therefore, the active power dissipation can be minimized and the switching speed can be fast.

In the conventional scaled CMOS, the required gate voltage swing to turn the device on and off is limited by the subthreshold swing of $SS \approx 60$ mV/dec at room temperature. To achieve a current ON/OFF ratio above 10^4, a minimum gate voltage swing is estimated to be about 0.25 V, which limits the voltage scaling in CMOS. On the contrary, the operation of the transpinor relies on the modulation of the magnetic moment rather than the charge current flow in the DMS channel; therefore, the gate voltage swing could be much smaller. To estimate the gate voltage swing for the transpinor, Fig. 6d simulates the gate-dependent carrier densities in a Ge MOS structure with a p-type doping of $N_A = 10^{18}$ cm^{-3}, flat band voltage of $V_{FB} = 0$ V, and 2 nm-thick Al_2O_3 as the gate dielectric. In the accumulation regime, the gate voltage is only changed by less than 0.15 V to increase the carrier density from $p_s = 10^{18}$ cm^{-3} to reach $p_{th} \approx 10^{19}$ cm^{-3}, and the swing can be further reduced by using other high-κ gate dielectrics with a higher dielectric constant (such as HfO_2). Thus a much lower V_{th} can be achieved.

It should be pointed out that one of the key challenges to integrate spintronic devices into CMOS circuits is to achieve efficient conversion between a spin signal (magnetization) and a charge signal (current/voltage), and the concatenability (device fan out). Similarly, in order to cascade the transpinor devices for logic computation, several strategies can be adopted to transfer information from one transpinor device to another, as shown in Fig. 6e. Possible solutions may involve magnetic tunnel junctions (MTJs), spin valves, or the spin Hall effect [56]. For instance, CoFe/MgO/Fe(Mn_5Ge_3) MTJ or CoFe/Cu/Fe(Mn_5Ge_3) metallic spin valve can be integrated on top of the nanomagnets to read out and also manipulate their magnetizations. Alternatively, the recently discovered inverse spin Hall effect and VCMA effect may also be used to read/write the magnetization information in the nanomagnets [57]. In this way, the control and clocking schemes will be carefully designed to enable the switching of the next stage. The read out voltage signal can be used to drive the switching of the next transpinor stage, which ideally requires a very small range of gate voltage in the vicinity of paramagnetic-to-ferromagnetic phase transition.

The proposed transpinor device can be readily used to explore new energy-efficient, nonvolatile circuits and architectures for very high fan-in digital logic. The computation progresses in stages by cascading novel dynamically controlled spin logic nanowires through electric control to provide nonvolatile implicit latching of spin state at the output nodes. One example is the high fan-in dynamic transpinor NAND logic (DTL-NAND), as shown in Fig. 7. It is a magnetic nanowire circuit that has an output nanomagnet whose magnetization is controlled through a DMS nanowire channel gated with multiple transpinors. Logic inputs are connected to these transpinors. A preset transpinor enables the control of dynamic operation as follows: first, the output nanomagnet magnetization is preset (Fig. 9b) using voltage-controlled (V_{pre}) phase transition through the *Preset* transpinor. Next, inputs connected to the other Transpinor devices are evaluated by injecting a spin-polarized current from the opposite end of the nanowire through the *Evaluate* nanomagnet. The output nanomagnet is reversed (Fig. 7c) only if all input voltages are "**1**" (where "**1**" is defined as the gate voltage that causes the ferromagnetic-to-ferromagnetic phase transition for the DMS nanowire channel in a transpinor); a single "**0**" voltage prevents further propagation of the spin information through the DMS nanowire channel, retaining instead previously latched preset state (Fig. 7d). This operation fulfills a NAND function, and other logic functions may also be possible. One of the shortcomings is the need to use front-end process versus all-metal devices, the latter of which low-temperature back-end processing suffices.

Arbitrary spin logic can be further realized by cascaded NAND-NAND schemes with electric control—a transpinor crossbar could be a possible fabric organization [58]. Cascading between high fan-in DTL elements requires the conversion of a output magnetism to a gate voltage (to be used as input to the next stage of transpinor NAND logic). As discussed above (see Fig. 6e), various techniques can be employed, including spin valve, inverse spin Hall effect, or MTJ devices. Figure 7e shows a possible cascade of DTL-NAND logic using a MTJ device, which is connected to each output nanomagnets. In this solution, the read-out

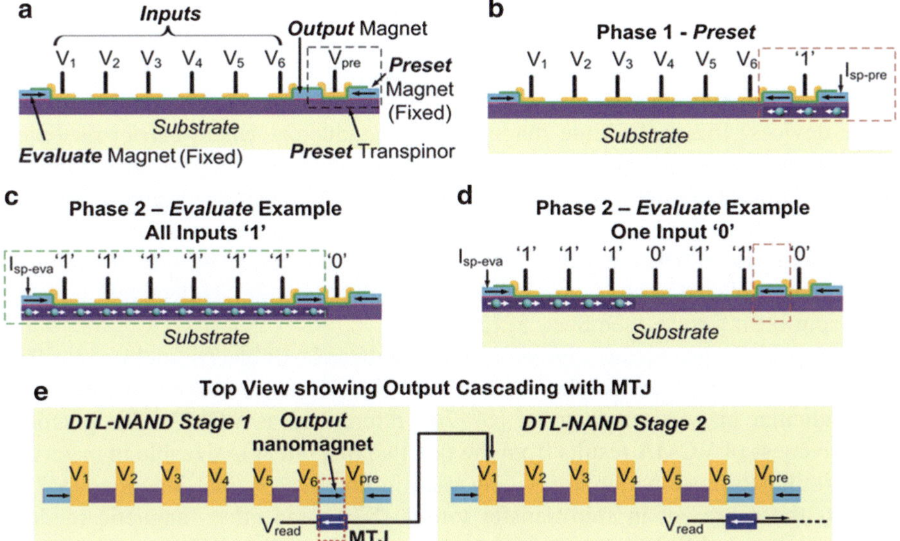

Fig. 7 Dynamic Transpinor Logic-NAND (DTL-NAND). (**a**) Side-view of a 6-input DTL-NAND gate. Dynamic circuit operation is described as follows: (**b**) Phase 1—*Spin Preset* where the output magnetization is aligned with the *Preset* nanomagnet; (**c**) Phase 2—*Spin Evaluate* example where all inputs are "**1**", then the output magnetization is reversed to align with the *Evaluate* nanomagnet and can be interpreted as output logic "**0**" using an adjacent MTJ; (**d**) Phase 2—*Spin Evaluate* example where one of the inputs is "**0**", then the output magnetization retains its *Preset* state and can be interpreted as logic "**1**" using an adjacent MTJ; (**e**) Top-view schematic showing cascaded DTL-NAND stages for arbitrary logic, where the cascading to stage 2 is achieved with an MTJ placed adjacent to the output nanomagnet in stage 1. Credit: Discussion with Csaba Andras Moritz within Western Institute of Nanoelectronics

voltage is clocked synchronously with the *Preset* and *Evaluate* operations. The spin-based switching and high fan-in—without the same performance degradation as in CMOS since the fan in is spin controlled rather than driving increasingly large load capacitance—implies potential for ultra low-energy operation as well as a very compact implementation (fewer logic levels) that saves area and improves performance. Based on the above assessment, the transpinor is envisioned to be an emerging low-power logic device that can possibly replace or complement CMOS in the future.

3 Voltage Control of Magnetic Anisotropy in Metallic Spintronic Devices

Several approaches have been proposed to realize electric-field-controlled magnetic devices and circuits, driven primarily by the potential of such solutions to achieve dramatically lower energy dissipation compared to their current-controlled counterparts. Some of the prominent examples include single-phase multiferroic materials [59, 60], multiferroic tunnel junctions having ferroelectric tunnel barriers [61], multiferroic heterostructures where magnetic and electrical properties are coupled through strain (see, e.g., Sect. 6.1 where this approach is applied to spin wave devices) [62, 63], dilute magnetic semiconductors where carrier-mediated ferromagnetism is controlled by a gate voltage (see Sect. 2) [28], and ultrathin metallic films where the Curie temperature (i.e., phase transition) is controlled by a voltage (see Sect. 4) [64, 65]. Yet other approaches to reduce power dissipation are based on the use of new types of STTs, which utilize the large spin-orbit coupling in heavy metals placed in close proximity to magnetic films (i.e., collectively referred to as spin-orbit torques, see Sect. 5) [56, 66].

An especially promising approach for the realization of electric-field-controlled spintronic devices relies on the electric field (i.e., voltage) control of interfacial perpendicular magnetic anisotropy [67–74], referred to as VCMA. The practical attractiveness of VCMA results from the fact that this effect is sizeable in materials very similar to those which are widely used in magnetic tunnel junctions (MTJ), such as those present in spin-transfer torque (STT) and other magnetic random access memory (MRAM) elements. As a result, VCMA devices are friendly to CMOS integration and processing, and benefit significantly from the existing know-how, processing tools and manufacturing infrastructure of STT-MRAM. This type of VCMA devices will be the focus of this section.

The discussion is organized as follows: Section 3.1 presents a short review of the motivation behind the development of VCMA-based devices, from the perspective of nonvolatile memory and logic applications. This is followed in Sect. 3.2 by an introduction to the VCMA effect in magnetic/non-magnetic interfaces. Section 3.3 discusses a single-domain model of VCMA-induced switching in nanomagnets, such as those used in magnetoelectric memory (MeRAM) elements. Section 3.4 discusses the temperature dependence of the VCMA effect, which can have important implications for practical applications. Section 3.5 presents examples of using the VCMA effect in switching of nanoscale memory devices, and how such memory elements can be integrated into circuit arrays, as well as their practical bit density considerations. Finally, Sect. 6 will discuss the scalability of this type of magnetoelectric memory and compare it to STT-MRAM.

3.1 Voltage-Controlled Magnetism: Impact on Memory and Logic

The scaling of CMOS technology over the past several decades, while resulting in increasingly faster and more powerful microprocessors [1, 75], is quickly approaching a limit due to increased static (leakage) power dissipation at small technology nodes. While scaling (i.e., reducing the dimensions of transistors on a chip) generally reduces the dynamic power dissipation during their switching, it also results in larger leakage power. Given that CMOS transistors, and logic or memory circuits based on them, require power to be continuously applied to them in order to retain their information (i.e., CMOS circuits are volatile, meaning that data is lost in case the supply voltage is removed), this standby power dissipation becomes a limiting factor, potentially eliminating other gains obtained by reducing device dimensions. The integration of a fast, energy-efficient nonvolatile memory technology with CMOS can alleviate this problem, by eliminating the need for a continuously applied power source (and hence the associated leakage current). NVL circuits, combining volatile CMOS logic and nonvolatile magnetic memory can thus allow for continued scaling with improved energy efficiency, by substantially reducing the static power dissipation [76].

Spintronic devices (i.e., magnetic memories, generally referred to as MRAM) are the leading contenders for such a paradigm shift towards NVL. This is due to their unique combination of attributes: The compatibility of their material stacks with standard CMOS back-end of line (BEOL) processing [77–80], the possibility of fast (<1 ns) read and write times, and their potentially unlimited write cycle endurance ($>10^{16}$, which is absolutely essential for logic operations, where very frequent re-writes of a single bit typically occur during circuit operation). Read-out is performed using the tunneling magnetoresistance (TMR) [81] effect. For the writing of information, earlier generations (i.e., toggle MRAM) utilized magnetic fields generated by currents flowing through adjacent metal lines, to switch the magnetization of memory cells. More recent generations utilize spin-transfer torque (STT) instead, where current directly passes through the memory bit [80, 82–84]. However, these strategies face limitations in terms of energy efficiency (due to the large Ohmic dissipation) and density (due to the large currents needed to accomplish switching, resulting in wide transistors). The manipulation of magnetic moments by electric voltages overcomes these shortcomings.

For ultralow-power spintronic memories going beyond STT-MRAM, at least two approaches are presently pursued: First, using current-induced spin-orbit torques (SOTs), such as the giant spin Hall effect (SHE), as an alternative mechanism to reduce the switching currents needed to write information [56, 82]. Secondly, a complete departure from current-controlled write mechanisms, replacing them by voltage-controlled (i.e., magnetoelectric) effects, resulting in Magnetoelectric RAM (MeRAM) circuits [85–87]. In the following we will review the physical principles involved in this approach.

3.2 *Voltage-Controlled Magnetic Interface Anisotropy*

Interfaces of ferromagnetic thin films and nonmagnetic materials can exhibit a large perpendicular magnetic anisotropy (PMA) [88–92], which itself is a function of electric fields applied perpendicular to the interface [67–69, 71–74, 93–99]. This effect has been attributed to the change (due to the applied voltage) of the relative occupancy of different atomic orbitals at the interface (e.g., 3d orbitals in the case of

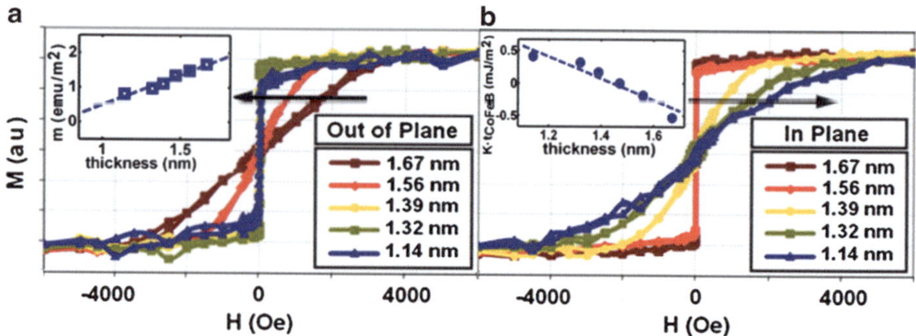

Fig. 8 Measured magnetization as a function of the applied magnetic field, measured with (**a**) out-of-plane and (**b**) in-plane external magnetic fields on films with different CoFeB thickness. The inset in (**a**) shows the magnetic moment as a function of film thickness, while the *inset* in (**b**) shows the dependence of the PMA on thickness. Reprinted with permission from [90]. Copyright 2011, AIP Publishing LLC

Fe and Co, which are commonly used in MTJs), which, combined with spin-orbit coupling, results in a change of the magnetic anisotropy energy [68, 69, 71, 99]. More generally, VCMA can be described in terms of the effect of Rashba spin-orbit coupling on the magnetic anisotropy of a ferromagnetic thin film sandwiched between non-magnetic media, which is a function of electric fields normal to the interfaces of such a layered structure [100]. If MgO is used as the dielectric barrier providing the above-mentioned interface, the resulting VCMA effect can be incorporated into MTJ devices, hence allowing for TMR readout similar to typical MRAM cells [81, 96, 101–103, 105].

The interfacial PMA in an MgO/Co$_{20}$Fe$_{60}$B$_{20}$/Ta material structure is illustrated in Fig. 8, where an increasing total perpendicular anisotropy is observed as the Co$_{20}$Fe$_{60}$B$_{20}$ layer thickness is reduced [90]. The inset in Fig. 8b shows the measured total perpendicular anisotropy energy as a function of the film thickness, which indicates a spin reorientation transition at ~1.5 nm, where the effective perpendicular anisotropy vanishes. This effect has been used to realize high-TMR perpendicular magnetic tunnel junctions [88, 89, 106, 107], reduce switching current in in-plane MTJs while maintaining their thermal stability [90], increase output power and achieve high quality factors in spin torque nano-oscillators [108–110], as well as to enhance the detection sensitivity of spin torque diode microwave detectors [111, 112]. Near the transition thickness, however, the magnetization configuration is particularly sensitive to voltages applied to the device [70, 96, 97, 113, 114], and can therefore be used to realize electric-field-controlled magnetic memory devices [85, 86, 96, 97, 102, 103, 105, 113].

The interfacial VCMA effect can be measured using a number of methods. One is using magnetometry measurements as a function of applied electric field bias [71], or (equivalently) in the case of MTJs, to measure magnetoresistance curves of the device under different bias conditions [73, 115]. Figure 9 illustrates the corresponding experiment as described in [115]. The fixed layer is in-plane in

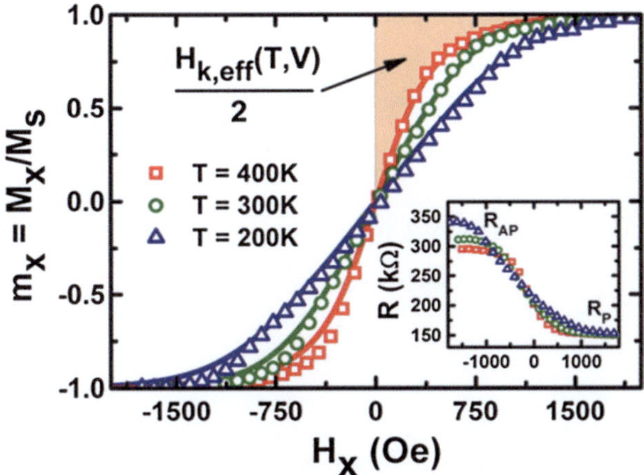

Fig. 9 Measurement of the perpendicular magnetic anisotropy using magnetoresistance loops in a magnetic tunnel junction [73, 115]. The inset shows the measured magnetoresistance loops, from which the magnetization versus field loop can be calculated. The area enclosed by the magnetoresistance curve varies with both temperature [115] and voltage [73], which allows for measurement of the temperature dependence of the PMA, as well as its voltage dependence (i.e., the VCMA strength). Reprinted with permission from Ref. [115]. Copyright 2014, AIP Publishing LLC

this case, while the free layer thickness is such that its magnetization is pointing out-of-plane, resulting in a hard-axis behavior for in-plane applied magnetic fields. The area enclosed by the curves in Fig. 9 can be used to infer the perpendicular anisotropy [73], and is a function of the voltage bias applied to the device, as well as of temperature (see Sect. 3.4).

In [93], the anisotropy modulation by electric fields is similarly measured using the anomalous Hall effect. A $Co_{40}Fe_{40}B_{20}$ free layer composition is used in this case, without the presence of a fixed layer. Magneto-Optical Kerr Effect (MOKE) measurements were used in [95] for the case of $Co_{20}Fe_{80}$ thin films, also in the absence of a fixed layer. The interfacial VCMA effect has also been studied by using microwave VCMA-FMR measurements [72–74, 94].

Most experimental results for the commonly used CoFe/MgO material system to date have reported VCMA magnitudes of ~30–40 fJ/V-m, in general agreement with theoretical predictions [67]. Significant effects of the capping or seed metallic material on both PMA and VCMA have been reported both in computational and experimental works [116, 117], as well as by using different materials for the magnetic free layer [70, 118]. Similarly, the use of dielectrics with high dielectric constant has been proposed as a way to enhance charge accumulation per unit electric field at the device interface, thereby enhancing the VCMA effect [119]. A magnetoelectric coefficient close to the one observed in CoFeB/MgO systems has also been observed for the FePt/MgO system [120]. For practical applications in nonvolatile memory, typically a value of >200 fJ/V-m will be required to simultaneously achieve high thermal stability and low switching voltage, as well as to

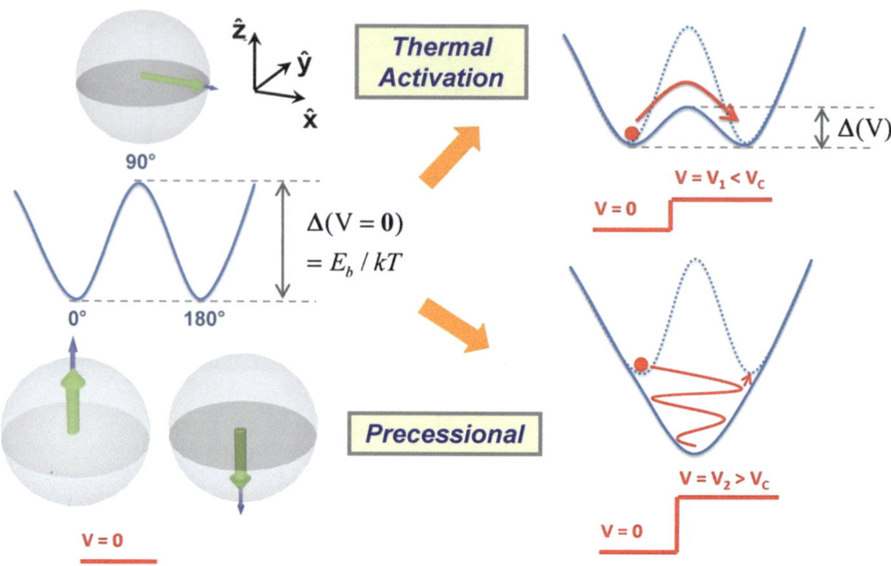

Fig. 10 Illustration of thermal and precessional electric-field-induced switching processes in perpendicular magnetic elements. At equilibrium ($V = 0$), the energy barrier separates the two stable states (0° and 180°), where the energy becomes maximum for in-plane magnetization (90° or 270°). In order to switch, the VCMA effect is used to lower the energy barrier. If the energy barrier is only partially lowered, thermal activation can result in a switching event. (A preferred switching direction can be induced in this case by additional application of a magnetic field, as illustrated in Figs. 11, 12, and 13). On the other hand, if the barrier is eliminated, a new energy minimum will form in the in-plane direction. By timing the resulting precession, switching to the opposite state can hence be obtained. The critical voltage, which separates these two regimes, is defined as the required voltage to lower the energy barrier to zero (Reprinted from [121])

ensure scalability to smaller dimensions, as discussed below in Sect. 3.3. While a number of works have reported such VCMA values [70, 116], realizing this requirement while simultaneously maintaining a high TMR ratio for readout remains a challenge and requires further investigation.

3.3 Voltage-Induced Switching of Magnetic Memory Bits

The anisotropy modulation due to VCMA can be used to induce (or assist) the switching of magnetic memory devices [82, 87, 96, 97, 102, 105, 113, 114]. Compared to STT-induced switching, VCMA can significantly reduce power dissipation, and enhance bit density by eliminating the need for large drive currents (hence large access devices).

The VCMA effect can be used to induce both thermally-activated and precessional types of switching, as illustrated in Fig. 10: (1) When ultra-fast (typically

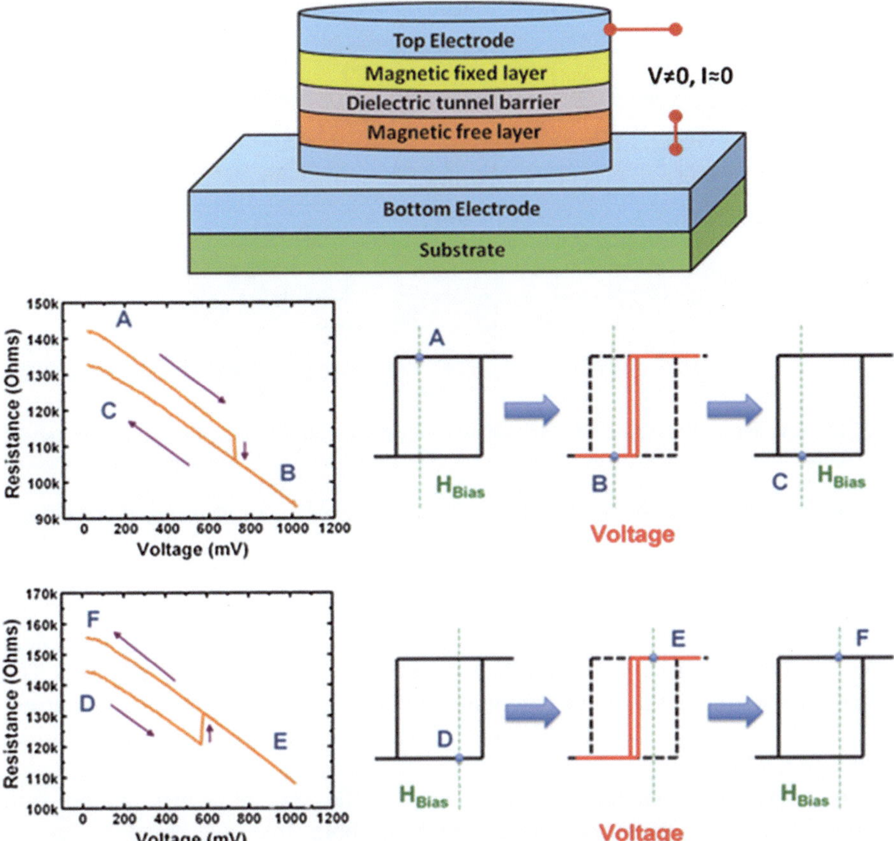

Fig. 11 Schematic illustration of a voltage-controlled magnetic tunnel junction (top) [82, 113, 114]. The bottom panel shows the voltage-induced switching of a nanoscale MTJ device in the thermally-activated regime, assisted by an external magnetic field which also determines the switching direction. Note that the same voltage polarity is used for switching in both directions. Reprinted with permission from Ref. [82]. Copyright 2013, Institute of Physics

<1 ns) voltage pulses are applied, the resulting voltage-induced precession can be timed to create full switching to the opposite orientation [96, 105]. This is illustrated in Fig. 10 (bottom) for the case of a bit with out-of-plane stable states (i.e., a perpendicular MTJ), where the application of a pulsed voltage reduces the perpendicular anisotropy, inducing a switching to the opposite state if the pulse is timed to half a precession cycle [96, 102, 105]. (2) In a second approach, the VCMA effect can be used to modulate the coercivity of the MTJ free layer, allowing for thermally-activated electric-field-assisted switching when a magnetic field is applied to the device [97, 103]. As illustrated in Fig. 10 (top) and Fig. 11, in this case the applied voltage induces a reduction in the free layer coercivity due to the modification of the interface anisotropy, resulting in thermally-activated switching. When a magnetic field is applied to a device under such an electric field bias, there

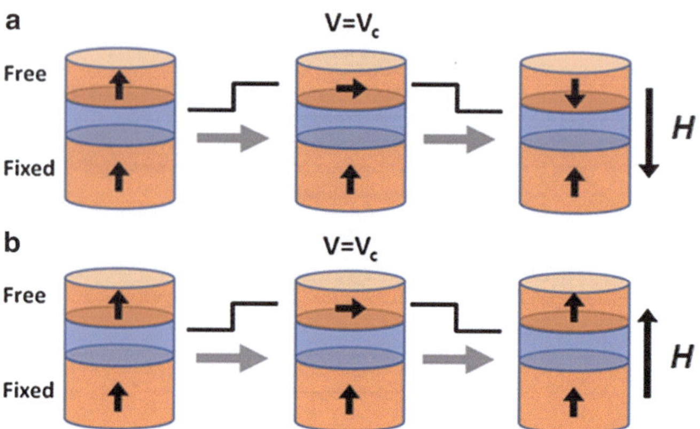

Fig. 12 Schematic illustration of perpendicular magnetoelectric tunnel junctions switched by the combination of an applied voltage and external magnetic field [86]. The final state depends on the direction of the overall magnetic field acting on the free layer, and can be controlled by applying an external field. The need for the varying external field can be eliminated in precessional switching, or when current-induced torques are used in addition to VCMA [85, 96, 103, 105]. Reprinted with permission from Ref. [86]. Copyright 2013, AIP Publishing LLC

will be only a single available state for a given applied bias magnetic field (see Fig. 11), which in turn determines the switching direction [82, 97, 103]. The role of the external field can be potentially replaced by allowing a non-zero current to pass though the device, thereby providing additional spin torque or field-like torque [87, 114].

To analyze the switching voltage, consider the case of a memory bit with perpendicular (out-of-plane) magnetization in both free and fixed layers, as shown in Fig. 12. The thermal stability factor of this bit is given by $\Delta = E_b/k_B T$, where k_B is the Boltzmann constant, and T is temperature. Similar to the case of other magnetic memories [80, 122, 123], Δ determines the retention (i.e., dwell) time of the device. E_b is the energy barrier between the two free layer states (up or down), which is given by $E_b = H_{k,eff}^{\perp}(V)M_s At/2$, where M_s is the free layer saturation magnetization, A is its area, and t is the free layer thickness. The effective perpendicular anisotropy field is given by

$$H_{k,eff}^{\perp}(V) = h_{k,s}^{\perp}(V)/t - 4\pi M_s, \tag{4}$$

where $h_{k,s}^{\perp}(V)/t$ is the thickness-dependent PMA field [88, 90], which can be modulated by the applied voltage V [67, 71, 73, 93]. The energy barrier is a function of voltage, and the standby thermal stability factor (for $V = 0$) is thus determined by the value of $H_{k,eff}^{\perp}(0)$. Assuming a linear dependence of the perpendicular anisotropy field on voltage, which has been observed in most experimental reports, one can write the interfacial anisotropy as $h_{k,s}^{\perp}(V) = h_{k,s}^{\perp}(0)(1 - \zeta V)$. Positive voltages

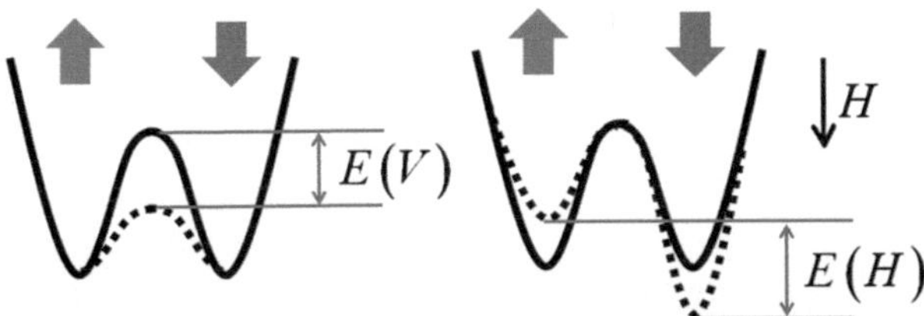

Fig. 13 Schematic representation of the effect of applied electric fields (*left*) and external magnetic fields (*right*) on the energy diagram of a bi-stable magnetic bit exhibiting VCMA. Reprinted with permission from Ref. [86]. Copyright 2013, AIP Publishing LLC

will thus reduce $H_{k,eff}^{\perp}(V)$ in this convention, contributing to thermally-activated switching by reducing the thermal stability of the perpendicular bit [124]. The case of a nonlinear VCMA effect follows similarly. The parameter ζ is proportional to the (linear) VCMA magnitude, and quantifies the sensitivity of the interfacial anisotropy to applied voltages, as determined experimentally by voltage-dependent magnetoresistance or ferromagnetic resonance measurements [71, 73, 74].

Using the voltage-dependent thermal stability factor of the magnetic bit $\Delta(V) = \Delta(0) - M_s A\left(\zeta V h_{k,s}^{\perp}(0) + tH\right)/2k_B T$, where $\Delta(0) = \left(h_{k,s}^{\perp}(0) - 4\pi M_s t\right) M_s A/2k_B T$, and where we have accounted for a bias field H applied externally to the bit, one can then determine the switching voltage as discussed in [86]. Note that the direction (i.e., sign) of H determines the switching direction, by increasing or decreasing $\Delta(V)$ depending on the initial state of the magnetic bit. The result is a voltage-dependent dwell time for the magnetic bit, allowing for thermally-activated switching on a time scale of $\tau(V) = \tau_0 \exp(\Delta(V))$, where τ_0 is the attempt time. The critical switching voltage V_c, which corresponds to $\tau(V_c) = \tau_0$, can then be obtained from the condition $\Delta(V_c) = 0$, and is given by

$$V_c = \left(h_{k,s}^{\perp}(0) - (4\pi M_s + H)t\right)/\zeta h_{k,s}^{\perp}(0). \tag{5}$$

Note that the contribution of the bias magnetic field is thus not only to determine the switching direction, but also to affect the critical switching voltage. The case of precessional switching follows from reducing the energy barrier to zero in the absence of an assisting field, i.e., for $H = 0$, as shown in Fig. 10 (bottom). In the limit where VCMA is negligible ($\zeta \approx 0$), Eq. (5) reduces to a critical switching field of $H_c = H_{k,eff}^{\perp}(0)$ as expected.

The external field thus acts to reduce or increase the energy barrier, depending on its direction with respect to the initial state of the device, while the VCMA effect acts to reduce or increase the barrier depending only on the voltage polarity. This comparison is illustrated in Fig. 13.

3.4 Temperature Dependence of the VCMA Effect

In this section, we briefly discuss the temperature dependence of both the perpendicular magnetic anisotropy (PMA) and of the VCMA effect in nanoscale MgO|CoFeB| Ta-based magnetic tunnel junctions, following closely the discussion of [115].

The magnetic anisotropy in ferromagnetic materials is known to have a strong dependence on temperature [125]. In particular, Callen and Callen's theory [126] predicts that uniaxial anisotropies (such as PMA in our case) decrease with $M_s^3(T)$, where $M_s(T)$ is the temperature-dependent saturation magnetization of the material. Quantifying the temperature dependence of both PMA and VCMA is important for MeRAM applications where the operation temperature may be well above room temperature. Beyond practical considerations, the study of the temperature dependence of PMA and VCMA effects is also of interest as it can contribute to a better understanding of the underlying physics behind both phenomena.

We consider a multilayer stack with the composition of substrate | bottom electrode | PtMn (20) | $Co_{70}Fe_{30}$ (2.3) | Ru (0.85) | $Co_{60}Fe_{20}B_{20}$ (3) [fixed layer] | MgO (1.3) | $Co_{20}Fe_{60}B_{20}$ (1.5) [free layer] | Ta (5) | top electrode (thickness in nm), patterned into 125 nm × 50 nm elliptical nanopillars for measurements. The MgO tunneling barrier is thick enough (resistance-area (RA) product ~750 Ω-μm^2) to make current-induced spin-torque effects negligible. The thickness of the $Co_{20}Fe_{60}B_{20}$ free layer is chosen such that the stable magnetization state of the free layer is in the perpendicular (z) direction.

Figure 9 shows resistance (R) versus in-plane magnetic field (H_x) curves measured at three different temperatures for such a device. The hard-axis-like curves confirm that the free layer has a stable perpendicular magnetization. The temperature- and voltage-dependent effective anisotropy field $H_{k,eff}(T, V)$ (i.e., the field required to saturate the perpendicular free layer in the in-plane direction) can then be obtained using a procedure described in [73, 115].

The dependence of the saturation magnetization on temperature $M_s(T)$ was independently measured in a separate test structure. Figure 14 shows the extracted data for M_s as a function of T in the range from 10 K to 400 K. The data is found to

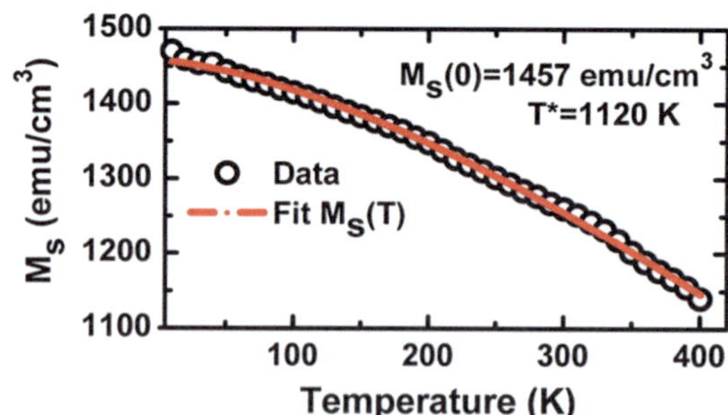

Fig. 14 The measured dependence of M_s as a function of temperature from SQUID magnetometry. The curve is found to fit well to Bloch's law with fitting parameters of $T^* = 1120\,K$ for the Curie temperature and $M_s(0) = 1457$ emu/cm^3. Reprinted with permission from Ref. [115]. Copyright 2014, AIP Publishing LLC

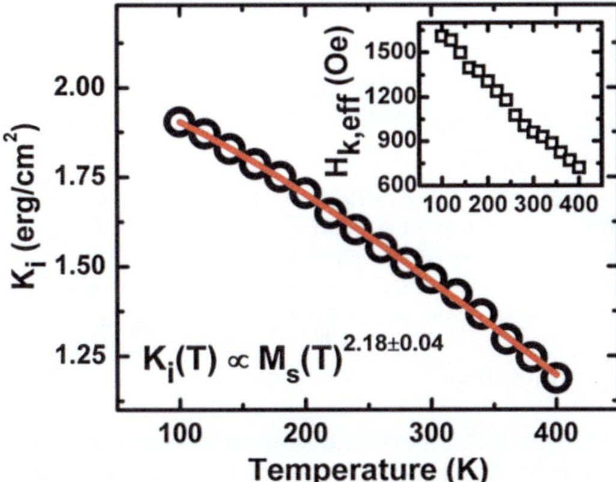

Fig. 15 The inset shows the extracted dependence of the effective anisotropy field as a function of temperature at zero bias voltage. Using the extracted values, the interface anisotropy is calculated and the temperature dependence is found to fit well to a power law of $M_s(T)$ with a power law exponent of $\gamma = 2.18 \pm 0.04$. The power law exponent combines contributions from the MgO|CoFeB and CoFeB|Ta interfaces. Reprinted with permission from Ref. [115]. Copyright 2014, AIP Publishing LLC

fit well to Bloch's law [127] $M_s(T) = M_s(0)\left(1 - \left(\dfrac{T}{T^*}\right)^{3/2}\right)$, due to the large Curie temperature T_C of CoFeB compared to the range of temperatures under consideration.

One can next quantify the dependence of the interfacial PMA energy K_i on temperature at equilibrium, i.e., without applying any voltages to the device. The inset in Fig. 15 shows the extracted values for the effective anisotropy field at equilibrium $H_{k,eff}(T, V = 0)$. The PMA energy K_i can then be directly calculated from $H_{k,eff}$ [115]. Figure 15 shows the extracted values for $K_i(T, V = 0)$ calculated in this manner, where one obtains a value of 1.45 mJ/m^2 for the PMA at room temperature ($T = 300$ K), in good agreement with previous reports for the MgO|Co$_{20}$Fe$_{60}$B$_{20}$|Ta system [88, 90]. The dependence of K_i on temperature is found to fit well to a power law of $M_s(T)$, i.e.,

$$K_i(T) = K_i(0)\left(\frac{M_s(T)}{M_s(0)}\right)^{\gamma} \qquad \gamma = 2.18 \pm 0.04, \tag{6}$$

where the PMA at zero temperature is $K_i(0) = 2.02\,\text{erg/cm}^2$.

Next the temperature dependence of the VCMA effect can be quantified, by looking at the temperature dependence of the material-level VCMA coefficient ξ (not to be confused with the differently-defined device-level VCMA parameter ζ in Sect. 3.3), which describes the change of interfacial anisotropy energy per unit electric field, i.e., $\xi = \Delta K_i / (\Delta V / d_{MgO})$, where d_{MgO} is the thickness of the MgO barrier. Figure 16a shows the change of the effective anisotropy field $\Delta H_{k,eff}$ as a function of the voltage V for four different temperatures, where

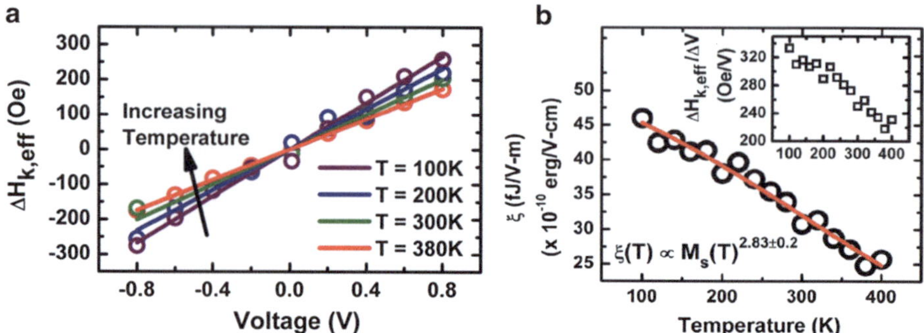

Fig. 16 (**a**) Measured change of the effective anisotropy field as a function of applied voltage for different temperatures. The VCMA effect is observed to remain linear over the temperature range of study, but its magnitude is reduced when temperature increases. (**b**) The inset shows the extracted values of $\Delta H_{k,eff}/\Delta V$ as a function of temperature. By using these values, the VCMA coefficient ξ (i.e., the change of interface anisotropy energy per unit electric field) is calculated and also found to fit well to a power law of $M_s(T)$, but with an exponent of $\gamma' = 2.83 \pm 0.2$, different from the PMA. Reprinted with permission from Ref. [115]. Copyright 2014, AIP Publishing LLC

$\Delta H_{k,eff} = H_{k,eff}(V) - H_{k,eff}(V = 0)$. Our results show a linear dependence of $\Delta H_{k,eff}$ on voltage over the wide temperature range under study. The inset in Fig. 16b shows the extracted values for $\Delta H_{k,eff}/\Delta V$ as a function of temperature during the experiment. Consequently, the temperature-dependent VCMA coefficient is calculated as $\xi(T) = M_s(T)\big(\Delta H_{k,eff}(T)/\Delta V\big)d_{MgO}t_{CoFeB}/2$ [115]. The obtained values are plotted in Fig. 16b. The VCMA coefficient at room temperature ($T = 300\,\mathrm{K}$) is found to be 31 fJ/V-m, in good correspondence with previous works [73]. Further, the temperature dependence of the VCMA coefficient is also found to fit well to a power law of $M_s(T)$, but notably with a different exponent compared to $K_i(T)$,

$$\xi(T) = \xi(0)\left(\frac{M_s(T)}{M_s(0)}\right)^{\gamma'} \qquad \gamma' = 2.83 \pm 0.2, \tag{7}$$

with $\xi(0) = 48.9\,\mathrm{fJ/V\text{-}m}$ as the VCMA coefficient at zero temperature.

Interestingly, while both $K_i(T)$ and $\xi(T)$ follow power laws of $M_s(T)$, they are described by different exponents. For the present material system, the measured PMA of the CoFeB free layer is an overall quantity that includes contributions from both the MgO|CoFeB and CoFeB|Ta interfaces, i.e., $K_i = K_{i,\mathrm{MgO/CoFeB}} + K_{i,\mathrm{CoFeB/Ta}}$. However, in principle one can postulate that only the MgO|CoFeB interface should be sensitive to the applied electric field, and therefore the VCMA coefficient may be primarily related to this interface. The perpendicular anisotropy contribution from the MgO|CoFeB interface has been attributed to hybridization of Fe d-orbitals with O p-orbitals, whereas the VCMA effect is due to the modification of the occupancy between the different hybridized orbitals [71]. Therefore, the fact that the localized orbitals (moments) modified by voltage follow more closely the conditions assumed in Callen-Callen's law might explain why the VCMA coefficient power law exponent is dominated by a $M_s^3(T)$ contribution. On the other hand, the exponent of $K_i(T)$ obtained in this work has a leading $M_s^2(T)$ contribution, which may be related to the influence of the high spin orbit coupling Ta, and a smaller $M_s^3(T)$ contribution, which could be accounted for by the MgO|CoFeB interface. In fact, a previous work [89] demonstrated that the CoFeB|Ta interface plays a key role in the strong PMA of the MgO|CoFeB|Ta stack. It should be stressed, however, that the exact role of Ta (and in general, of the high spin orbit coupling metal interfacing with the magnetic material used in the MTJ) is more complicated, involving its role both as a sink of boron atoms during annealing, as well as promotion of (001) crystalline orientation in the resulting CoFe(B) layer. A more detailed understanding of the exact contribution of each mechanism to explain the different power law exponents will require first principles calculations.

In terms of device applications where the VCMA effect is used to switch MTJs, a key figure of merit is the ratio between the VCMA coefficient and the PMA, i.e., $\xi(T)/K_i(T)$ [86]. Given that $\xi(T)$ decreases faster as a function of temperature than $K_i(T)$ does, $\xi(T)/K_i(T)$ decreases with temperature, and therefore larger voltages will be needed to achieve the electric-field-induced switching for a constant value of the interfacial anisotropy [115].

3.5   *Circuit Implementation of VCMA-Controlled Memory*

In this section we discuss a possible circuit implementation of electric-field-controlled MeRAM using diode-based arrays, following closely the discussion presented in [87]. STT-MRAM normally uses a 1-Transistor/1-MTJ cell structure with CMOS transistors as the access devices. However, the large currents required for STT switching of MTJ bits require large transistors to drive them [87]. As a result, the density of STT-MRAM arrays is not limited by the MTJ size, but rather by the switching current of the magnetic bits, which in turn dictates the MTJ separation. Moreover, the use of three-terminal CMOS transistors also imposes a layout-based minimum size limit of ~8 F^2 on the 1-T/1-MTJ cells. By contrast, in principle, crossbars are the densest memory arrays possible (with a 4 F^2 cell size, assuming circular bits with perpendicular magnetization). Due to its bipolar switching scheme, however, STT-MRAM does not lend itself easily to a crossbar implementation. MeRAM, on the other hand, has a unipolar switching behavior, and hence the realization of a diode-controlled MeRAM array is possible, and can greatly increase the density and scalability of the overall memory. Additionally, the crossbar architecture allows for 3D stacking of multiple diode-MTJ memory layers in the CMOS back-end-of-line (BEOL) fabrication steps, potentially doubling the effective density with each layer [87].

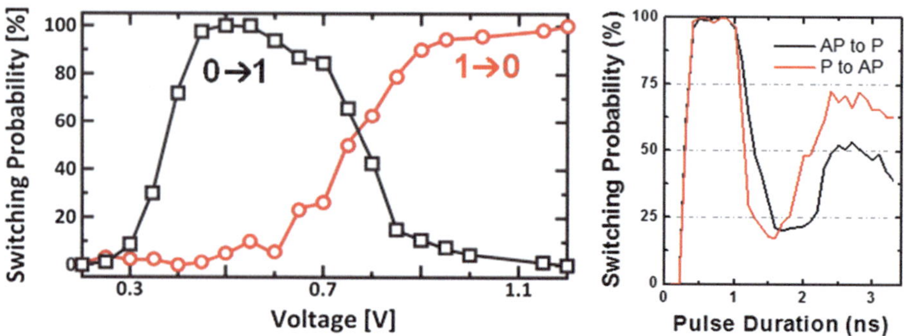

Fig. 17 (*Left*) Measured probability of switching curves for VCMA-controlled MTJs with STT-assisted switching, in the thermally-activated regime [87, 114]. The combination of VCMA and STT effects allows for a unipolar set/reset switching scheme. The switching direction is controlled by the input pulse amplitude, and the write process is thermally activated. © 2013 IEEE. Reprinted, with permission, from Ref. [87]. (*Right*) Measured probability of switching, as a function of the input pulse voltage length, for an 80 nm perpendicular MTJ, where writing is performed by precessional switching (see illustration in Fig. 10). Switching in both directions can be achieved by the same pulse shape in this case, and no STT is involved in the process

Figure 17a shows how voltage pulses of the same polarity, but different amplitudes, can be used to switch an MTJ device using VCMA in opposite directions (thermally activated switching) [87, 114]. For comparison, Fig. 17b shows the switching probability of a magnetic bit precessionally switched by VCMA, for a fixed voltage amplitude, as a function of the pulse width. It can be seen that voltage pulses timed to half the precession of the free layer can achieve full switching. Importantly, both types of thermal and precessional switching illustrated in this figure use only one polarity of voltage to achieve switching in both directions (i.e., the polarity which reduces the perpendicular interface anisotropy) [87, 96, 97, 102, 103, 105]. This allows for integration of such devices into a 1-Diode/1-MTJ crossbar array, resulting in an improved bit density.

Figure 18 shows the schematic and layout view for a high-density crossbar memory array using voltage-controlled MTJs. The unipolar set/reset write scheme of the devices allows for a diode to be integrated in series without any loss in functionality. The series diode also eliminates sneak currents present in traditional crossbar arrays, improving the read performance.

The following description corresponds to the thermally-assisted VCMA switching illustrated in Fig. 17a, but in principle a similar scheme can be applied to precessional writing as well. During both reading and writing, unaccessed bit-lines (BLs) are grounded, while unaccessed source-lines (SLs) are pulled to V_{DD} (1.5 V), thus reverse biasing the series diode for unaccessed bits. During the write operation, the target SL is pulled to ground, while the target BL is pulsed with the appropriate set/reset voltage. During the read operation, the target SL is pulled to ground and the target BL is connected to a sense amplifier. To prevent disturbing the state of the desired bit cell, a small sensing voltage of 0.2 V is used, which is

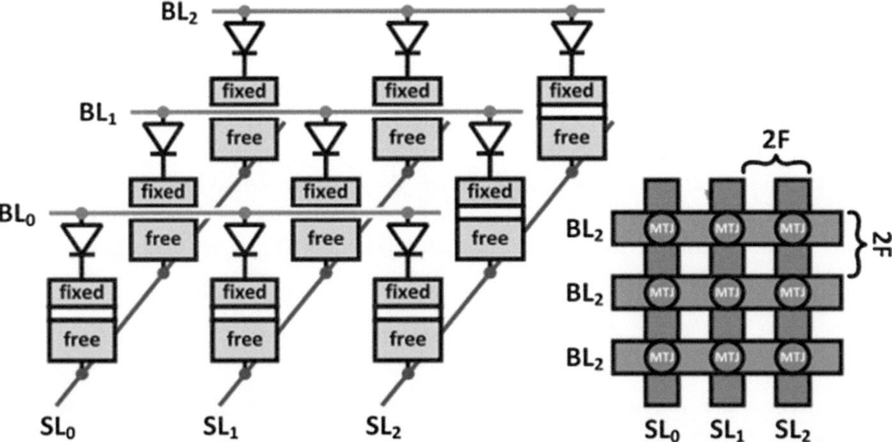

Fig. 18 Schematic representation and layout (*top view*) of a crossbar array structure, with integrated diodes, for high-density voltage-controlled MeRAM. © 2013 IEEE. Reprinted, with permission, from Ref. [87]

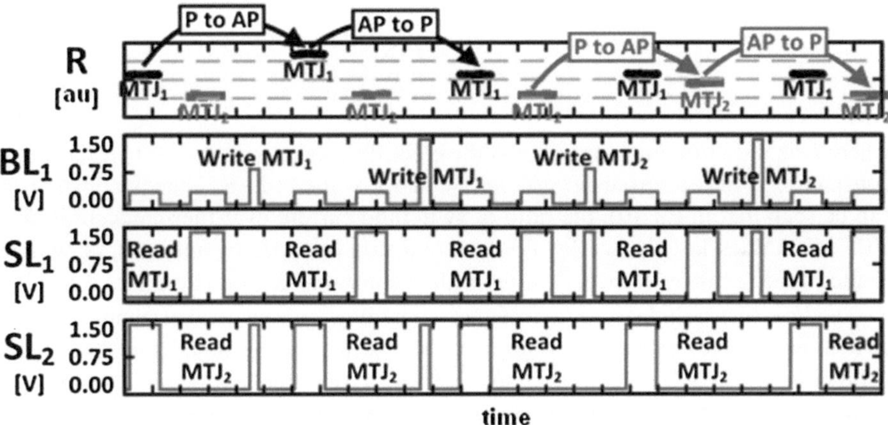

Fig. 19 Measured transient waveforms for reading and writing of addressed memory cells in a MeRAM crossbar array, demonstrating that two different cells can be written and read selectively with the approach shown in Fig. 17 (*left*). © 2013 IEEE. Reprinted, with permission, from Ref. [87]

sufficiently low to prevent unintended writes. Larger arrays can be built if the resistance of the MTJs is increased or if the leakage current of the diodes is reduced.

Figure 19 shows experimental transient waveforms demonstrating the functionality of the crossbar array concept. MTJ_1 and MTJ_2 are first initialized into the P state. MTJ_1 is then first switched from P to AP, and back to P, without disturbing the value stored in MTJ_2. Similarly, MTJ_2 is then switched from P to AP, and back to P, without disturbing the state of MTJ_1.

3.6 *Scalability of VCMA-Based Magnetoelectric RAM*

One important consideration to integrate a specific spintronic device into a chip is the scalability of the device. The goal of scaling is to provide better performance and/or lower cost with higher density.

A critical parameter which determines the scaling behavior for nonvolatile MRAM is the ratio of switching voltage (or current) over the thermal stability factor, which should ideally be minimized. For perpendicular MeRAM this is given by

$$\frac{V_c}{\Delta} = \frac{2k_B T}{\zeta M_s H_{k,s}^{\perp}(0)A}.$$

(8)

As with other magnetic memories, scaling to smaller bit areas requires an increase in $H_{k,s}^{\perp}(0)$ and/or M_s to maintain a constant thermal stability Δ (hence nonvolatile retention time) of the bits. However, Eq. (8) shows that this does not necessarily lead to an increase of the switching voltage V_c, provided that the VCMA parameter can be adjusted accordingly. It is interesting to compare this to the corresponding scaling parameter for perpendicular STT-MRAM, which is given by [88, 122, 128]

$$\frac{I_c}{\Delta} = \frac{4e\alpha k_B T}{\hbar \eta},$$

(9)

where I_c is the switching current, e is the electron charge, α is the Gilbert damping constant, η is the spin-transfer efficiency, and $\hbar$ is the reduced Planck constant. It can be seen that in Eq. (9), the ratio of switching current over thermal stability for STT-MRAM is largely set by fundamental constants or by parameters with a limited tuning range. Hence, scaling with a constant- Δ rule (by increasing $H_{k,s}^{\perp}$ (0) and/or M_s) will lead to a constant switching current (rather than constant switching voltage) across technology nodes. (It should be noted that Eq. (9) is valid only in a single-domain switching behavior, and may not be valid when sub-volume excitations and magnetic textures are present, as has been pointed out in a number of recent works [129]. However, this does not affect the overall validity of our discussion and we will not consider these effects in the present text.)

This presents a fundamental problem for the scaling of current-controlled STT-MRAM, as the transistor widths needed to drive this constant switching current will not significantly shrink with successive technology nodes, hence hitting a current-drive-limited barrier on transistor width (hence cell area). By contrast, MeRAM switching voltages are associated with very small leakage currents through the device, which allow for the use of minimum-sized transistors at each technology node, hence imparting a growing density advantage with progressive scaling to smaller bit dimensions. This is illustrated in Fig. 20, which compares the transistor area required for a 1-transistor/1-MTJ MeRAM cell to that of a 1-transistor/1-MTJ STT-MRAM cell, for three different values of STT-MRAM

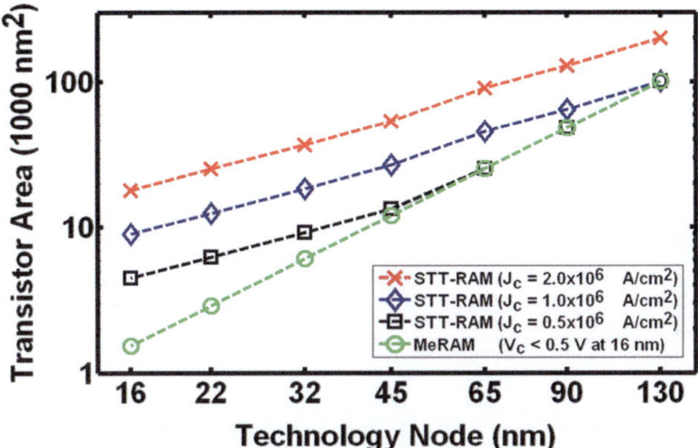

Fig. 20 Comparison of transistor area scaling for 1-transistor/1-MTJ MeRAM cells and 1-transistor/1-MTJ STT-MRAM cells across technology nodes down to 16 nm, for three different values of STT-MRAM switching current densities. While voltage-controlled MeRAM allows for using minimum transistor widths at each technology node, thereby maximizing density, STT-MRAM bit stability dictates a constant switching current even at scaled bit dimensions (see Eq. (9)), leading to a saturation behavior in the transistor width. While the development of STT-MRAM structures with reduced switching current densities can delay the onset of this scaling dilemma, even a switching current density as low as 0.5×10^6 A/cm^2 would result in a three times larger transistor size compared to the MeRAM case at the 16 nm node

switching current densities. Estimates on CMOS operating voltage and current drive capability were obtained from Ref. [131]. Figure 20 shows that, while lower switching current densities can delay the onset of the scaling dilemma for STT-MRAM, even switching current densities as low as 0.5 MA/cm^2 represent a sizeable penalty in terms of area at scaled technology nodes.

A similar scaling advantage is possible in terms of energy efficiency. While MeRAM already provides an advantage over STT-MRAM (see Sect. 3.3) in terms of switching energy per bit, this advantage grows quickly as bit dimensions are scaled. This is due to the fact that, assuming a constant write time t_w across technology nodes, STT-MRAM bits retain an approximately constant write energy $E_w = RI_c^2 t_w$, provided that the MTJ's resistance-area product (RA) can be reduced sufficiently (i.e., MgO barrier can be made thin enough) to prevent the resistance R, and hence the write voltage RI_c from increasing. This reduction of RA with scaled device dimensions is in fact necessary to maintain compatibility with CMOS read and write circuits in STT-MRAM. In addition to not allowing for reduction of the write energy by scaling bit dimensions, this may eventually lead to reliability issues due to the increasingly thinner MgO barriers needed for STT-MRAM scaling.

In the case of MeRAM, the voltage-controlled switching mechanism does not impose such a constraint to reduce RA (hence MgO thickness) with scaled device dimensions. Hence, the write energy (where C is the capacitance of the MTJ) reduces with each technology node, as the resistance increases and C decreases due to the smaller bit area A. Note that, due to the high resistance of voltage-switched MTJs, one may in general need to consider the parasitic capacitance C in

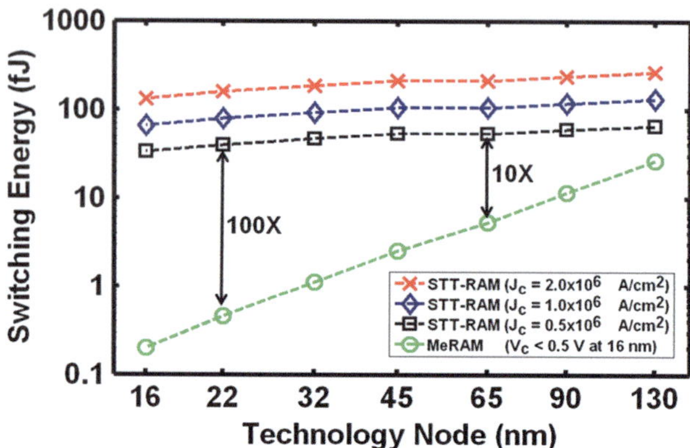

Fig. 21 Comparison of switching energy (per bit) scaling for 1-transistor/1-MTJ MeRAM cells and 1-transistor/1-MTJ STT-MRAM cells across technology nodes down to 16 nm, for three different values of STT-MRAM switching current densities. While voltage-controlled MeRAM allows for reduced switching energies as the bit dimensions are reduced, STT-MRAM bit stability dictates a constant switching current even at scaled bit dimensions (see Eq. (9)), leading to a saturation behavior in the energy efficiency. The slight reduction of STT-MRAM write energies in this plot is due to the reduced operating voltages at smaller nodes, which dictate a lower write power dissipation. However, this comes at the cost of drastically reducing the RA product of MTJ cells, which may itself pose a limitation on scaling. The gap in energy efficiency increases with reduced bit dimensions

calculations of switching energy, although in devices such as those presented in this work (with resistances <200 KΩ) the energy dissipation is still dominated by the resistance R. This energy scaling comparison is illustrated in Fig. 21, which compares the projected write energy per bit for MeRAM and STT-MRAM (assuming three different values of STT-MRAM switching current densities), for a write time of 1 ns. It shows that, while even at present technology nodes MeRAM would represent a significant improvement of energy efficiency, the efficiency gap would grow further at scaled technology nodes. These projections indicate that MeRAM would be able to address memory applications where energy efficiency and/or density are major concerns (e.g., embedded SRAM Cache and DRAM in mobile applications), where STT-MRAM would face a scaling issue. It should be noted, however, that both Figs. 20 and 21 are based on considering only a 1-transistor/1-MTJ cell, with no particular assumption on the memory array size. Hence, obtaining precise values of energy as well as array density would require a full simulation of a MeRAM array, which is beyond the scope of this discussion.

The unipolar write scheme in MeRAM also confers an indirect advantage on it in terms of the readout process. In particular, given that the free layer switching in MeRAM is only sensitive to voltages of one polarity, voltages of the opposite polarity can be used for a read-disturbance-free readout. This also means that the device can be read out using voltage levels similar to those used for writing, but with opposite polarity, (unlike STT-MRAM, which requires much smaller read voltages to prevent read disturbance), hence allowing for a fast readout despite the increased resistance of the device.

Successful realization of these potential advantages will require additional development of improved MeRAM bits. In particular, it is important to maximize the VCMA effect (i.e., increase ζ) through materials optimization, while maintaining high TMR for readout. In addition, high-speed bidirectional switching of MTJs using techniques similar to those described in Sect. 3.3 will be needed for the realization of fast MeRAM arrays. Once these challenges are addressed, it is expected that MeRAM can then be a candidate for energy-efficient, dense, and fast nonvolatile memory with better scalability than previous types of MRAM.

4 Electric Field Controlled Magnetism in Metallic Films Beyond VCMA

Previously, we have introduced the VCMA effect in the ferromagnetic thin films, where electric field modifies the magnetic anisotropy of the thin film. Actually, electric field can induce modifications in materials' magnetism in terms of other aspects, such as domain wall, phase transition, Curie temperature and so on. In this section, we first introduce electric field controlled ferromagnetic phase transition in ultrathin metallic films beyond VCMA effect, and then discuss a specific parameter, Curie temperature, which can be controlled by external voltage.

4.1 Electric Field Controlled Phase Transition in Thin Metallic Films

Theory of electric-field-controlled surface ferromagnetic transition has also been proposed by Ovchinnikov and Wang [132]. It is widely believed that in metals, unlike in the diluted magnetic semiconductors (DMS), the electric field control of ferromagnetic order is hardly achievable. There are two arguments to support this opinion: (a) the carrier density ρ in metals is extremely high, the variation of T_C induced by the change of surface charge density σ is negligible; (b) the electric field cannot penetrate the surface, since it is screened by the surface atomic layer due to high ρ. However, we will show both arguments are incorrect in the following part.

The mechanism for electric field control of magnetism in the DMS is originated from the control of local (bulk) Curie temperature T_C^b, which is directly related with the number of itinerant carriers, by the electric field existing in the penetrating region of DMSs. Since the variation of T_C^b is relatively small, it is necessary to make operation temperature T_0 of the device close to the T_C^b, $\tau = (T_0 - T_C^b)/T_C^b \rightarrow 0$.

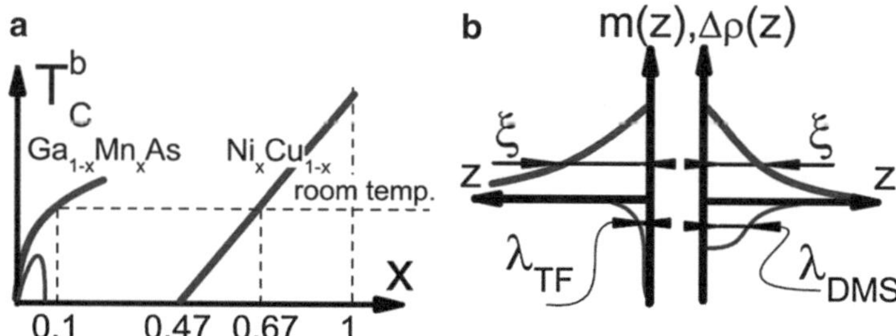

Fig. 22 (**a**) T_C^b for Ga$_{1-x}$Mn$_x$As from mean-field considerations and Ni$_x$Cu$_{1-x}$ from experimental studies. The thin line represents T_C^b of the DMS from Monte Carlo simulations. (**b**) Even though the width of the injected charge distribution, $\Delta\rho(z)$, (*thin lines*) is higher in the high-T_C DMS (*right side*), $\lambda_{TF} < \lambda_{DMS}$, the magnetization profile, $m(z)$, (*thick lines*) being determined by the spin-spin-correlation length, ξ, is wider in metals (*left side*). Reprinted with permission from Ref. [132]. Copyright 2009 by the American Physical Society

The dependence of general Curie temperature T_C on the itinerant carrier density is determined by $\Delta T_C \approx \partial_\rho T_C \cdot \Delta\rho$, where $\partial_\rho T_C$ is the material-specific T_C sensitivity. So, it is not like $\Delta T_C/T_C \propto \Delta\rho/\rho$, which indicates the influence of changing surface carrier density is very small at high ρ. Figure 22a gives the T_C^b of two specific materials: Ni$_x$Cu$_{1-x}$ and Ga$_{1-x}$Mn$_x$As, which shows that at room temperature, they have the similar sensitivity. Therefore, there should be a similar behavior of electric field controlled phase transition in metallic films, and thus argument (a) is not correct.

Due to high ρ, the Thomas-Fermi screening length λ_{TF} of metals is much smaller than DMS, which indicates the electric field can only exist at the surface atomic layer and modify the carrier density of surface atoms (see Fig. 22b). Therefore, the phenomenon of electric field controlled phase transition in metals should be viewed as a surface transition, rather than local bulk transition. The width of this surface transition is much larger than the λ_{TF}, and is called spin-spin-correlation length, ξ, which for metals is wider than DMS as shown in Fig. 22b. An additional surface integral term, $c\int_S m^2$, where m is the spatial density of the magnetization and c is the surface enhancement. The effects of the surface-enhanced magnetization [71] and the so-called magnetic "dead" layers [133] are the direct indications that $c > 0$ and $c < 0$, respectively. Therefore, with the temperature-surface enhancement, the electric field control of ferromagnetic phase transition in metals is possible. In the previous section, the VCMA effect is one manifest of this theory. The detailed theory is given in Ref. [132], and within in this theoretical framework, in principle, dependence of any magnetic properties of metallic films on the electric field can be analyzed. In next section, one important parameter, Curie temperature, of ferromagnetic metals controlled by the electric field is discussed.

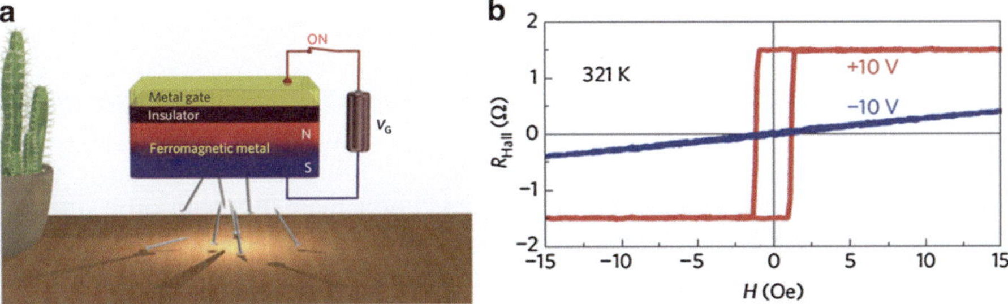

Fig. 23 Electric field induced ferromagnetic phase transition near room temperature in the ultrathin Cobalt (Co) film. Reprinted by permission from Macmillan Publishers Ltd: Nature Materials Ref. [64], copyright 2011. (**a**) Illustrative picture of a ferromagnetic phase transition device, which can change from ferromagnet to normal metal by applying voltage. The device for the transport measurements consists of a metal gate (Au/Cr), an insulator layer (HfO_2), and an ultrathin Co layer. (**b**) Switch between ferromagnetic phase and paramagnetic phase: at +10 V gate voltage, the Co film is in the ferromagnetic phase, and at −10 V gate voltage, the Co film is in the paramagnetic phase

4.2 An Example: Electric Field-Controlled Curie Temperature

Change of the Curie temperature of metallic films induced by electric fields has been theoretically proposed [65] and experimentally demonstrated in cobalt [64]. Typically, Curie temperature is used to characterize a ferromagnet, below which the ferromagnet has spontaneous magnetization. The spontaneous magnetization phase transition from nonzero value to zero is described by $M \approx M_0(1 - T/T_c)^{\beta}$, where β is a critical exponent and it only depends on the system dimensionality and order parameter (here is magnetization) degree of freedom. There are two possible ways to tune the Curie temperature: one is changing the magnetic anisotropy, which is VCMA described above, and the other is changing the magnetization interaction between, which is typically described by a parameter, exchange interaction energy J. In principle, change of carrier density will induce change in average J, and because in two-dimensional system, adjustability of carrier density in the two-dimensional system is larger than the three-dimensional system. Therefore, in ultrathin ferromagnetic metals, the voltage sensitivity of Curie temperature could be noticeable.

Chiba et al. [64] demonstrated that one can use the electric field effect to switch a thin metal film between ferromagnetic phase (ferromagnet) and paramagnetic phase (normal metal) (see Fig. 23). Compared with the traditional electromagnet, this kind of new electromagnet has advantage in terms of device scalability and power consumption. Since it doesn't require a conducting wire and a large current within it, it has a better scalability and is more energy efficient. Moreover, the gate controlled ferromagnetic phase transition is like a switch, which can turn on and off the magnetic properties of the channel, and hence it can potentially be integrated into the spintronic devices.

5 Spin-Orbitronic Devices

Spin-orbit interaction, describing the coupling between the orbital and spin degrees of freedom provides yet another avenue to control magnetism electrically. The devices that exploit spin-orbit interaction for controlling the ferromagnetic order are termed here as the spin-orbitronic devices. Fundamentally, spin-orbit interaction is a relativistic effect as explained in the following. The relativistic equation describing electrons in an atom is the well-known Dirac equation, which when approximated up to second order in v/c (with v and c being the electron's velocity and the speed of light), generates (amongst others) the following Pauli spin-orbit coupling term [134]:

$$H_{SO} = -\frac{\hbar}{4m_0^2 c^2}\sigma \cdot p \times \nabla V_0 .\tag{10}$$

Here $\hbar$ is the reduced Planck's constant, m_0 is the electron's mass, σ is the Pauli spin matrix vector, p is the momentum operator and V_0 is the atomic potential. Typically, this relativistic effect is more important for the core electrons, which are closer to the nucleus and are thus moving at much higher velocities. Similarly, the heavier elements have a larger spin-orbit coupling. In a crystal, the valence electrons move in a periodic "pseudopotential", i.e., the atomic potential, due to a periodic arrangement of nuclei, renormalized by the core electrons. In this case, these valence electrons are described by the so called Bloch bands and the effect of spin-orbit coupling, entering through the pseudopotential, is conveniently parameterized by a material dependent effective parameter, which can be obtained from the band structure [134]. The consequence of a spin-orbit term is that the orbital degrees of freedom, which can be controlled electrically, become coupled to the spin-degrees of freedom, which in turn can interact with magnetic order, and hence provide the sought mechanism for electric control of magnetism.

The requirement of having ferromagnetic order and high spin-orbit interaction, for constructing spin-orbitronic devices, is typically met either by interfacing a heavy metal with ferromagnets (devices of this type are discussed in Sects. 5.2 and 5.3) [56, 135] or by doping materials possessing high spin-orbit coupling with magnetic impurities (devices of this type are discussed in Sect. 5.1) [20, 136]. Before discussing the particular examples in details, we discuss a general symmetry based phenomenology for understanding the various possibilities offered by the introduction of the spin orbit term in controlling magnetic order.

All spin-orbitronic devices discussed here work below the Curie temperature, where the magnitude fluctuations in the magnetization are suppressed and the information is stored in the orientation of magnetization. The equation of motion conserving the magnitude of magnetization can then be written within the Landau Lifshitz Gilbert (LLG) phenomenology as [137, 138]:

$$\partial_t m = -\gamma m \times H(m, I, V) + \alpha m \times \partial_t m .\tag{11}$$

Here, α is the Gilbert damping parameter, γ is the gyromagnetic ratio and m is a unit vector oriented along the magnetization. The effective field $H(m, I, V) \equiv H_m + H_I$ $+H_V$ includes the following contributions: H_I and H_V represent the electrically controlled terms allowed via spin orbit interaction (and the subject of the present section) while H_m represents the sum of externally applied and the internal effective magnetic fields, which in its minimalistic form is derived from exchange, dipolar and crystalline anisotropy energy. The crystalline anisotropy energy is the first example of a term that arises due to spin orbit interaction. Intuitively, in the presence of spin orbit interaction the magnetization, i.e., the spin degrees of freedom, becomes "aware" of the crystalline structure, i.e., the orbital degrees of freedom, and hence prefers to orient along certain crystalline axis, known as the easy axis. In this sense, even in the absence of any electrical control, spin-orbit interaction influences the orientation of magnetization. In fact, for computational devices this crystalline anisotropy term is crucial, as by restricting the magnetization to point in opposite directions along the easy axis, the information becomes binary. Moreover, the barrier to overcome the anisotropy energy makes this information non-volatile.

To arrive at the form of allowed terms in H_I and H_V one can either start from the microscopic picture or utilize the symmetry principle. In the following we adopt the latter approach which states that the equation of motion should remain invariant under the transformations which respect the symmetry of the system under consideration. The typical symmetries of the devices discussed here are (see Fig. 27a):

(a) Rotation invariance about the growth axis (chosen to be oriented along the z-axis): the hetero-structures considered here are typically amorphous or polycrystalline films, thus having no favored direction in the film plane.

(b) Mirror symmetry: In all devices the mirror symmetry about the x-y plane is broken intentionally by having different material environment on top and bottom of the ferromagnet, leaving the structure mirror symmetric only about the y-z and x-z plane. In Sect. 5.2 we will also break the mirror symmetry about the x-z plane to introduce additional useful spin-orbit terms.

Imposing the above mentioned symmetries on the equation of motion and knowing that magnetization transforms as a pseudo-vector, the leading order terms allowed in H_I and H_V can be written as (some of the terms appearing at the same order as below are suppressed for simplicity, interested readers are referred to the Refs. [139, 140]):

$$H_I = H_D m \times (i \times z) + H_F(i \times z), \qquad (12)$$

$$H_V = H_E m_z z. \qquad (13)$$

Here, i and z are the unit vectors along the current and the normal to the film plane, respectively. The H_I terms and corresponding torques, i.e., $m \times H_I$, are known as the current-induced spin-orbit fields and spin-orbit torques, respectively. The first term on the right hand side of H_I is known as the damping (or anti-damping)-like term, owing to its non-conservative nature with the corresponding torque changing sign under time reversal. While the second term is referred to as the field-like term, due to its conservative nature with the corresponding torque preserving sign under time reversal. The strengths of the damping-like and the field-like terms are in turn parameterized by H_D and H_I. Microscopically, two primary effects are typically associated with the spin-orbit torques: the so called spin-Hall [141] and Rashba/Edelstein effect [142, 143]. The spin-Hall effect is the flow of spin current transverse to charge current which in the diffusive metal systems arises due to inequivalent scattering of up and down spin electrons in the presence of spin-orbit interaction. This spin current can then interact with the magnetization and thus apply torque on it. While, the Rashba/Edelstein effect is an interfacial effect where, due to change in the bandstructure at the interface, large internal electric fields are typically present. These electric fields via spin-orbit interaction in turn lead to a net spin polarization which upon interacting with the magnetization can also result in the above mentioned spin orbit torques. It is important to note that the current-induced spin-orbit torques vanish, even in the presence of spin orbit interaction, if all the mirror-symmetries are present.

Similarly, the H_V terms and corresponding torques, i.e., $m \times H_V$, are the voltage-induced fields and torques. Microscopically one phenomenon responsible for the voltage torques is the so called voltage control of magnetic anisotropy (VCMA) [71]. As was mentioned above, the magnetic anisotropies arise due to spin-orbit interaction, which are sensitive to electric field due to charging effect at the interface. As a result, the magnetic anisotropies can be tuned via application of gate-voltage, which in turn reorients magnetization by application of voltage-induced torques.

Thus based on the allowed terms, a basic element of a spin-orbitronic device could be a three terminal device. Information can be stored in the orientation of magnetization, which can in turn be manipulated by a current, a voltage via a gated structure or both. Such a spin-orbitronic device was experimentally demonstrated recently in a heavy metal/ferromagnet/oxide heterostructure. A lateral current above a critical threshold, flowing in the plane of the heavy-metal film, was shown to switch a magnet oriented both perpendicular and parallel to the interface [56, 144]. Of particular importance is the case of switching perpendicularly oriented magnets as they allow for packing information more densely than their in-plane counterparts. For the case of in-plane oriented magnet a magnetic tunnel junction was also fabricated to read the orientation of magnetization thus demonstrating a spin-orbit torque memory [56]. Additionally, for these in-plane devices gate control of the critical threshold current via VCMA has been demonstrated. However, the critical currents required for the switching are still notoriously high resulting in high power consumption due to joule heating. One of the primary goals of current research is to find ways to reduce these critical currents for low power applications. In the next section we discuss one such material system with required critical currents about three orders of magnitude lower than those in the heavy metal system.

Another desirable property of spinorbitronic devices is the ability to operate them without use of any external magnetic fields. The denser perpendicular spinorbitronic devices demonstrated so far, however require application of in-plane external magnetic fields in addition to the current-induced spin orbit torque. In Sect. 5.2 we demonstrate a strategy to overcome this challenge and achieve zero-field switching of perpendicular bits purely by current-induced spin orbit torques. Finally, in Sect. 5.3, taking advantage of the multi-terminal nature of the spinorbitronic devices, use of spin orbit torques for logic devices is discussed.

5.1 Current-Induced Magnetization Switching in Topological Insulators

One route to increasing the strength of spin-orbit torques and hence decreasing the critical currents for magnetization switching is to look for material system with high spin-orbit interaction. Recently, topological insulators(TI) have emerged as a new class of materials where spin orbit interaction is strong enough to cause the so called topological phase transition [145]. The non-trivial topology, as characterized by a topological invariant constructed from the band structure, results in an insulating state in the bulk while conduction via spin-polarized surface states on the surface of a topological insulator. Thus topological insulators provide an ideal playground for exploring spin-orbit torques. However, typical TIs are not magnetic and interfacing them with metallic ferromagnets will not serve the purpose as most of the current will be "shunted" by the metallic ferromagnet, where the high spin-orbit interaction is absent. Thus, one of the routes to exploring spin-orbit torques in TI is provided by doping TI with magnetic impurities. Recently such a magnetic TI state was achieved by doping Bi_2Se_3 (a typical TI) with Chromium [146]. The easy axis of the resultant ferromagnet is oriented along the perpendicular (to the film plane) axis. Interestingly this magnetic state can also be controlled by a voltage from top gate. Having achieved both magnetism and its proximity to a high spin orbit interaction system the final requirement of breaking mirror symmetry, for observing non-zero spin orbit torques, is fulfilled by making a heterostructure of TI/magnetic TI (as illustrated in Fig. 24). In the particular example presented in Ref. [136] the magnetic TI consisted of six quintuple layers (~ 6 nm) while TI was three quintuple layers thick. To avoid the aforementioned current shunting problem both TI and magnetic TI were additionally doped with Sb such that the conductance of TI and magnetic TI layers were approximately same.

The switching of magnetization due to current-induced spin orbit torque is clearly observed in this material system. It should be noted for the case of perpendicular oriented magnetization an additional externally applied in-plane magnetic field oriented along the current direction is necessary for obtaining deterministic switching by spin-orbit torque of form Eq. (12) (this will be explained in details in Sect. 5.2). The strength of critical currents required to achieve this switching can be obtained by constructing a phase diagram of possible magnetic states in the presence of current and the external in-plane magnetic field as shown in Fig. 25. The maximum level of current density required in this material system

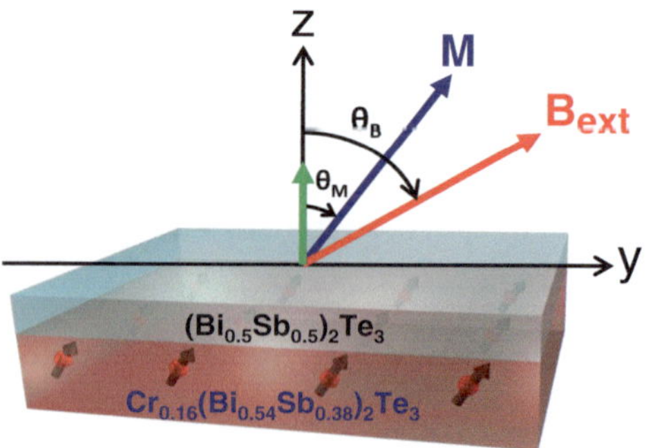

Fig. 24 Illustration of the bilayer TI/magnetic-TI hetero-structure: The magnetic TI consist of a Chromium doped (the dopants and their magnetization vector are depicted in the bottom layer) topological insulator. For characterization of current induced torque, an in-plane current is passed along the y axis along with application of an external field at an angle θ_B with respect to normal. The resultant overall magnetization of the magnetic TI is indicated by M. Reprinted by permission from Macmillan Publishers Ltd: Nature Materials Ref. [136], copyright 2014

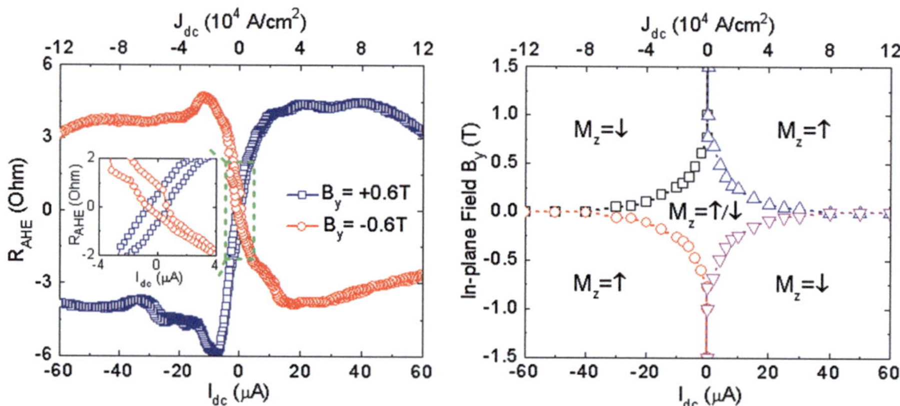

Fig. 25 Observation of spin orbit torque induced switching: Magnetization orientation as inferred from the measured anomalous Hall resistance (R_{AHE}) as a function of current is shown *left panel* for a field applied parallel (*squares*) and anti-parallel (*circles*) to the y axis. Inset shows the zoomed in region near zero current. The measured phase diagram of states of magnetization allowed in presence of current and in-plane field is shown in the *right panel*. The symbols mark the boundary between different phases. The states with positive (negative) component of magnetization along the z axis are represented by ↑ (↓). Reprinted by permission from Macmillan Publishers Ltd: Nature Materials Ref. [136], copyright 2014

($\sim 10^4$ A/cm^2) is about three orders of magnitude smaller than that required for heavy metallic system, indicating the spin-orbit interaction is extremely efficient in generating spin-orbit torques. This efficiency is quantified in terms of the so called spin-Hall tangent defined as: $\eta = 2eM_sH_It/\hbar J$, where e is the electron's charge, M_s is the saturation magnetization, t is the thickness of the ferromagnet, J is the current

Fig. 26 Quantitative measurement of spin-orbit torque in TI/magnetic-TI heterostructure: The effective field extracted from second-harmonic generation experiment for three different orientations of external magnetic field, as measured by the angle it makes with the z axis (θ_B), and current density. Reprinted by permission from Macmillan Publishers Ltd: Nature Materials Ref. [136], copyright 2014

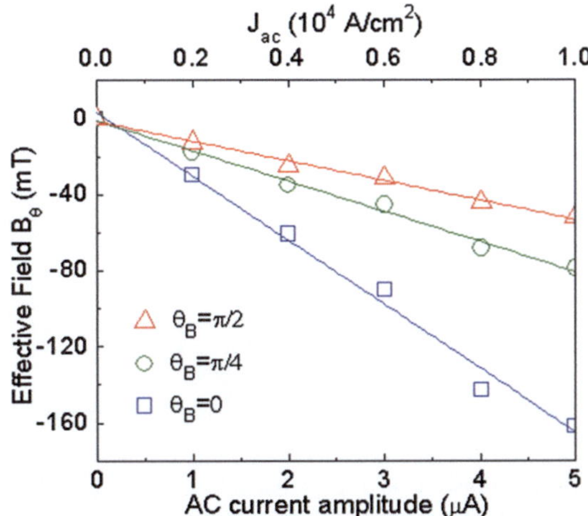

density and H_I is the strength of the effective current induced spin-orbit field . This spin-Hall tangent can be directly measured by measuring the strength of current-induced spin orbit fields via the technique of harmonic generation [147].

Figure 26 shows the result of harmonic generation experiment resulting in the spin Hall tangents of ~425, three orders of magnitude larger than that of the heavy metal system. It is likely that the spin-polarized surface states might have a role in producing spin orbit torque however the microscopic mechanism resulting in the colossal spin orbit torques would require further experimentation. Irrespective of the microscopic mechanism, TIs could provide the necessary pathway towards ultra-low power spinorbitronic devices for computing. The key challenge to overcome would be to increase the Curie temperature of magnetic TIs, which in the present case ~10 K, and/or interface TIs with magnetic insulators.

5.2 *SOT-Induced Zero-Field Magnetization Switching in Perpendicular Devices*

As mentioned in Sect. 5.1 current-induced spin orbit torque was used to switch both in-plane and perpendicular magnets in thin film heterostructures with broken mirror symmetry about the film plane. However, the form of current induced spin orbit torques for these structures necessitates use of external in-plane magnetic fields along the current direction in order to switch the perpendicular magnets (as explained later). This in-plane magnetic field can in principle be integrated into the device by using additional ferromagnetic layers at the cost of lowering the thermal stability advantage of the perpendicular bits. Thus for computational spin-orbitronic devices it is desirable to find strategies to switch perpendicular magnets without the external magnetic field. In order to devise a strategy for switching in absence of external magnetic field we discuss first why an external field is required

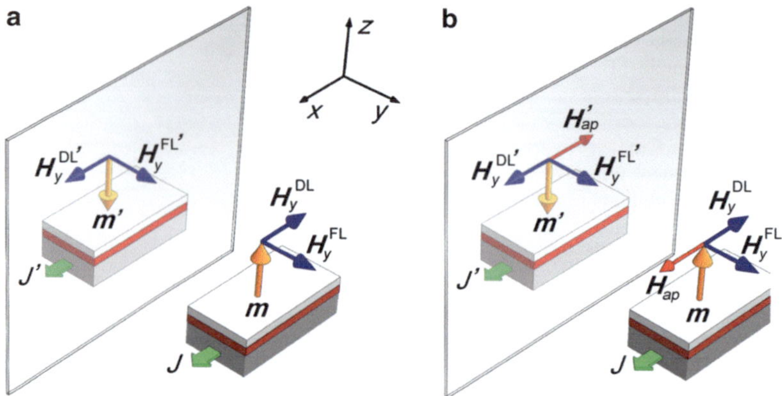

Fig. 27 Illustration of requirement of external field to achieve current induced switching for structures with symmetry broken only along the xy plane. (**a**) The figure shows a state along with its mirror transformation about the xz plane in absence of external field. H_y^{DL} and H_y^{FL} represents current (J) induced damping-like and field-like spin orbit fields. Mirroring in the xz plane reverses the magnetization state for the same direction of current hence resulting in both up and down states allowed. (**b**) By fixing the direction of external field H_{ap} one can choose between up state or down state for a given direction of current. Reprinted by permission from Macmillan Publishers Ltd: Nature Nanotechnology Ref. [139], copyright 2014

for switching perpendicular magnets in heterostructures with broken symmetry about the film plane.

The requirement of the external field can be explained by using symmetry principles [139]. Structures with broken mirror symmetry about the film plane are depicted in Fig. 27a. Without the external magnetic field if a perpendicular state with positive magnetization is allowed for a current flowing along the +x axis, by performing a mirror transformation in the xz plane (and remembering magnetization transforms as a pseudo-vector) we see that the same current should allow for a state with magnetization along −z axis. Thus a unique magnetization state is not selected by the current and hence no deterministic switching occurs. However once an external magnetic field is applied along the current direction, the mirror transformation also flips the external magnetic field (Fig. 27b). Hence, by fixing the external magnetic field a particular magnetization state is chosen for a particular direction of current, resulting in deterministic switching. In essence, the role of magnetic field is to break the mirror symmetry about the xz plane. Thus if a method can be devised to break mirror symmetry about the xz plane without using external magnetic fields, a deterministic switching can be achieved in spin-orbitronic devices purely via spin orbit torques. This also reflects in the equation of motion for magnetization, derived by the method used in Sect. 5. Breaking mirror symmetry about the xz plane, in addition to breaking the mirror symmetry about the xy plane, allows for two new leading order terms, namely $H_F^z(I)z$ and $H_D^z(I)m \times z$. The term $H_F^z(I)z$ acts like a current induced effective spin orbit field in the perpendicular direction which can thus facilitate switching of the perpendicular magnetization in the absence of external magnetic field.

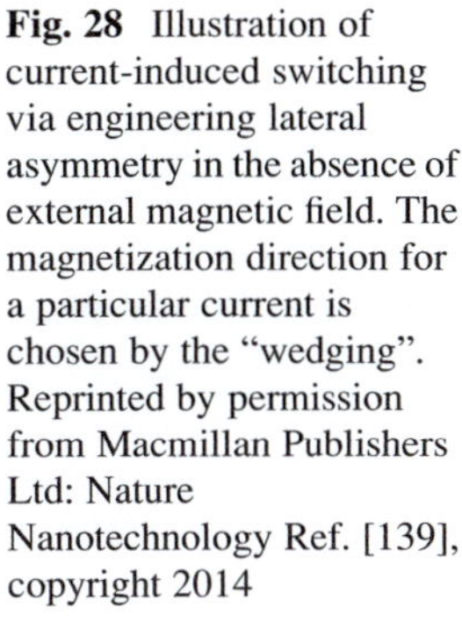

Fig. 28 Illustration of current-induced switching via engineering lateral asymmetry in the absence of external magnetic field. The magnetization direction for a particular current is chosen by the "wedging". Reprinted by permission from Macmillan Publishers Ltd: Nature Nanotechnology Ref. [139], copyright 2014

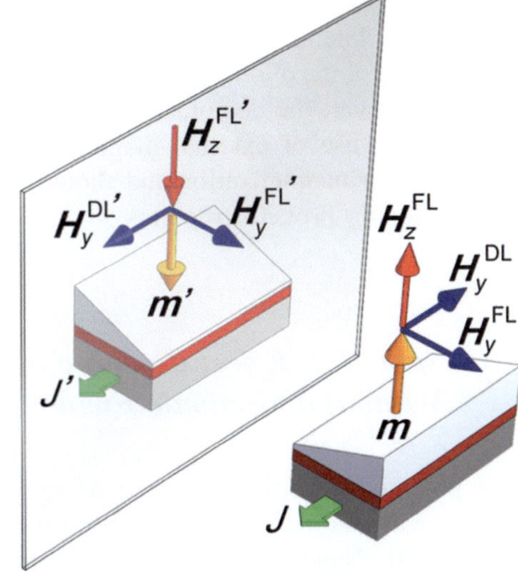

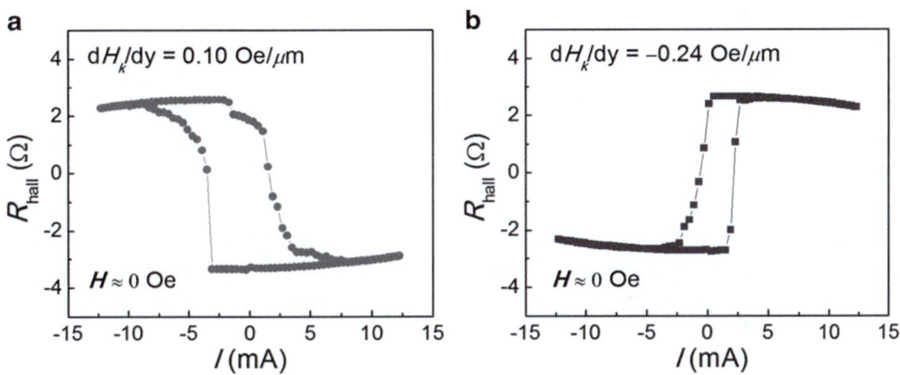

Fig. 29 Experimental observation of current-induced switching of magnetization as measured via anomalous Hall resistance for samples with additional lateral asymmetry (**a**, **b**) depicts switching in samples with "positive" and "negative" lateral asymmetry as measured by the slope of variation in anisotropy along the lateral direction, i.e., $\dfrac{dH_k}{dy}$, respectively. Reprinted by permission from Macmillan Publishers Ltd: Nature Nanotechnology Ref. [139], copyright 2014

One method for breaking mirror symmetry about the xz plane was demonstrated in Ref [139] by engineering a lateral asymmetric structure via depositing a wedge shaped oxide layer as schematically depicted in Fig. 28. Resultant current induced switching of the magnetization in the absence of external magnetic field was subsequently observed as shown in Fig. 29. The strength of the new z-directed perpendicular spin orbit field, as also measured for single domain magnets via harmonic generation technique, was found comparable to the regular spin orbit fields produced due to symmetry breaking about the xz plane. This is a somewhat surprising result given that the symmetry breaking about the xz plane is abrupt, i.e., takes place on atomic length scales, as compared to more gradual symmetry breaking by lateral oxide gradients. This along with the microscopic mechanism responsible for the observed new field deserves further experimental and theoretical effort. However, the demonstration of zero field switching provides a pathway to alleviate the use of external magnetic fields in the spin-orbitronic devices with perpendicular magnetization and should encourage the search for additional practical symmetry broken structures.

5.3 Spin Hall Effect Clocking of Nanomagnetic Logic Without a Magnetic Field

The potential of spin-orbit torques in reducing power consumption for logic devices was also demonstrated recently for the so called scheme of nanomagnetic logic [148]. Nanomagnetic logic is a spin-based computing scheme where the information, stored in the orientation of nano-scale magnets, is manipulated based on the dipole-dipole interaction among the nanomagnets. Typically, the nanomagnetic logic elements consist of a chain of dipole coupled nanomagnets, where the state of the chain (as described by the orientations of each element) can be changed by the application of a "clocking" field as follows. The chain of magnets, sitting in one of the many available equilibrium configurations, is put into a metastable configuration by the application of the clocking field. On turning the clocking field off, the system settles into the desired low energy configuration minimizing the magnetic dipole energy, based on the input bits. The advantage offered by such a scheme is the possibility of lower power consumption than that used by transistors, as the reconfiguration caused by dipole interaction dissipate orders of magnitude low energy than that caused by joule heating due to reconfiguration of charges in transistors. However, so far, one of the major bottlenecks in utilizing this promise of nanomagnetic logic has been the requirement of external magnetic fields for clocking the magnets. The power consumed in the external circuitry to generate these external magnetic fields overwhelms the advantage offered by nanomagnetic logic. This challenge was overcome recently utilizing spin orbit torques [149].

Bhowmik et al. showed a chain of perpendicularily oriented nanomagnets of CoFeB clocked in the absence of external magnetic fields (as shown in Fig. 30). Spin orbit torques from a current flowing in a Tantalum underlayer replace the role of external magnetic field. The above chain can be stabilized into one of the two possible states, namely (1) up down up and (2) down up down, depending upon the orientation of an additional input magnet (which usually is designed so that its state is unaltered by the clocking pulse). The typical nanomagnetic logic scheme would change the state of the chain (for e.g., from (1) to (2)) by the application of a power hungry external magnetic field clock applied in-plane, strong enough to put the chain into a metastable position with the magnetization oriented in the plane. However, as shown in Ref. [149], the same objective can be achieved by SOT

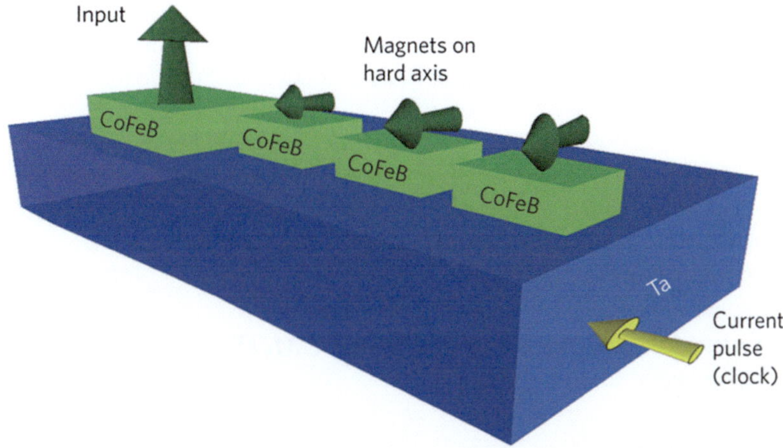

Fig. 30 Schematics of the device fabricated for demonstration of spin-Hall clocking. A current pulse is passed through the heavy metal underlayer of Tantalum. This "clock" pulse puts the magnetization in three identical bits of magnetic elements along the plane of the fim. The output after the pulse is decided by a fourth input bit having larger perpendicular anisotropy (engineered such that it is not affected by the clock pulse). Reprinted by permission from Macmillan Publishers Ltd: Nature Nanotechnology Ref. [149], copyright 2014

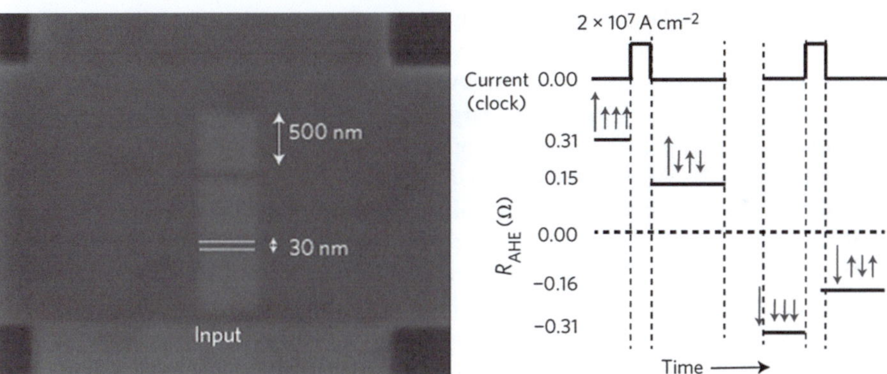

Fig. 31 Experimental structure for demonstration of spin-Hall clocking is depicted on the *left panel*. The measured state (as inferred from the anomalous Hall resistance) of the four bits before and after the application of spin-Hall clock pulse is shown on the *right panel*. The ↑ (↓) represents the state with magnetization along positive (negative) z axis. Reprinted by permission from Macmillan Publishers Ltd: Nature Nanotechnology Ref. [149], copyright 2014

field induced by symmetry breaking in the growth direction (i.e., Eq. (12)), which can be understood as follows. In the limit when SOT is the only term in Eq. (12), the equilibrium orientation is the case when magnetization is pointing in the plane, i.e., the same metastable position obtained by the application of external magnetic fields. Once the spin orbit current pulse is turned off the chain goes to a state decided by the orientation of input magnet. This functionality is shown in Fig. 31.

Moreover, it was shown that the current level required by such a SOT clocking should consume about three orders of lower power than traditional methods of current induced Oersted fields thus providing an attractive alternate for nanomagnetic logic applications.

6 Applications and Perspectives on Spintronic Computation

6.1 Nonvolatile Circuits with Hybrid CMOS and Spintronic Memory

While memory is the most straightforward application area of MTJs, nanomagnetic devices also offer opportunities for integration of their inherent nonvolatility and reconfigurability for application of circuit functions. Magnetic non-volatile logic provides the possibility of reducing standby power consumption significantly, for energy-efficient nonvolatile systems.

We will discuss the potential integration of MTJ based spintronic devices for several levels of applications or different system levels from the circuit point of view: replacing SRAM, complementing logic circuits and eventually improving performance at gate and transistor levels. Nowadays, the minimum switching energy achieved for standard STT-MRAM cells is on the order of ~100 fJ, which is about 2–3 orders of magnitude larger than that of CMOS. Relatively high dynamic switching power limits the applications of STT-MRAM. For the implementation of STT-MRAM for caches (in replacing or complementing SRAM), according to Smullen et al. [150]'s results from the simulation of several different configurations at the 32 nm node (see Fig. 32), the energy efficiency can be improved by a factor of about five by relaxing the nonvolatility and the memory density can be increased by using a single MTJ cell to replace six-transistors based SRAM. As the MTJ size is reduced, to keep the thermal stability factor Δ as a constant, higher (perpendicular) magnetic anisotropy is required for long term storage, which typically requires 10 years retention time. However, for logic applications, 1 hour retention time is enough in most cases and hence the relaxation of the thermal stability requirement is reasonable here.

To achieve realistic nonvolatile electronics, the integration needs to go down to small circuits or logic level and eventually transistor level. Therefore, the switching energy has to be scaled down further to be compared to or better than that of CMOS. Alternatively, a more energy efficient mechanism is required for information processing and propagation, which is to be explored in the next spin wave logic section. For hybrid CMOS/MTJ circuits, one approach is using MTJs as both memory cells and functional inputs to latch data, which is also referred to be a logic-in-memory (LIM) architecture. A recent study that evaluates the energy performance of the MTJ-based LIM architecture (see Fig. 33) in comparison with

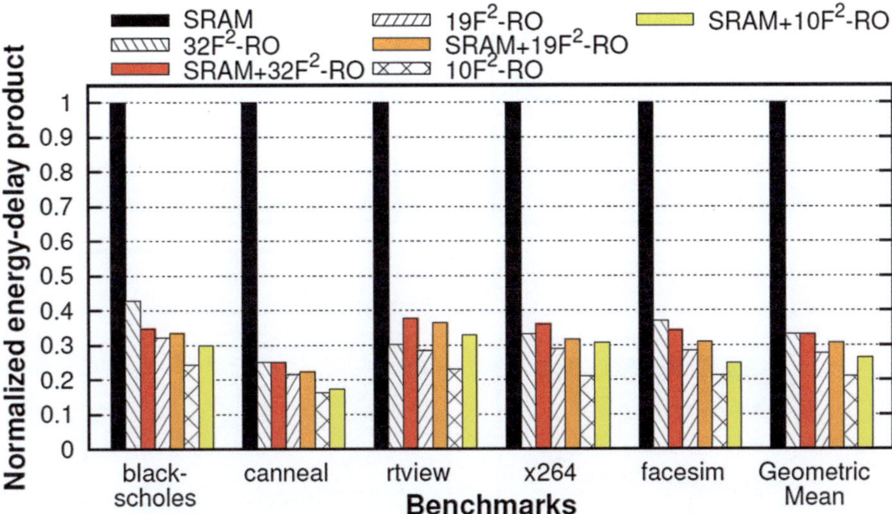

Fig. 32 Energy-efficiency of hybrid SRAM and STT-MRAM cache hierarchies. XX F^2 is the area of the STT-MRAM (F is the technology node, half-pitch), and RO represents that read-optimized STT-MRAMs are used. The horizontal axis stands for different workloads. The use of the hybrid architecture with a relaxed nonvolatility requirement reduces energy-plot product as well as increasing density. © 2011 IEEE. Reprinted, with permission, from Ref. [150]

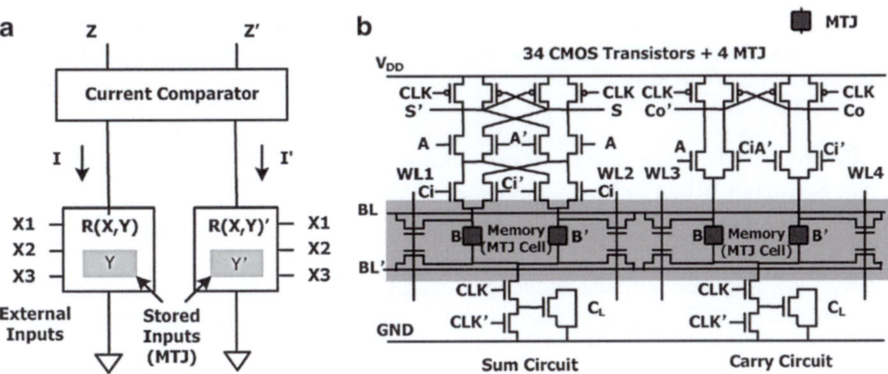

Fig. 33 (**a**) Schematic of MTJ-based LIM architecture; (**b**) a hybrid CMOS-MTJ dynamic current-mirror-logic 1-bit full adder. © 2010 IEEE. Reprinted, with permission, from Ref. [151]

static and dynamic CMOS implementations, shows that MTJ-based LIM architecture has no tangible advantage in energy performance over its equivalent CMOS design. This is due to the fact that the write energy in STT devices is still too high compared to CMOS.

To dramatically improve the energy efficiency of magnetic memory and make it suitable for integration with CMOS for nonvolatility at the gate level, we need to

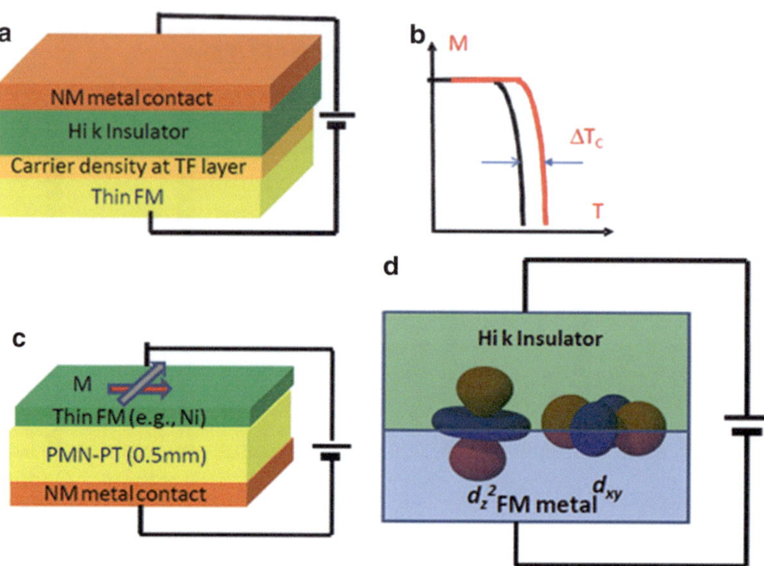

Fig. 34 Principles of electric field control of magnetism. (**a**) Control of ferromagnetic phase transition through the Thomas-Fermi (TF) layer at the interface where extraordinary temperature-surface enhancement occurs. The result of electric field induced Curie temperature shift is shown on the *right* (**b**). (**c**) Multiferroic heterostructure (magnetoelectric cell), where the magnetic anisotropy is affected by the electric field induced strain via piezoelectric material. (**d**) Interfacial magnetic anisotropy controlled by an electric field. For example, occupancies of the different 3d orbitals at the interface are changed by an applied electric field. In turn the spin-orbit interaction results in a reorientation of the magnetic moments. Reprinted with permission from Ref. [13]. Copyright 2012, World Scientific

further reduce the write energy of nonvolatile elements. One approach is using the giant Spin Hall Effect in heavy metals as discussed in the spin-orbitronics section, which can significantly reduce the switching current due to large spin Hall angle and adjustable ratio of heavy metal cross-section to MTJ area [56, 82]. Alternatively, one may use voltage-induced switching of magnetization, as opposed to current-driven STT switching. Electric field control of magnetism, similar to that of MOS structure as illustrated in Fig. 34, can lead to an ultra-low power nonvolatile magnetic memory.

Figure 34 gives several principles of electric field control of ferromagnetism: (1) control the phase transition and hence control the Curie temperature (Fig. 34a, b); (2) control the magnetic anisotropy, like VCMA effect (Fig. 34d); (3) control magnetic anisotropy by magnetostriction of a thin magnetic film with strain (Fig. 34c). The first two mechanisms have already been discussed in the Sects. 3 and 4, respectively. In this section, we focus on magnetostriction and the use of multiferroic materials. The multiferroic materials include single-phase and synthetic multiferroic materials [59]. One example is to use synthetic multiferroic heterostructures consisting of piezoelectric and ferromagnetic materials to realize

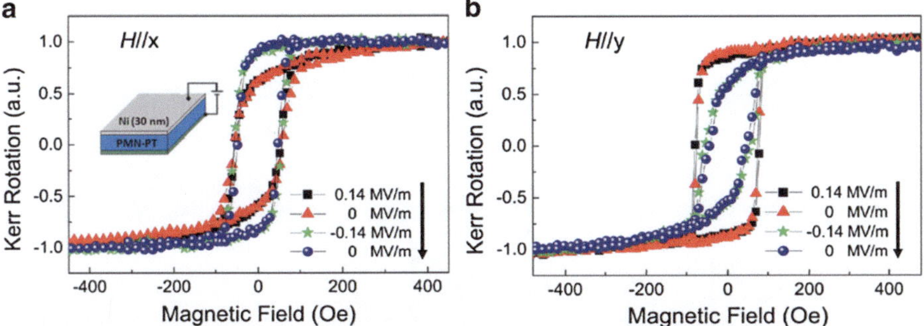

Fig. 35 Magnetization rotation and switching by electric field in 30 nm Ni films on a piezoelectric PMN-PT substrate. Permanent and reversible voltage-induced 90° switching of the magnetic easy axis is achieved in this system, as demonstrated by MOKE measurements for different electric fields. The magnetic moment is along the x-direction with no bias; it switches to the y-direction when a bias of 0.14 MV/cm is applied. (**a**) MV loop at different biases with the magnetic field along the x-axis; (**b**) with the magnetic field along the y-axis. *Arrows* indicate the order in which the electric field is applied in the experiment in order to switch the magnetization. Reprinted with permission from Ref. [63]. Copyright 2011, AIP Publishing LLC

voltage control of magnetization as illustrated in Fig. 34c. In this case, a voltage applied to the material stack generates a mechanical strain in the piezoelectric material. Due to the magnetostrictive property of the adjacent magnetic film, this strain can lead to a reorientation of magnetization as a result of the magnetoelectric (ME) effect.

Figure 35 gives the experimental evidence of voltage control of magnetization for a 30 nm magnetostrictive Ni film on a piezoelectric PMN-PT substrate. By measuring the state of the magnetic film using the magneto-optical Kerr effect (MOKE), it is shown that a 90° reorientation of the easy axis can be observed with field ∼0.14MV/m (or 1.4 kV/cm). This experiment demonstrates voltage-controlled reorientation of magnetization, as well as non-volatility—requiring no continuously applied voltage to keep the magnetization in the reoriented state, thus demonstrating an electric field controlled non-volatile magnetic memory operation. Furthermore, the state of such a memory bit can be readout, for example, by fabricating the magnetostrictive layer into a MTJ structure to allow for resistive TMR readout.

6.2 Spin Wave Logic

Besides hybrid CMOS/MTJ circuits, there are a number of promising ideas emerging for spin-based logic. We will limit our discussion to mostly our spin wave bus (SWB) approach. Other approaches, such as nanomagnetic logic based on dipole-coupled magnetic cellular automata, can be found in related literatures [4, 13, 148]. Figure 36 illustrates the SWB logic approach and its basic operation principle. A

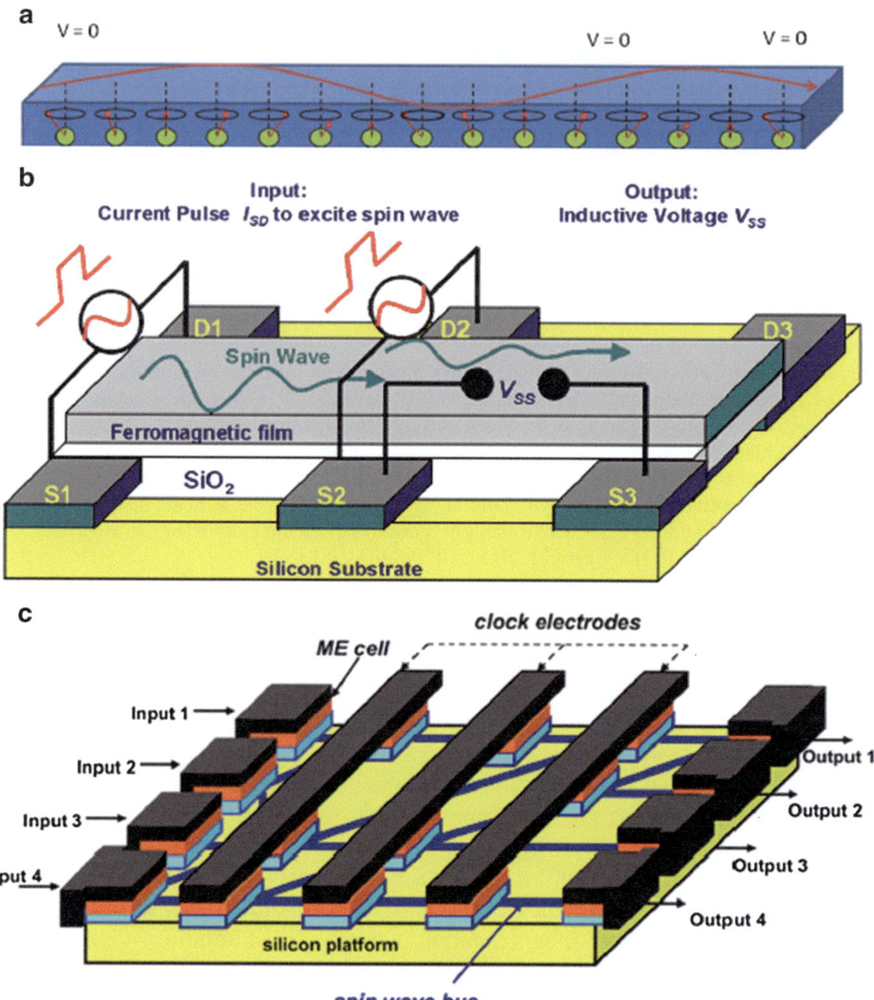

Fig. 36 (**a**) A spin wave, viewed as a collective excitation of exchange-coupled spins in a ferromagnetic waveguide. (**b**) Illustration of a device using two input elements where spin waves are excited inductively using electrical currents passing below the SWB, and are detected inductively by reading out a voltage at the output element. (**c**) Schematic of a logic circuit based on SWB and ME gates. The ME devices are used for input and output functions, interfacing with spin waves propagating in the SWB. Reprinted with permission from Ref. [13]. Copyright 2012, World Scientific

spin wave is a collective excitation of spins due to exchange interaction, which can also be viewed as a magonic wave as shown in Fig. 36a. SWB uses spin wave interference for various gate operations; the result of the operations can be readout by a nonlinear switch (a ME gate, similar to a MeRAM cell) in the final stage as illustrated in the output side of Fig. 36c.

Using SWB, we have proposed and developed the concept of magnonic logic circuits [7, 10, 152], where an applied in-plane magnetic field is used to control the spin wave propagation frequency and dispersion characteristics. The basic structure uses a magnetic film as a spin conduit of wave propagation as SWB shown in Fig. 36a, where the information can be coded into the phase of a propagating spin wave and the logic functions are performed using spin wave interference in the bus. A prototype device is illustrated in Fig. 36b, showing two microstrip lines (S1 and S2) acting as the spin wave input with a third microstrip (S3) used to readout the signal. The amplitude and phase of propagating waves can be modulated by an applied or an effective magnetic field, that latter of which can be provided by a ME gate as in the case of MeRAM devices. In this case, the modulation of spin wave is done via electric field as discussed previously through the change of magnetic anisotropy. Nonvolatile circuits can be constructed using SWB structures and ME gates as illustrated in Fig. 36c. Spin wave generation and detection at the input and output can also be achieved by ME gates for higher energy-efficiency. This approach combines several advantages: (a) information transmission is accomplished without electron transport enabling one to minimize power dissipation in the interconnects; (b) ability to use wave superposition in the SWB to enhance functionality, while switching is done at readout; (c) a number of spin waves with different frequencies can be simultaneously transmitted among a number of spin-based devices (i.e., frequency multiplexing) [153]. There are other approaches, which are based on spin wave Mach-Zehnder-type interferometer proposed in recent years [154, 155], where spin wave amplitude was used to define the logic state of the output. In contrast, in SWB, information is coded into the phase of the spin wave signal. Two logic states 0 and 1 are assigned to two phases of the spin waves having the same amplitude. In this approach, data processing is accomplished via the change of the phase of the propagating spin waves. Compared with the spin wave Mach-Zehnder-type interferometer, the main advantage of this design is that the length of the circuit is no longer limited by the phase accumulation length [154, 155]. This translates in the intriguing possibility of scaling down the length of the whole circuit to the wavelength of the spin wave (10–100 nm).

Using the latter approach, a prototype majority gate has been demonstrated. Figure 37a shows a photo of the test chip used in the experimental study of a prototype majority (MAJ) gate. Each of the five wires can be used as an input or an output port for which microwave coplanar structures are used, similar to the basic structure in Fig. 37b. In order to demonstrate a three-input one-output MAJ gate, three of the five wires were used as input ports, and two other wires were connected in a loop to detect the inductive voltage produced by the spin wave interference. The plot in Fig. 37c shows that the output inductive voltage detected from different combinations of spin wave phases. In this experiment, an electric current passing through each wire generates an Oersted magnetic field, which, in turn, excites spin waves in the ferromagnetic layer. The direction of the current flow (the polarity of the apply voltage) defines the initial spin wave phase. The relative phases between the input waves may be 0 or π, for the same or opposite direction of the current. Figures 37c, d shows the results of the inductive voltage as a function of time for

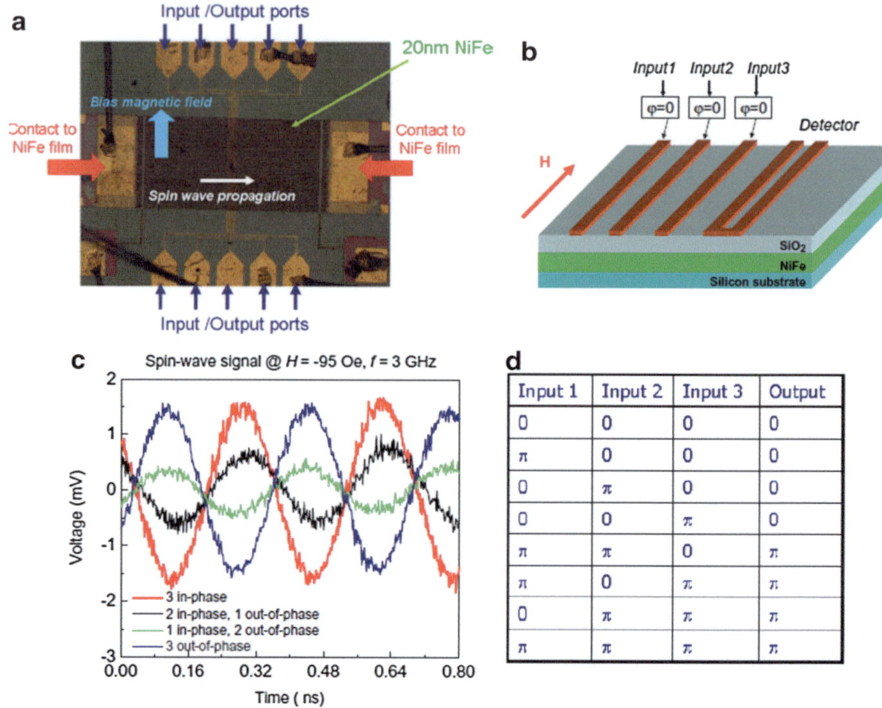

Input 1	Input 2	Input 3	Output
0	0	0	0
π	0	0	0
0	π	0	0
0	0	π	0
π	π	0	π
π	0	π	π
0	π	π	π
π	π	π	π

Fig. 37 Prototype majority (MAJ) gate. The image of the prototype four-terminal spin wave device for MAJ logic demonstration is shown on *top left* (**a**) and the schematic of the central device on the *top right* (**b**). The experimental data illustrating device operation is given in the *bottom left* (**c**) with the logic table on the *bottom right* (**d**). The operation frequency is 3 GHz. Reprinted with permission from Ref. [13]. Copyright 2012, World Scientific

different combinations of the input spin wave phases. The phase of this voltage corresponds to the majority of phases of the interfering spin waves. The data are taken for 3 GHz excitation frequency and at bias magnetic field of 95 Oe (perpendicular to the spin wave propagation). Note that this bias magnetic field may be replaced with a ME gate. This simple demonstration of the concept verifies the feasibility of the spin wave MAJ logic using phase, and further reduction of energy consumption requires the ME cell, which will be discussed in coming section.

6.3 Electric-Field-Induced Spin Wave Generation Using Multiferroic Magnetoelectric Cells

To maintain low energy consumption of magnonic devices, however, the spin wave generation and detection elements need to be energy-efficient as well, so as to minimize the dynamic power dissipation. However, the commonly used current-driven transducers such as inductive antennas [156] and STT [157] elements require

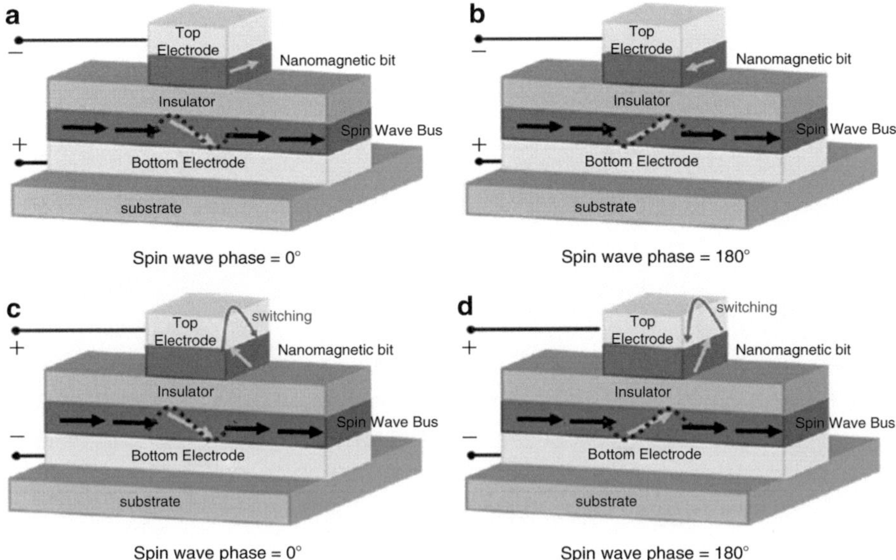

Fig. 38 Schematic of input (**a**, **b**) and output (**c**, **d**) cell operation in the all-spintronic spin-wave-based nonvolatile logic scheme. Input voltage pulses excite spin wave signals, which propagate through the SWB. The phase of the generated spin wave is determined by the state of the input memory cell via dipole or exchange interaction. The ME gates excite and store spin wave information. At the output end, the ME gate is switched by the arriving spin wave, with the switching direction being determined by the spin wave phase. A bias voltage is applied by the clock to reduce the energy barrier to allow for the spin wave to switch the state of the ME cell. Reprinted with permission from Ref. [13]. Copyright 2012, World Scientific

large currents to operate, resulting in high power dissipation and scaling issues due to intrinsic ohmic losses. To reduce the power dissipation with magnetization excitation in spin wave logic, we have proposed and demonstrated spin wave transducers where spin waves are generated by alternating voltages (i.e., electric fields) rather than currents. We refer to such devices as ME cells, and their general pictures are given in Fig. 38.

In the illustration, we show that ME gates with the SWB can realize spin wave excitation and storage of the final computational results. The only nonmagnetic input required for the circuit is a clock applied to the ME gates for triggering the switching at the output end and the spin wave generation ate the input bit. The direction of switching is determined by the spin wave phase at the output cell. The voltage pulses are converted into spin waves, then, the data transmission and processing within the SWB is accomplished through the ME gate lick the memory devices: (a) output data from each computational step is stored in a ME memory cell, which can be switched by the spin wave under the ME gate, assisted by a pulsed bias voltage. The bias is used to reduce the energy barrier for facilitating the switching by the spin wave. The direction of switching is determined by the spin wave phase; (b) a ME cell generates a spin wave for the next computational step

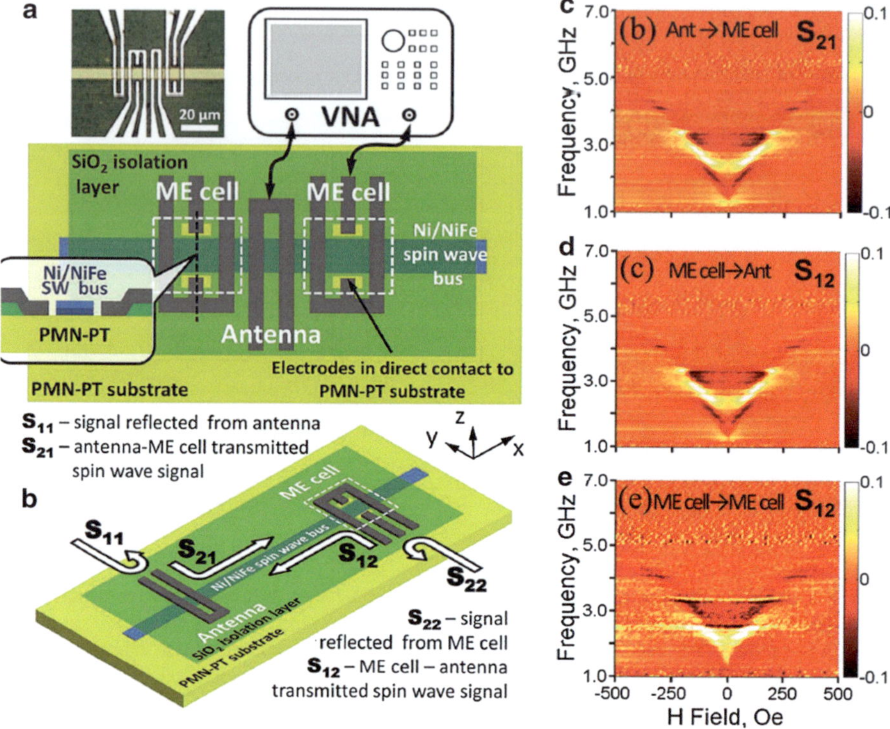

Fig. 39 (**a**) Schematic of the studied device: spin wave generation and propagation measurements using a vector network analyzer were performed on the 5 μm wide Ni/NiFe bus lithographically defined on a PMN-PT piezoelectric substrate. Inset shows cross-section view of the ME cell. (**b**) The schematic of two-port measurements of transmission (S_{21} and S_{12}) and reflection (S_{11} and S_{22}) measurements between conventional loop antennas and voltage-driven magnetoelectric cells. (**c–e**) Two-port measurements between antenna and ME cell. (**c**) spin wave signal detected by the ME cell, while being generated by the loop antenna. (**d**) transmission signal generated by the ME cell and detected by the loop antenna. (**e**) transmission signal recorded between two ME cells. Reprinted with permission from Ref. [158]. Copyright 2014, AIP Publishing LLC

with the next clock cycle and the spin wave pulse is determined by its memory state through exchange and/or dipole coupling to the SWB. Logic functionality is achieved as a sequence of the ME switching events, where each ME gate changes its state according to the magnetization of the preceding cells via propagating and interfering spin waves as assisted by the ME gate bias.

Energy-efficient electric-field-induced generation of spin waves is a key requirement of this NVL scheme to work. For ME gates or cells, we have recently successfully demonstrated this effect using a multiferroic heterostructure, showing spin waves generation by voltage in a capacitve ME gate and propagating over distances of up to 40 μm [158]. Figure 39a is the schematic of ME cells and antenna, where the field-frequency two-dimensional study results of spin wave transmission between ME cells and antenna are given Fig. 39c–e.

6.4   Spintronics for Special Task Data Processing

General computing requires a complete set of Boolean operations. For spintronics, these operations have already been experimentally demonstrated in Refs. [154, 155, 159]. However, in addition to general computing functions, spin waves may be particularly suited to a certain subset of problems that are difficult or inefficient to be solved in standard CMOS logic. One example is magnonic cellular neural networks (MCNN). Cellular neural networks (CNN) have a specialized ability for data processing [160, 161]. MCNN has been proposed and numerically studied [152]. A MCNN consists of two-dimensional arrays of ME cells (see Fig. 40a), which can excite spin waves and achieve voltage/magnetization conversion. In the arrays, each ME cell is a MAJ output, the value of which is determined by the surrounding ME cells due to exchange coupling or so called spin wave interference. Here, we give an example of image processing to show the advantage of MCNN in special task data processing. An image data (see Fig. 40b, A) are input through the ME cells at the rectangular boundaries of the MCNN as shown in Fig. 40a. The initial magnetization states of the MCNN represent the input data, which is an

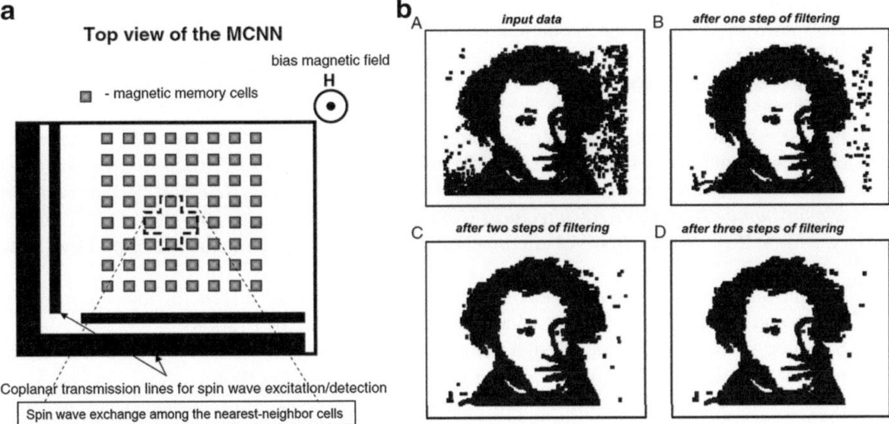

Fig. 40 (a) Schematic of the MCNN for special type data processing. The core of the MCNN consists of a two-dimensional array of magnetic memory cells sharing the common magnetic layer—spin wave bus. The elementary cell is a bi-stable element with two preferable magnetic polarizations. The cells communicate via spin waves propagating in the spin wave bus. Each cell changes its state according to the sum of all incoming spin waves. External magnetic field is used as a global parameter to govern network operation. (b) The results of the numerical modeling illustrating the image processing by the MCNN on the mesh consisting of 100×51 magneto-electric cells. The black marks depict ME cells polarized along one direction. ME cells polarized opposite to the direction are shown in white color. (*A*) input image; (*B*) input image after one step of filtering; (*C*) after two steps of filtering; (*D*) after three steps of filtering. Reprinted from Ref. [152], Copyright 2010, with permission from Elsevier

obscure picture. To execute a filter function, we only need to apply the majority logic once for each ME cell, and after the first step of filtering, the result is shown in Fig. 40b–B. The steps of the filtering operation can be optimized by the visual choice. Figure 40b–D shows the picture after three steps of filtering, which gives better contrast compared with the original picture. MCNN also has other important properties, in particular reconfigurable function, which is controlled by an external magnetic or electric field. In Ref. [152], a detailed analysis and more numerical simulations based on MCNN are given.

References

1. G.E. Moore, Cramming more components onto integrated circuits. Electron. Mag. **38**, 114–117 (1965)
2. International technology roadmap for semiconductors, (ed.), 2005
3. T.N. Theis, P.M. Solomon, In quest of the "next switch": prospects for greatly reduced power dissipation in a successor to the silicon field-effect transistor. Proc. IEEE **98**, 2005–2014 (2010)
4. S.A. Wolf et al., Spintronics: a spin-based electronics vision for the future. Science **294**, 1488–1495 (2001)
5. D.A. Allwood et al., Magnetic domain-wall logic. Science **309**, 1688–1692 (2005)
6. R.P. Cowburn, M.E. Welland, Room temperature magnetic quantum cellular automata. Science **287**, 1466–1468 (2000)
7. A. Khitun et al., Magnonic logic circuits. J. Phys. D **43**, 264005 (2010)
8. F.B. Ren, D. Markovic, True energy-performance analysis of the MTJ-based logic-in-memory architecture (1-bit full adder). IEEE Trans. Electron Devices **57**, 1023–1028 (2010)
9. P. Shabadi et al., Spin wave functions nanofabric update, in *2011 IEEE/ACM International Symposium on Nanoscale Architectures (NANOARCH)* (2011), pp. 107–113
10. A. Khitun, K.L. Wang, Non-volatile magnonic logic circuits engineering. J. Appl. Phys. **110**, 034306 (2011)
11. S. Matsunaga et al., Fabrication of a nonvolatile full adder based on logic-in-memory architecture using magnetic tunnel junctions. Appl. Phys. Exp. **1** (2008)
12. S.Y. Lee et al., A full adder design using serially connected single-layer magnetic tunnel junction elements. IEEE Trans. Electron Devices **55**, 890–895 (2008)
13. K.L. Wang, P.K. Amiri, Non-volatile spintronics: perspectives on instant-on nonvolatile nanoelectronic systems. Spin **02**, 1250009 (2012)
14. B. Behin-Aein et al., Proposal for an all-spin logic device with built-in memory. Nat. Nanotechnol. **5**, 266–270 (2010)
15. T. Dietl, H. Ohno, Dilute ferromagnetic semiconductors: physics and spintronic structures. Rev. Mod. Phys. **86**, 187–251 (2014)
16. T. Dietl, A ten-year perspective on dilute magnetic semiconductors and oxides. Nat. Mater. **9**, 965–974 (2010)
17. H. Ohno, Making nonmagnetic semiconductors ferromagnetic. Science **281**, 951 (1998)
18. Y. Ohno et al., Electrical spin injection in a ferromagnetic semiconductor heterostructure. Nature **402**, 790–792 (1999)
19. D. Chiba et al., Magnetization vector manipulation by electric fields. Nature **455**, 515–518 (2008)
20. A. Chernyshov et al., Evidence for reversible control of magnetization in a ferromagnetic material by means of spin-orbit magnetic field. Nat. Phys. **5**, 656–659 (2009)

21. C. Gould et al., Tunneling anisotropic magnetoresistance: a spin-valve-like tunnel magneto-resistance using a single magnetic layer. Phys. Rev. Lett. **93**, 117203 (2004)
22. M. Yamanouchi et al., Current-induced domain-wall switching in a ferromagnetic semiconductor structure. Nature **428**, 539–542 (2004)
23. H. Ohno et al., (Ga, Mn)As: a new diluted magnetic semiconductor based on GaAs. Appl. Phys. Lett. **69**, 363–365 (1996)
24. H. Munekata et al., Diluted magnetic III–V semiconductors. Phys. Rev. Lett. **63**, 1849–1852 (1989)
25. D. Ferrand et al., Carrier-induced ferromagnetic interactions in p-Doped $Zn_{1-x}Mn_xTe$ epilayers. J. Cryst. Grow. **214–215**, 387–390 (2000)
26. Y.D. Park et al., A group-IV ferromagnetic semiconductor: Mn_xGe_{1-x}. Science **295**, 651–654 (2002)
27. F. Xiu et al., Electric-field-controlled ferromagnetism in high-curie-temperature $Mn_{0.05}Ge_{0.95}$ quantum dots. Nat. Mater. **9**, 337–344 (2010)
28. H. Ohno et al., Electric-field control of ferromagnetism. Nature **408**, 944 (2000)
29. T. Dietl et al., Zener model description of ferromagnetism in Zinc-Blende magnetic semiconductors. Science **287**, 1019–1022 (2000)
30. Y. Wang et al., Direct structural evidences of $Mn_{11}Ge_8$ and Mn_5Ge_2 clusters in $Ge_{0.96}Mn_{0.04}$ thin films. Appl. Phys. Lett. **92**, 101913 (2008)
31. T. Nie et al., Quest for high-Curie temperature Mn_xGe_{1-x} diluted magnetic semiconductors for room-temperature spintronics applications. J. Cryst. Growth, in press, 2015, doi: 10.1016/j.jcrysgro.2015.01.025
32. J. Tang et al., Spin transport in Ge nanowires for diluted magnetic semiconductor-based nonvolatile transpinor. ECS Trans. **64**, 613–623 (2014)
33. J. Tang et al., Electrical spin injection and detection in $Mn_5Ge_3/Ge/Mn_5Ge_3$ nanowire transistors. Nano Lett. **13**, 4036–4043 (2013)
34. International Technology Roadmap of Semiconductors (ITRS) http://www.itrs.net, 2013 Edition
35. I. Žutić et al., Spintronics: fundamentals and applications. Rev. Mod. Phys. **76**, 323–410 (2004)
36. M. Johnson, Bipolar spin switch. Science **260**, 320–323 (1993)
37. D.J. Monsma et al., Room temperature-operating spin-valve transistors formed by vacuum bonding. Science **281**, 407–409 (1998)
38. D.J. Monsma et al., Perpendicular hot electron spin-valve effect in a new magnetic field sensor: the spin-valve transistor. Phys. Rev. Lett. **74**, 5260–5263 (1995)
39. M.E. Flatté, G. Vignale, Unipolar spin diodes and transistors. Appl. Phys. Lett. **78**, 1273–1275 (2001)
40. J. Fabian et al., Magnetic bipolar transistor. Appl. Phys. Lett. **84**, 85–87 (2004)
41. M.E. Flatté et al., Theory of semiconductor magnetic bipolar transistors. Appl. Phys. Lett. **82**, 4740–4742 (2003)
42. D.E. Nikonov, G.I. Bourianoff, Spin gain transistor in ferromagnetic semiconductors-the semiconductor Bloch-Equations approach. IEEE Trans. Nanotech. **4**, 206–214 (2005)
43. S. Datta, B. Das, Electronic analog of the electro-optic modulator. Appl. Phys. Lett. **56**, 665–667 (1990)
44. S. Sugahara, M. Tanaka, A spin metal-oxide-semiconductor field-effect transistor using half-metallic-ferromagnet contacts for the source and drain. Appl. Phys. Lett. **84**, 2307–2309 (2004)
45. M. Johnson, R.H. Silsbee, Interfacial charge-spin coupling: injection and detection of spin magnetization in metals. Phys. Rev. Lett. **55**, 1790 (1985)
46. A. Hirohata, K. Takanashi, Future perspectives for spintronic devices. J. Phys. D Appl. Phys. **47**, 193001 (2014)
47. H.C. Koo et al., Control of spin precession in a spin-injected field effect transistor. Science **325**, 1515–1518 (2009)

48. D. Osintsev et al., Temperature dependence of the transport properties of spin field-effect transistors built with InAs and Si channels. Solid-State Electron. **71**, 25–29 (2012)
49. Y. Saito et al., Spin injection, transport, and read/write operation in Spin-Based MOSFET. Thin Solid Films **519**, 8266–8273 (2011)
50. T. Inokuchi et al., Reconfigurable characteristics of spintronics-based MOSFETs for nonvolatile integrated circuits. in *2010 Symposium on VLSI Technology (VLSIT)* (2010), pp. 119–120
51. T. Marukame et al., Read/write operation of spin-based MOSFET using highly spin-polarized ferromagnet/MgO tunnel barrier for reconfigurable logic devices. in *2009 I.E. International Electron Devices Meeting (IEDM)* (2009), pp. 1–4
52. S. Sugahara, J. Nitta, Spin-transistor electronics: an overview and outlook. Proc. IEEE **98**, 2124–2154 (2010)
53. S. Salahuddin, S. Datta, Interacting systems for self-correcting low power switching. Appl. Phys. Lett. **90**, 093503 (2007)
54. D. Nikonov, G. Bourianoff, Operation and modeling of semiconductor spintronics computing devices. J. Supercond. Novel Mag. **21**, 479–493 (2008)
55. S. Das Sarma et al., How to make semiconductors ferromagnetic: a first course on spintronics. Solid State Commun. **127**, 99–107 (2003)
56. L. Liu et al., Spin-torque switching with the giant spin hall effect of tantalum. Science **336**, 555–558 (2012)
57. P.K. Amiri, K.L. Wang, Voltage-controlled magnetic anisotropy in spintronic devices. Spin **02**, 1240002 (2012)
58. M. Rahman et al., Experimental prototyping of beyond-CMOS nanowire computing fabrics. in *2013 IEEE/ACM International Symposium on Nanoscale Architectures (NANOARCH)* (2013), pp. 134–139
59. W. Eerenstein et al., Multiferroic and magnetoelectric materials. Nature **442**, 759–765 (2006)
60. R. Ramesh, N.A. Spaldin, Multiferroics: progress and prospects in thin films. Nat. Mater. **6**, 21–29 (2007)
61. M. Gajek et al., Tunnel junctions with multiferroic barriers. Nat. Mater. **6**, 296–302 (2007)
62. G. Srinivasan et al., Magnetoelectric bilayer and multilayer structures of magnetostrictive and piezoelectric oxides. Phys. Rev. B **64**, 214408 (2001)
63. T. Wu et al., Electrical control of reversible and permanent magnetization reorientation for magnetoelectric memory devices. Appl. Phys. Lett. **98**, 262504 (2011)
64. D. Chiba et al., Electrical control of the ferromagnetic phase transition in cobalt at room temperature. Nat. Mater. **10**, 853–856 (2011)
65. I.V. Ovchinnikov, K.L. Wang, Voltage sensitivity of Curie temperature in ultrathin metallic films. Phys. Rev. B **80**, 012405 (2009)
66. C.-F. Pai et al., Spin transfer torque devices utilizing the giant spin Hall effect of tungsten. Appl. Phys. Lett. **101**, 122404–4 (2012)
67. M.K. Niranjan et al., Electric field effect on magnetization at the Fe/MgO(001) interface. Appl. Phys. Lett. **96**, 222504 (2010)
68. C.-G. Duan et al., Surface magnetoelectric effect in ferromagnetic metal films. Phys. Rev. Lett. **101**, 137201 (2008)
69. J.P. Velev et al., Multi-ferroic and magnetoelectric materials and interfaces. Philos. Trans. R Soc. A: Math. Phys. Eng. Sci. **369**, 3069–3097 (2011)
70. F. Bonell et al., Large change in perpendicular magnetic anisotropy induced by an electric field in FePd ultrathin films. Appl. Phys. Lett. **98**, 232510 (2011)
71. T. Maruyama et al., Large voltage-induced magnetic anisotropy change in a few atomic layers of iron. Nat. Nanotechnol. **4**, 158–161 (2009)
72. T. Nozaki et al., Voltage-induced perpendicular magnetic anisotropy change in magnetic tunnel junctions. Appl. Phys. Lett. **96**, 022506 (2010)
73. J. Zhu et al., Voltage-induced ferromagnetic resonance in magnetic tunnel junctions. Phys. Rev. Lett. **108**, 197203 (2012)

74. T. Nozaki et al., Electric-field-induced ferromagnetic resonance excitation in an ultrathin ferromagnetic metal layer. Nat. Phys. **8**, 492–497 (2012)
75. M. Mayberry, *Emerging Technologies and Moore's Law: Prospects for the Future*. http://csg5.nist.gov/pml/div683/upload/Mayberry_March_2010.pdf
76. K.L. Wang, P. Khalili Amiri, Nonvolatile spintronics: perspectives on instant-on nonvolatile nanoelectronic systems. J. Spin. **2**, 1250009 (2012)
77. C.J. Lin et al., 45nm low power CMOS logic compatible embedded STT MRAM utilizing a reverse-connection 1T/1MTJ cell. in *2009 I.E. International Electron Devices Meeting (IEDM)* (2009), pp. 1–4
78. S. Assefa et al., Fabrication and characterization of MgO-based magnetic tunnel junctions for spin momentum transfer switching. J. Appl. Phys. **102**, 063901 (2007)
79. Y.M. Huai et al., Observation of spin-transfer switching in deep submicron-sized and low-resistance magnetic tunnel junctions. Appl. Phys. Lett. **84**, 3118–3120 (2004)
80. E. Chen et al., Advances and future prospects of spin-transfer torque random access memory. IEEE Trans. Magn. **46**, 1873–1878 (2010)
81. S.S.P. Parkin et al., Giant tunnelling magnetoresistance at room temperature with MgO (100) tunnel barriers. Nat. Mater. **3**, 862–867 (2004)
82. K.L. Wang et al., Low-power non-volatile spintronic memory: STT-RAM and beyond. J. Phys. D **46**, 074003 (2013)
83. P.K. Amiri et al., Low write-energy magnetic tunnel junctions for high-speed spin-transfer-torque MRAM. IEEE Electr. Device Lett. **32**, 57–59 (2011)
84. Y. Huai, Spin-transfer Torque MRAM (STT-MRAM): challenges and prospects. AAPPS Bulletin **18**, 33–40 (2008)
85. P. Khalili Amiri, K.L. Wang, Voltage-controlled magnetic anisotropy in spintronic devices. Spin **2**, 1240002 (2012)
86. P.K. Amiri et al., Electric-field-induced thermally assisted switching of monodomain magnetic bits. J. Appl. Phys. **113**, 013912 (2013)
87. R. Dorrance et al., Diode-MTJ crossbar memory cell using voltage-induced unipolar switching for high-density MRAM. EEEE Electr. Device Lett. **34**, 753–755 (2013)
88. S. Ikeda et al., A perpendicular-anisotropy CoFeB-MgO magnetic tunnel junction. Nat. Mater. **9**, 721–724 (2010)
89. D.C. Worledge et al., Spin torque switching of perpendicular Ta | CoFeB | MgO-based magnetic tunnel junctions. Appl. Phys. Lett. **98**, 022501 (2011)
90. P.K. Amiri et al., Switching current reduction using perpendicular anisotropy in CoFeB-MgO magnetic tunnel junctions. Appl. Phys. Lett. **98**, 112507 (2011)
91. S. Yakata et al., Influence of perpendicular magnetic anisotropy on spin-transfer switching current in CoFeB/MgO/CoFeB magnetic tunnel junctions. J. Appl. Phys. **105**, 07D131 (2009)
92. W.-G. Wang et al., Rapid thermal annealing study of magnetoresistance and perpendicular anisotropy in magnetic tunnel junctions based on MgO and CoFeB. Appl. Phys. Lett. **99**, 102502 (2011)
93. M. Endo et al., Electric-field effects on thickness dependent magnetic anisotropy of sputtered MgO/Co(40)Fe(40)B(20)/Ta structures. Appl. Phys. Lett. **96**, 212503 (2010)
94. S.S. Ha, Voltage induced magnetic anisotropy change in ultrathin $Fe_{80}Co_{20}$/MgO junctions with Brillouin light scattering. Appl. Phys. Lett. **96**, 142512 (2010)
95. Y. Shiota et al., Voltage-Assisted Magnetization Switching in Ultrathin $Fe_{80}Co_{20}$ Alloy Layers. Appl. Phys. Exp. **2**, 063001 (2009)
96. Y. Shiota et al., Induction of coherent magnetization switching in a few atomic layers of FeCo using voltage pulses. Nat. Mater. **11**, 39–43 (2012)
97. W.-G. Wang et al., Electric-field-assisted switching in magnetic tunnel junctions. Nat. Mater. **11**, 64–68 (2012)
98. A.J. Schellekens et al., Electric-field control of domain wall motion in perpendicularly magnetized materials. Nat. Commun. **3**, 847 (2012)

99. C.G. Duan et al., Tailoring magnetic anisotropy at the ferromagnetic/ferroelectric interface. Appl. Phys. Lett. **92**, 122905 (2008)
100. S.E. Barnes et al., Rashba spin-orbit anisotropy and the electric field control of magnetism. Sci. Rep. **4**, 4105 (2014)
101. S. Yuasa et al., Giant room-temperature magnetoresistance in single-crystal Fe/MgO/Fe magnetic tunnel junctions. Nat. Mater. **3**, 868–871 (2004)
102. Y. Shiota et al., Pulse voltage-induced dynamic magnetization switching in magnetic tunneling junctions with high resistance-area product. Appl. Phys. Lett. **101**, 102406 (2012)
103. J.G. Alzate et al., Voltage-induced switching of nanoscale magnetic tunnel junctions. IEDM Tech. Digest **29**(5) 1–4 (2012)
104. W. Feng et al., Intrinsic spin Hall effect in monolayers of group-VI dichalcogenides: a first-principles study. Phys. Rev. B **86**, 165108 (2012)
105. S. Kanai, Electric field-induced magnetization reversal in a perpendicular-anisotropy CoFeB-MgO magnetic tunnel junction. Appl. Phys. Lett. **101**, 122403 (2012)
106. M.T. Rahman et al., Reduction of switching current density in perpendicular magnetic tunnel junctions by tuning the anisotropy of the CoFeB free layer. J. Appl. Phys. **111**, 07C907 (2012)
107. M. Gajek et al., Spin torque switching of 20 nm magnetic tunnel junctions with perpendicular anisotropy. Appl. Phys. Lett. **100**, 132408 (2012)
108. Z.M. Zeng et al., Ultralow-current-density and bias-field-free spin-transfer nano-oscillator. Sci. Rep. **3**, 1426 (2013)
109. Z. Zeng et al., High-power coherent microwave emission from magnetic tunnel junction nano-oscillators with perpendicular anisotropy. ACS Nano (2012)
110. H. Maehara et al., High Q factor over 3000 due to out-of-plane precession in nano-contact spin-torque oscillator based on magnetic tunnel junctions. Appl. Phys. Exp. **7**, 023003 (2014)
111. B. Fang et al., Giant spin-torque diode sensitivity at low input power in the absence of bias magnetic field. arXiv:1410.4958
112. S. Miwa et al., Highly sensitive nanoscale spin-torque diode. Nat. Mater. **13**, 50–56 (2014)
113. J.G. Alzate et al., Voltage-induced switching of CoFeB-MgO magnetic tunnel junctions. in *56th Conference on Magnetism and Magnetic Materials* Scottsdale, Arizona (2011), pp. EG-11
114. J.G. Alzate et al., Voltage-induced switching of nanoscale magnetic tunnel junctions. in *Presented at the IEEE International Electron Devices Meeting (IEDM)*, San Francisco, CA (2012)
115. J.G. Alzate et al., Temperature dependence of the voltage-controlled perpendicular anisotropy in nanoscale MgO|CoFeB|Ta magnetic tunnel junctions. Appl. Phys. Lett. **104**, 112410 (2014)
116. A. Rajanikanth et al., Magnetic anisotropy modified by electric field in V/Fe/MgO(001)/Fe epitaxial magnetic tunnel junction. Appl. Phys. Lett. **103**, 062402 (2013)
117. T. Liu et al., Thermally robust Mo/CoFeB/MgO trilayers with strong perpendicular magnetic anisotropy. Sci. Rep. **4**, 5895 (2014)
118. P.V. Ong et al., Electric field control and effect of Pd capping on magnetocrystalline anisotropy in FePd thin films: a first-principles study. Phys. Rev. B **89**, 094422 (2014)
119. K. Kita et al., Electric-field-control of magnetic anisotropy of $Co_{0.6}Fe_{0.2}B_{0.2}$ oxide stacks using reduced voltage. J. Appl. Phys. **112**, 033919 (2012)
120. T. Seki et al., Coercivity change in an FePt thin layer in a Hall device by voltage application. Appl. Phys. Lett. **98**, 212505 (2011)
121. J.G. Alzate, Voltage-controlled magnetic dynamics in nanoscale magnetic tunnel junctions. Ph.D., Electrical Engineering, University of California, Los Angeles (2014)
122. J.A. Katine, E.E. Fullerton, Device implications of spin-transfer torques. J. Magn. Magn. Mater. **320**, 1217–1226 (2008)
123. K. Lee, S.H. Kang, Development of Embedded STT-MRAM for Mobile System-on-Chips. IEEE Trans. Magn. **47**, 131–136 (2011)

124. Note, 'however', that the validity of the present analysis does not depend on the sign of the VCMA effect
125. C.A.F. Vaz et al., Magnetism in ultrathin film structures. Rep. Progr. Phys. **71**, 056501 (2008)
126. H.B. Callen, E. Callen, The present status of the temperature dependence of magnetocrystalline anisotropy, and the l(l+1)/2 power law. J. Phys. Chem. Solids **27**, 1271–1285 (1966)
127. N.W. Ashcroft, N.D. Mermin, *Solid State Physics* (Saunders College, Philadelphia, 1976)
128. J.Z. Sun, Spin-current interaction with a monodomain magnetic body: a model study. Phys. Rev. B **62**, 570–578 (2000)
129. J.Z. Sun et al., Effect of subvolume excitation and spin-torque efficiency on magnetic switching. Phys. Rev. B **84**, 064413 (2011)
130. A. Sellai et al., Barrier height and interface characteristics of $Au/Mn_5Ge_3/Ge$ (111) Schottky contacts for spin injection. Semicond. Sci. Technol. **27**, 035014 (2012)
131. W. Zhao, Y. Cao, Predictive technology model for nano-CMOS design exploration. J. Emerg. Technol. Comput. Syst. **3**, 1 (2007)
132. I. Ovchinnikov, K. Wang, Theory of electric-field-controlled surface ferromagnetic transition in metals. Phys. Rev. B **79**, 020402 (2009)
133. S.K. Kim et al., Experimental observation of magnetically dead layers in Ni/Pt multilayer films. Phys. Rev. B **64**, 052406 (2001)
134. R. Winkler, Spin-orbit coupling effects in two-dimensional electron and hole systems. http://rave.ohiolink.edu/ebooks/ebc/10864040 (2003)
135. I.M. Miron et al., Current-driven spin torque induced by the Rashba effect in a ferromagnetic metal layer. Nat. Mater. **9**, 230–234 (2010)
136. Y. Fan et al., Magnetization switching through giant spin-orbit torque in a magnetically doped topological insulator heterostructure. Nat. Mater. **13**, 699–704 (2014)
137. L.D. Landau, E.M. Lifschitz, *Statistical Physics, Part 2* (Pergamon Press, Oxford, 1981)
138. T.L. Gilbert, Classics in magnetics: a phenomenological theory of damping in ferromagnetic materials. IEEE Trans. Magn. **40**, 3443–3449 (2004)
139. G. Yu et al., Switching of perpendicular magnetization by spin-orbit torques in the absence of external magnetic fields. Nat. Nanotechnol. **9**, 548–554 (2014)
140. S. Hoffman et al., Spin-torque ac impedance in magnetic tunnel junctions. Phys. Rev. B **86**, 214420 (2012)
141. J. Hirsch, Spin hall effect. Phys. Rev. Lett. **83**, 1834–1837 (1999)
142. Y.A. Bychkov, E.I. Rashba, Oscillatory effects and the magnetic susceptibility of carriers in inversion layers. J. Phys. C **17**, 6039–6045 (1984)
143. V.M. Edelstein, Spin polarization of conduction electrons induced by electric current in two-dimensional asymmetric electron systems. Solid State Commun. **73**, 233–235 (1990)
144. I.M. Miron et al., Perpendicular switching of a single ferromagnetic layer induced by in-plane current injection. Nature **476**, 189–193 (2011)
145. M.Z. Hasan, C.L. Kane, Colloquium: topological insulators. Rev. Mod. Phys. **82**, 3045–3067 (2010)
146. X. Kou et al., Interplay between different magnetisms in Cr-doped topological insulators. ACS Nano **7**, 9205–9212 (2013)
147. J. Kim et al., Layer thickness dependence of the current-induced effective field vector in Ta| CoFeB|MgO. Nat. Mater. **12**, 240–245 (2013)
148. A. Imre et al., Majority logic gate for magnetic quantum-dot cellular automata. Science **311**, 205–208 (2006)
149. D. Bhowmik et al., Spin hall effect clocking of nanomagnetic logic without a magnetic field. Nat. Nanotechnol. **9**, 59–63 (2014)
150. Smullen CW et al., Relaxing non-volatility for fast and energy-efficient STT-RAM caches. in *Presented at the IEEE 17th Int. Symp. High Performance Computer Architecture*, San Antonio, TX (2011)

151. R. Fengbo, D. Markovic, True energy-performance analysis of the MTJ-based logic-in-memory architecture (1-bit full adder). IEEE Trans. Electr. Devices **57**, 1023–1028 (2010)
152. A. Khitun et al., Magnetic cellular nonlinear network with spin wave bus for image processing. Superlattices Microstruct. **47**, 464–483 (2010)
153. A. Khitun, Multi-frequency magnonic logic circuits for parallel data processing. J. Appl. Phys. **111**, 054307 (2012)
154. T. Schneider et al., Realization of spin-wave logic gates. Appl. Phys. Lett. **92**, 022505 (2008)
155. M.P. Kostylev et al., Spin-wave logical gates. Appl. Phys. Lett. **87**, 153501-1-3 (2005)
156. A.A. Serga et al., YIG magnonics. J. Phys. D **43**, 264002 (2010)
157. S.I. Kiselev et al., Microwave oscillations of a nanomagnet driven by a spin-polarized current. Nature **425**, 380–383 (2003)
158. S. Cherepov et al., Electric-field-induced spin wave generation using multiferroic magneto-electric cells. Appl. Phys. Lett. **104**, 082403 (2014)
159. Y.N. Wu et al., A three-terminal spin-wave device for logic applications. J. Nanoelectron. Optoelectron. **4**, 394–397 (2009)
160. L.O. Chua, L. Yang, Cellular neural networks—theory. IEEE Trans. Circuits Syst. **35**, 1257–1272 (1988)
161. L.O. Chua, L. Yang, Cellular neural networks—applications. IEEE Trans. Circuits Syst. **35**, 1273–1290 (1988)

Advanced Perpendicular STT-MRAM Technologies for Power Reduction of High-performance Processors

Naoharu Shimomura, Shinobu Fujita, Keiko Abe, Hiroki Noguchi, and Hiroaki Yoda

1 Introduction

Although mobile devices such as smartphones are convenient in many respects, short battery lifetime remains an issue. Moreover, the energy consumption of the mobile processor is increasing as its performance improves. Accordingly, there are high expectations that the energy consumption of mobile devices will be reduced in order to prolong battery lifetime.

In order to reduce the energy consumption of a computer, the concept of normally-off computing was proposed in 2001 and an improved concept was subsequently presented [1]. The basic idea of normally-off computing is to shut down the power during the idle time while the device is on. For example, the computer is just waiting, wasting energy while you move your finger from one key on the keyboard to another. If the power of the computer is shut down and kept off during this waiting time and turned on abruptly just when you touch another key, most of the wasted energy can be saved.

We can apply this concept of power saving to a high-performance and low-power (hplp-) processor. The detail of this application is explained below. But the most important technology to realize this power saving is that of nonvolatile

N. Shimomura (✉) • S. Fujita • K. Abe • H. Noguchi
Advanced LSI Technology Laboratory, Corporate Research & Development Center,
Toshiba Corporation
e-mail: Naoharu.Shimomura@toshiba.co.jp

H. Yoda
Center For Semiconductor Research & Development, Toshiba Corporation

© Springer International Publishing Switzerland 2015
W. Zhao, G. Prenat (eds.), *Spintronics-based Computing*,
DOI 10.1007/978-3-319-15180-9_3

memory. The required features for nonvolatile memory in hplp-processors are low power, fast access speed and high endurance. Among nonvolatile memories, spin transfer torque (STT)-MRAM is the sole candidate for the hplp-processor.

2 Fundamentals and Development of Advanced Perpendicular STT-MRAM

2.1 Features of Existing Memories and MRAM's Target in the Memory Hierarchy

Figure 1 represents the capacity and the access time map of the memories [2]. The ideal memory that has large capacity, fast access speed, unlimited endurance and nonvolatility does not exist. The storage memories such as NAND FLASH and HDD are nonvolatile and have large capacity but the access time is slow. On the other hand, access time of work memories such as SRAM and DRAM is fast but their capacity is not large and they are volatile. Therefore, the work memories and storage memories have to work together complementarily in the computer, forming the hierarchical structure shown in Fig. 2a.

In the memory hierarchy in Fig. 2a, SRAM and DRAM are used for the work memories. The SRAM is used for a cache memory and the DRAM for a main memory. Since they are volatile memories, power consumption of the work memory is large. A leak current of SRAM and refresh energy of DRAM are big sources of power dissipation.

MRAM is a nonvolatile memory that has a possibility of replacing the work memories because of its potential for fast access time and high endurance. Replacing volatile work memories with nonvolatile memories opens up a

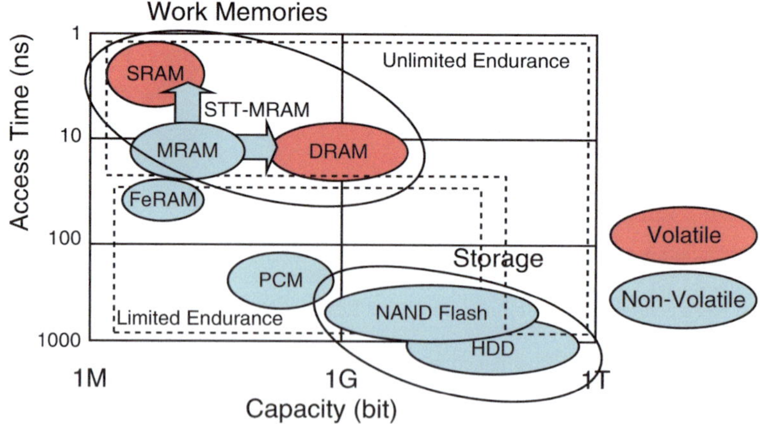

Fig. 1 Memory capacity and access time map

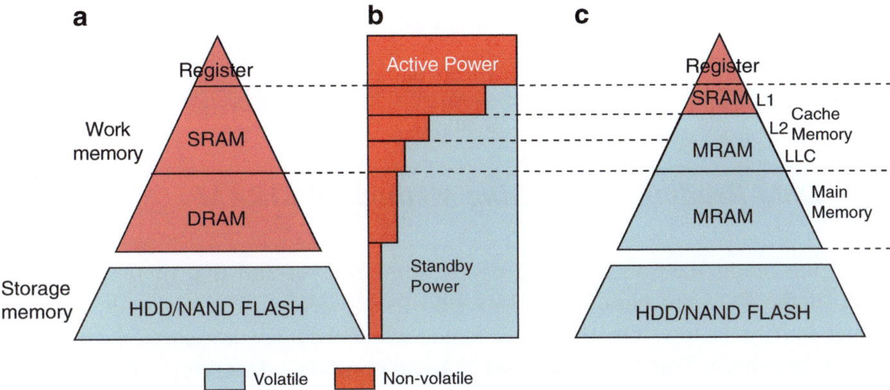

Fig. 2 (**a**) Current memory hierarchy diagram. (**b**) Power dissipation ratio between active power and standby power of the memories in the hierarchy. (**c**) Suitable memory hierarchy for reduction of power dissipation

possibility of eliminating this standby power and reducing the energy consumption of the memories.

However, simple replacement does not mean reducing the power consumption. Figure 2b shows ratios of active power to standby power for each memory level. Most of the power dissipation of a high-level memory such as an L1 cache memory is active power. Therefore, there is little room for reduction of energy consumption by replacing it with a nonvolatile memory. On the other hand, in the case of a low-level memory in the hierarchy, such as an L2 cache memory or a last-level cache (LLC) memory, most of the power dissipation is standby power. Therefore, a memory hierarchy suitable for reduction of energy consumption by using MRAM is the one illustrated in Fig. 2c. In this memory hierarchy, MRAM is used as the main memory and low-level cache memory such as L2 cache and LLC. However, SRAM is still used as the L1 cache.

2.2 Basic Structure of the MTJ Storage Element

The storage element of MRAM is Magnetic Tunnel Junction (MTJ) whose schematic diagram is shown in Fig. 3. Intrinsic parts of the MTJ consist of three layers: two magnetic layers and a tunnel barrier between them. One of the magnetic layers is a storage layer whose magnetization is variable. The other is a reference layer whose magnetization is fixed. A different magnetization direction of the storage layer is assigned to each binary digit value.

Fig. 3 Schematic diagram
of MTJ

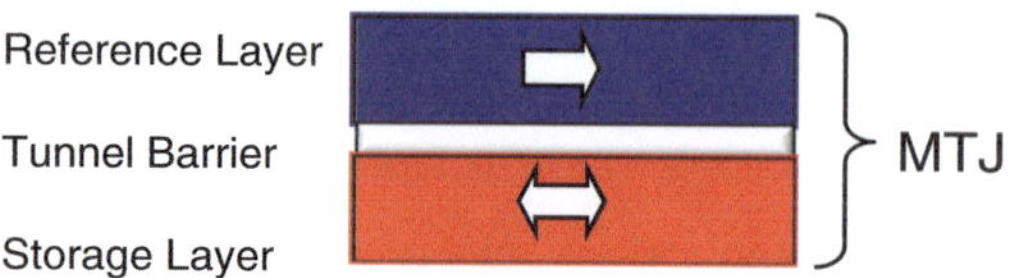

2.3 Data Reading and Writing Methods of MRAM

When the magnetization direction of the storage layer and that of the reference
layer of the MTJ are parallel (P) to each other, electrical resistance through the
tunnel barrier is low. On the other hand, when they are antiparallel (AP), the
resistance is high. This change in the electrical resistance depending on the relative
direction of these magnetic layers is called the tunnel magnetoresistance (TMR)
effect [3, 4]. Hence, measurement of the resistance of the MTJ makes it possible to
identify whether the MTJ is in a parallel or antiparallel state. This TMR effect is
used to read data from the MTJ in the MRAM. This TMR signal is amplified by the
sense amplifier connected to top and bottom electrodes and used as the output of
the MRAM.

When the spin-polarized current is injected into the storage layer of the MTJ, it
interacts with localized spin in the ferromagnetic material in the storage layer
exerting the spin transfer torque (STT) [5, 6]. The STT-MRAM uses STT switching
as a method of writing data to MTJ. It has been widely investigated because a
critical current of the switching decreases as the MTJ size shrinks. It means the
STT-MRAM is scalable memory, which is in contrast to the conventional field
writing MRAM whose critical switching field increases as the MTJ size shrinks.

2.4 Structure of a Unit Cell of the STT-MRAM

The schematic of a typical unit cell structure of the STT-MRAM is illustrated in
Fig. 4. The unit cell consists of an MTJ and a selection transistor that are connected
with each other. The other terminal of the MTJ is connected to a bitline and that of
the selection transistor is connected to another bitline. And a gate terminal of the
selection transistor is connected to a wordline as illustrated in Fig. 4.

2.5 Key Features to Outperform Conventional Memories

The biggest obstacle to the replacement of the work memory by the STT-MRAM is
the large write current of the STT-MRAM. The write current of the STT-MRAM is
limited by the drive current of the selection transistor. As the cell size shrinks, the
size of the selection transistor also shrinks and drivable write current decreases.

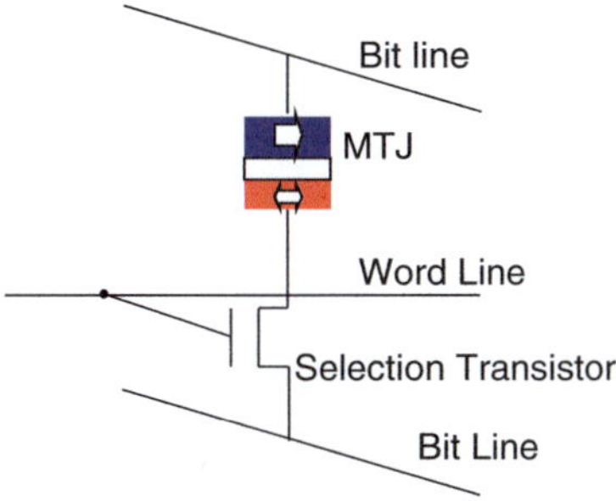

Fig. 4 Unit cell structure of the STT-MRAM

Therefore, the large write current of the STT-MRAM limits the scaling of the cell size. And furthermore, the large write current prevents the STT-MRAM-based cache memory from reducing the energy consumption of the processor, which is explained in detail below. Hence, the reduction of the write current of the STT-MRAM is critical for the MRAM work memory.

2.6 Perpendicular Magnetization MTJ for Reduction of the Write Current

A very effective strategy to reduce the write current of the STT-MRAM is to use a perpendicular magnetization MTJ (p-MTJ) whose magnetization direction is perpendicular to the plane.

Figures 5 and 6 show why the p-MTJ enables reduction of the write current of STT switching compared to that of the in-plane MTJ. Figure 5 explains the in-plane MTJ and Fig. 6 explains the p-MTJ. For simplicity we consider only a macrospin model where all the magnetization in the storage layer is aligned to the same direction. The in-plane MTJ has an anisotropic shape such as a rectangle or an oval. The anisotropic shape induces magnetic shape anisotropy because the demagnetization field varies depending on the direction of the magnetization. The in-plane MTJ is energetically the most stable when the magnetization is parallel to the long axis as indicated by A or A' in Fig. 5a. Therefore this axis is an easy axis. The hard axis of the in-plane MTJ lies within the plane and also is perpendicular to the easy axis, which is indicated by B and B' in Fig. 5a. When the magnetization is perpendicular to the plane, as indicated by C and C', energy of the in-plane MTJ becomes the highest because a large demagnetization field is induced and Zeeman energy becomes very large.

During the data retention period, the magnetization has to stay in one of the states along the easy axis, which is A or A'. In order to prevent unintended switching between them by thermal energy, a certain amount of energy barrier, typically around $60\,k_\mathrm{B}T$ for 10 years of retention time, is required. Here, k_B denotes the Boltzmann constant and T denotes absolute temperature. However, the magnetization switches when it receives the energy larger than this energy barrier. In this

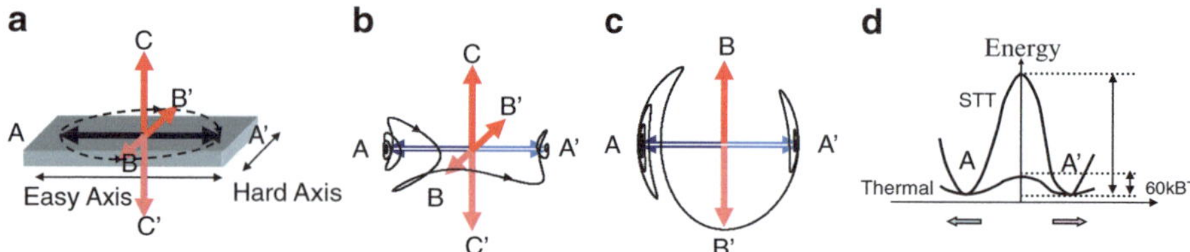

Fig. 5 Schematics of magnetization switching path of in-plane MTJ. (**a**) explains thermal switching. (**b**) and (**c**) show STT switching projected from a different side. (**c**) shows energy barriers of these two switching modes

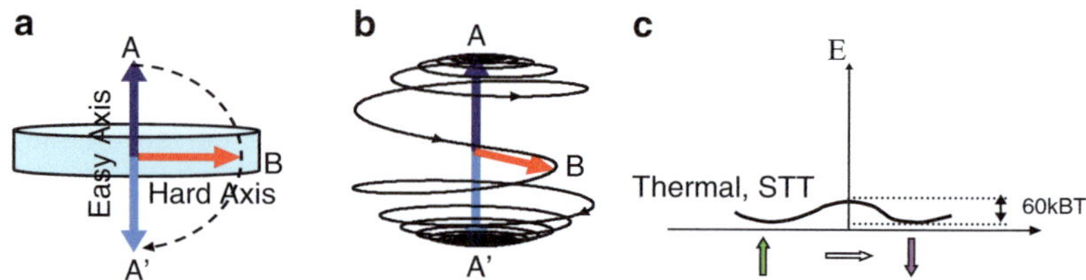

Fig. 6 Switching path of (**a**) thermal switching, (**b**) STT switching, and (**c**) energy barrier for the p-MTJ

situation, the magnetization takes the minimum energy path. It is the path within the plane that is indicated by the broken line from A through B to A' or that from A through B' to A' in Fig. 5a. Throughout the switching path, the magnetization can stay within the plane without having a perpendicular component. Therefore, induced Zeeman energy is small.

Contrary to this, an example of the magnetization path during the STT switching is shown in Fig. 5b, which is a macrospin simulation result. Figure 5c illustrates the same path, but projected from the C axis. As the STT switching is a process of increasing precession amplitude, the magnetization component along the C axis is not zero, which increases the Zeeman energy. Therefore, for the in-plane MTJ, much larger energy is required for the STT switching than that of the thermal switching. That is schematically illustrated in Fig. 5d.

On the other hand, the easy axis and the hard axis of p-MTJ are shown by A (or A') and B, respectively, in Fig. 6a. If the shape of the p-MTJ is cylindrical, it does not have shape anisotropy when the magnetization direction is along the hard axis. In that case, all the directions perpendicular to the easy axis are energetically identical. Therefore, we can choose any of these axes as the hard axis. In this case, the energy barrier for the magnetization to switch from A to A' is independent of its path.

An example of the path of the thermal switching of the p-MTJ is indicated by the broken arrow in Fig. 6a, which is the path from A through B to A'. An example of the STT switching path from A to A', which is precessional movement, is shown in Fig. 6b. In both cases, energy level becomes the highest when the magnetization lies in the hard axis direction. This energy level is not changed by azimuth. Therefore, the energy barrier heights of these two switching modes are the same. For example, if the energy barrier of the thermal switching is 60 k_BT, that of the STT switching is also 60 k_BT as explained in Fig. 6c.

Accordingly, less energy is required to overcome the energy barrier of the STT switching of the p-MTJ than that of the in-plane MTJ in order to keep the same amount of thermal switching energy barrier. Therefore, the p-MTJ enables reduction of the write current of STT switching compared to that of the in-plane MTJ.

2.7 *Trend of the Switching Current of p-MTJ*

The trend of the STT switching current is shown in Fig. 7. In the initial stage of the investigation of STT switching, most research involved the use of in-plane MTJs. The switching current of the in-plane MTJ was reduced from a milliampere to about 100 μA [7, 8]. However, further reduction in the switching current of the in-plane MTJ has yet to be achieved.

The first STT switching of the perpendicular magnetization was demonstrated with a giant magnetoresistance (GMR) element [9–11]. However, the switching current was milliamperes.

The first verification of STT switching of the p-MTJ was published in 2007 [12] and further reductions in switching current were subsequently reported. As shown in Fig. 8, the MTJ stack structure used in the first demonstration consisted of cap/TbCoFe(3 nm)/ CoFeB(1 nm)/ MgO (1 nm)/ CoFeB (2 nm)/TbCoFe (30 nm)/ substrate [12, 13]. TbCoFe (3 nm)/ CoFeB (1 nm) was a storage layer, CoFeB (2 nm)/TbCoFe (30 nm) was a reference layer and MgO (1 nm) was a tunnel barrier. Ferrimagnetic material was used for this p-MTJ. All the magnetic layers showed perpendicular anisotropy. The diameter was 130 nm and retention energy was 107 k_BT. MR through this p-MTJ was measured and is shown in Fig. 8b. STT

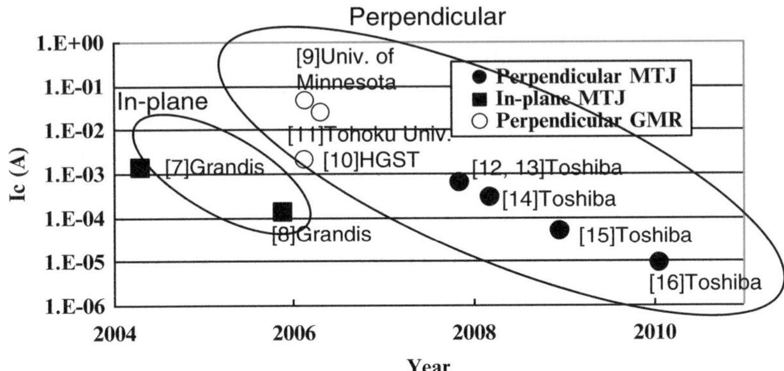

Fig. 7 Switching current reduction trend of in-plane MTJ, perpendicular GMR and perpendicular TMR

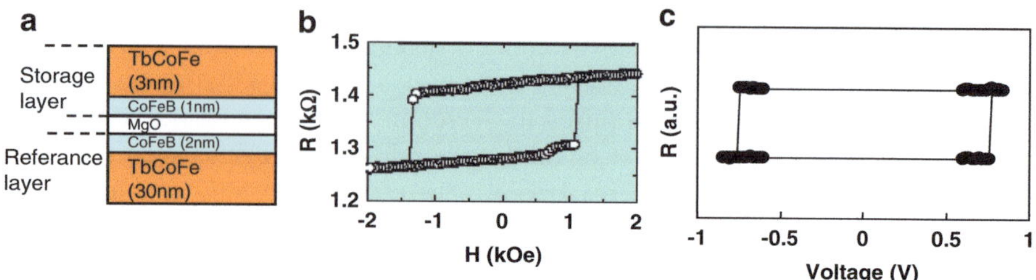

Fig. 8 The world's first demonstration of STT switching of p-MTJ. (**a**) Stack structure (**b**) RH-loop (**c**) VR plot of the p-MTJ that shows STT switching

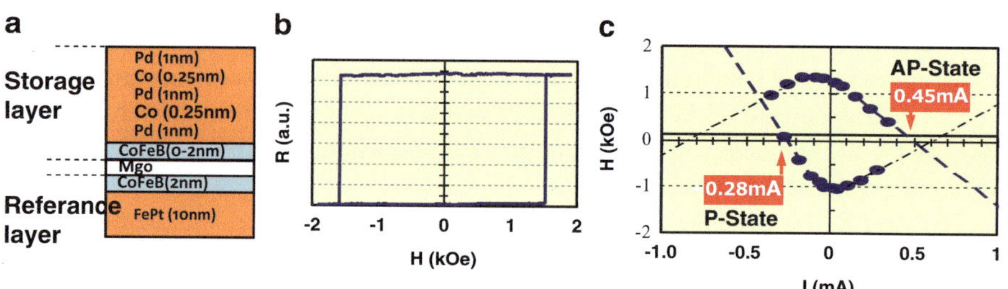

Fig. 9 Reduction of the switching current in the perpendicular CoFeB/MgO/CoFeB TMR with a Co-based multilayer

switching of the storage layer of this p-MTJ stack was successfully demonstrated as shown in Fig. 8c. However, the switching current from AP state to P state of this p-MTJ was about 0.62 mA and current density was 4.7 MA/cm^2, which was still very large.

Reduction of the STT switching current of p-MTJ was verified with the perpendicular CoFeB/MgO/CoFeB TMR with a Co-based multilayer [14]. As shown in Fig. 9a, a wedged CoFeB layer was used for the storage layer. The switching currents from AP to P switching and from P to AP switching were 0.28 mA and 0.45 mA, respectively, which are shown in Fig. 9c. The energy barrier height for the retention of this p-MTJ was 114 k_BT.

Figure 10 shows further reduction of the switching current on a p-TMR with an L1$_0$ ordered Fe-based alloy [15]. The p-MTJ layer structure consists of capping layer/perpendicular reference layer /MgO/L1$_0$-crystal storage layer /underlayer. A TEM image of this p-MTJ is shown in Fig. 10a, whose diameter was 55 nm. Figure 10b shows STT switching (R-V curve). The x-axis is a voltage imposed on the P-MTJ and the switching transistor connected to the P-MTJ. The switching current from AP to P switching was 49 µA and the retention energy was 56 k_BT.

Figure 11 shows a demonstration of the STT switching using p-MTJ, whose switching current is of the order of several microamperes [16]. Figure 11a shows a layer structure and Fig. 11b an R-I curve indicating the STT switching of this p-MTJ.

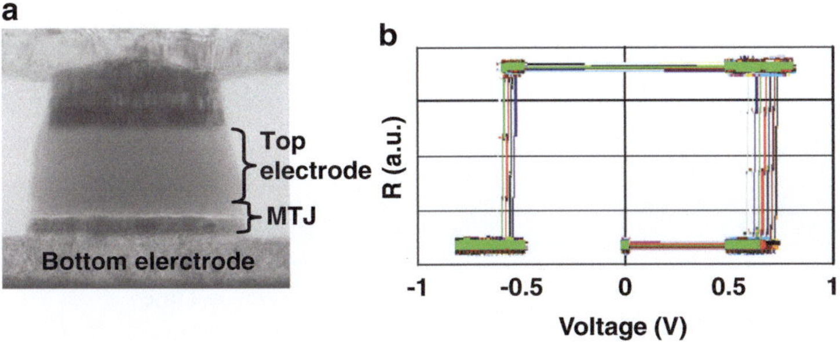

Fig. 10 Reduction of the switching current on a perpendicular TMR with an $L1_0$ ordered Fe-based alloy (**a**) TEM image, (**b**) STT switching (R-V curve)

Fig. 11 Demonstration of STT switching using p-MTJ, whose switching current is of the order of several microamperes. (**a**) p-MTJ layer structure, (**b**) STT switching (R-I curve)

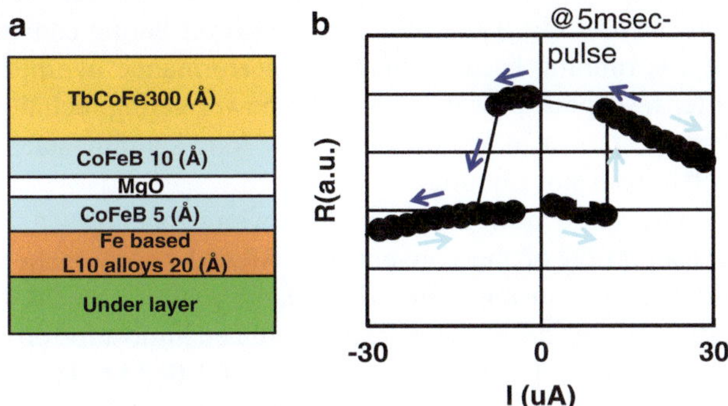

Figure 11b indicates that the switching current from AP state to P state of this p-MTJ is 9 µA and that of the other direction is 11 µA.

The switching current of this p-MTJ is reduced by an order of magnitude compared to that of the in-plane MTJ. Thus, it was proved by demonstration data that the p-MTJ has a big advantage over the in-plane MTJ for the reduction of write current. Subsequently, much research and many developments concerning STT-MRAM with p-MTJ have been reported [17, 18].

2.8 Reduction of Energy Consumption of a Mobile Processor by Nonvolatile Cache Memory

Figure 12 shows schematic diagrams that explain the reduction of the power consumption of the nonvolatile cache-based mobile processor [2]. Figure 12a, b illustrate the conventional SRAM-based cache memory and the STT-MRAM-based one, respectively. In order to reduce the leakage power, the power gating technique is used in the current processor for a long standby state, namely, when the waiting time is of the order of several milliseconds, the cache memory is shut down and no

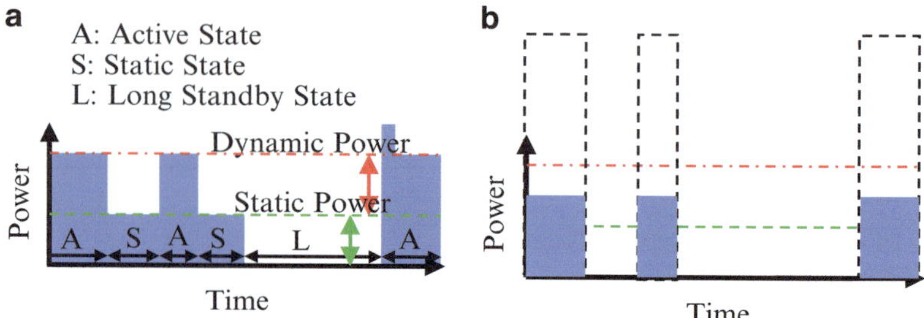

Fig. 12 Schematic diagrams of the power of the cache memory. (**a**) Conventional SRAM-based cache memory. (**b**) STT-MRAM-based normally-off processer

energy is consumed during that time, which is illustrated by the long standby state indicated in Fig. 12a. However, the power gating cannot be used while an application is running because of power/performance overhead and operation instability after fast speed power gating. The power gating technique enables reduction of the leakage power in SRAM during the long standby state, only when the application software is not running.

The static power in Fig. 12a, which occupies a large portion of the power consumption of the conventional SRAM-based cache memory, is mainly due to the leakage current of the SRAM. In contrast, as shown in Fig. 12b, the leakage power during a static state can be reduced almost to zero by using STT-MRAM cell. Furthermore, if write power of the STT-MRAM element is low enough, the power of the processor even in active states can be saved, which is indicated by the filled square in Fig. 12b. However, if the write power of the STT-MRAM element is large, the power of the processor in the active state increases, which is indicated by the dotted square in Fig. 12b. And consequently, that increases the total energy consumption of the processor rather than reducing it. Therefore, low-power STT-MRAM cell is important for reducing the energy consumption of the processor.

In order to realize the low-power STT-MRAM cell, fast and low-current-switching MTJ is necessary. In other words, write charge is a critical parameter for reducing power consumption of the processor [19]. Here, the write charge is defined as (write pulse width) × (write current), which has the dimension of the electric charge.

It should be noted is that the low-power STT-MRAM cell does not mean low-power switching of the storage element. A critical parameter is the write charge instead of write power applied to the storage element. The write power applied to the storage element is defined by (write pulse width) × (write current) × (voltage applied to the storage element). Reduction of the voltage without that of the write current does not help to reduce memory power consumption because voltage is dissipated not only in the storage element but also in connected transistors. The memory power dissipation is mainly dominated by (write pulse width) × (write current) × Vdd, where Vdd is the voltage applied to the memory, which is mainly determined by the CMOS transistor.

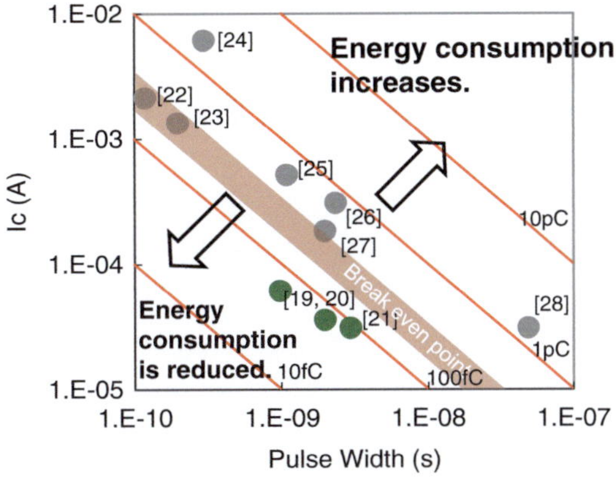

Fig. 13 Pulse width and switching current (Ic) map of the STT-MRAM. The cache memory with the STT-MRAM located in the lower left region reduces the energy consumption compared to a conventional one. However, that in the upper right region increases the energy

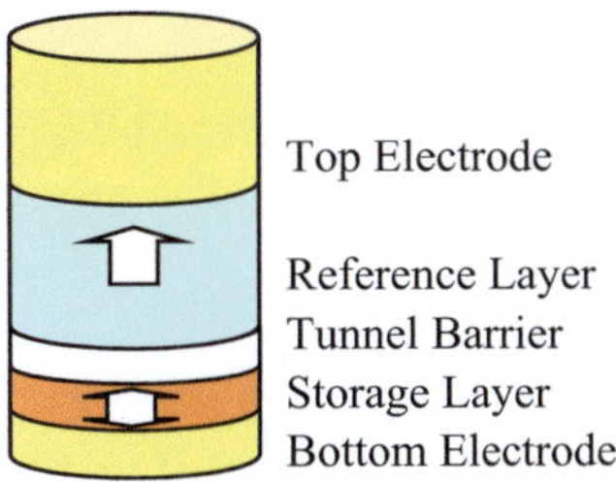

Fig. 14 Schematics of fast and low current switching p-MTJ

Figure 13 shows the pulse width and the switching current (Ic) map data of recently reported STT-MRAM elements. Data points on the same diagonal line in Fig. 13 have the same write charge. The STT-MRAM-based cache memory located in the lower left region in the map can reduce the energy consumption compared to a conventional SRAM-based cache memory. However, that in the upper right region increases the energy consumption. A diagonal stripe in the map indicates break-even points, where the energy consumption of the STT-MRAM-based cache memory is almost the same as that of the conventional SRAM-based cache memory. In order to reduce the energy consumption, the MTJ whose write charge is lower than this stripe is required. Most of the reported data are located in the upper right region, which is where energy is wasted. However, some of the data are located in the lower left region, which is where energy is saved.

The p-MTJ that shows the smallest write charge in Fig. 13 has the layer structure top electrode/reference layer/MgO/storage layer/bottom electrode as illustrated in Fig. 14 [20]. In order to achieve the low-current switching, the damping factor of the storage layer p-MTJ is reduced to 0.003.

The switching probability of this p-MTJ measured by the pulse current from 0.8 to 4 ns in pulse width as a function of the write current is shown in Fig. 15. Each plot in Fig. 15 is estimated from the results of a switching test performed 30 times. The switching was observed for the pulse shorter than 1 ns and with the current smaller than 100 µA.

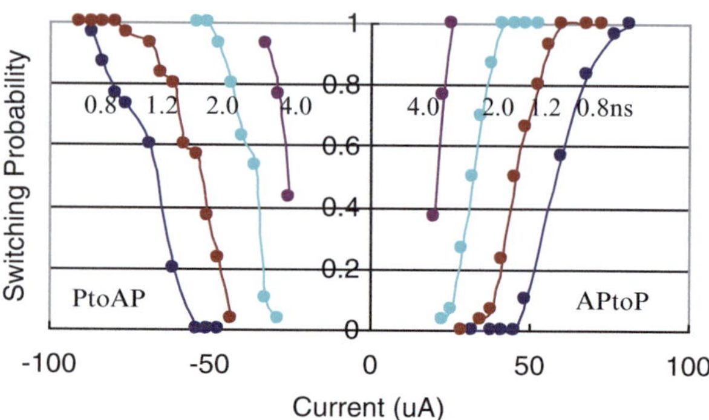

Fig. 15 Switching probability of p-MTJ measured by the pulse current with 0.8–4 ns in pulse width. The inset values indicate the pulse width in nanoseconds

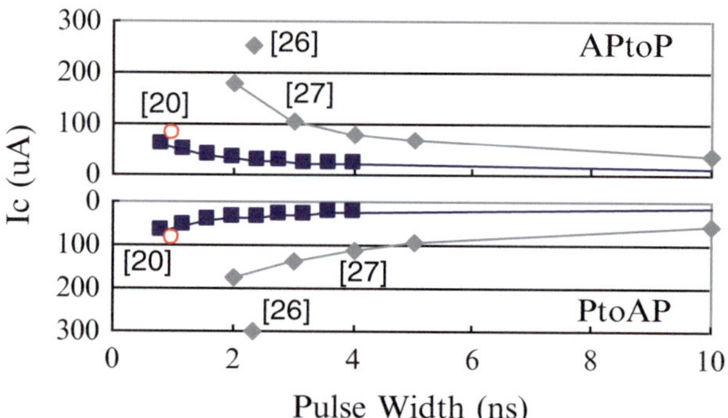

Fig. 16 Switching current Ic of MTJs with current pulse

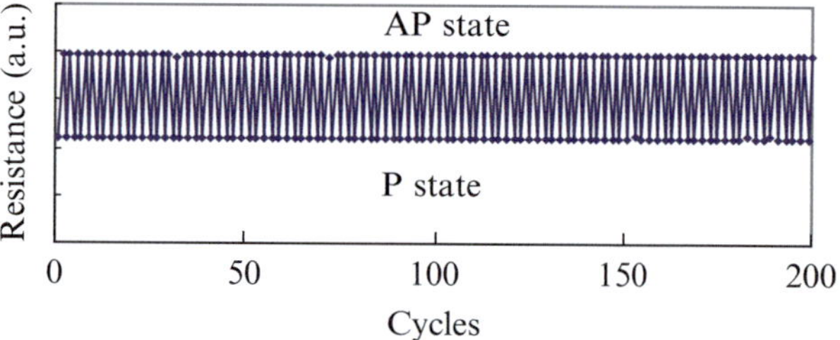

Fig. 17 Write cycle tests on p-MTJ with 1 ns pulse current. The applied currents were 90 and 80 μA for P to AP and AP to P switching, respectively

Figure 16 shows Ic with each pulse width. The Ic of this p-MTJ with 1 ns pulse is 60 μA for P to AP switching, and 50 μA for AP to P switching. Figure 17 shows results of a write cycle test performed 200 times on the p-MTJ with 1 ns pulse current. The applied current was 90 and 80 μA for P to AP switching and AP to P switching, respectively. The write pulse conditions are indicated by the blank circles in Fig. 16. These pulse currents were alternately applied to the p-MTJ and no switching error was observed.

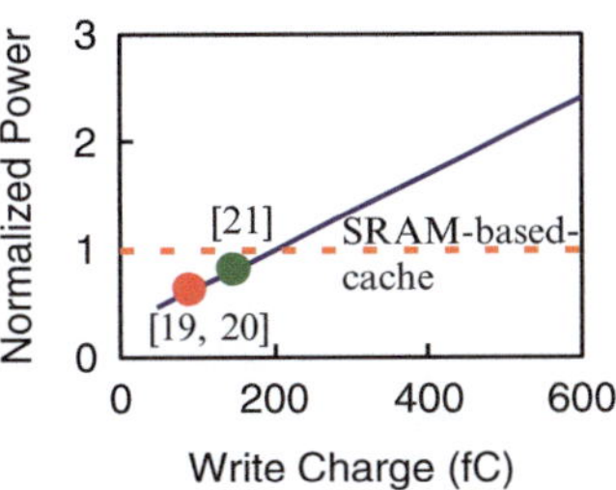

Fig. 18 The impact of the write charge on the power of the processor with STT-MRAM-based cache memory

The write charge of this p-MTJ is 1 ns $\times$ 90 µA = 90 fC. Figure 18 shows the power of the processor with the STT-MRAM-based cache memory normalized by that of the conventional SRAM-based cache memory as a function of the write charge, considering active-state profiling with high performance benchmarks such as h. 264 encoding and 3-D dynamics. The assumed STT-MRAM circuit topology in this simulation is based on the 2T-2MTJ cell [36]. Figure 18 proves that the impact of the write charge on the power is so critical that only the STT-MRAM that shows small write charge can achieve power saving. The advanced p-MTJs plotted in Fig. 18 can save energy even in active states.

3 High-Performance and Low-Power Circuit Designs Based on Advanced STT-MRAM

3.1 Combination of MRAM and SRAM

As illustrated in the previous section, replacing SRAM with the fast and low-current-switching advanced STT-MRAM cell can reduce the energy consumption of the processor. However, programming speed of SRAM is still faster than MRAM. Therefore, various cell-level combinations of MRAM and SRAM have been proposed to improve access time even though these have leakage current paths. To replace SRAM for LLC with STT-MRAM cell, there have been candidate memory cells such as one transistor (1T)-1MTJ (conventional MRAM [35]), cell-level combinations of MRAM and SRAM, 6T-2MTJ (Nonvolatile (NV-) SRAM [29]), 8T-2MTJ [30] and 4T-2MTJ [31] (other types of NV-SRAM), as shown in Fig. 19. Their features are listed in Table 1. In the case of conventional MRAM, the access speed is slow for cache memory applications and power is high compared with SRAM. 6T-2MTJ and 8T-2MTJ based on SRAM cell are SRAM/MRAM hybrid memories, operating as SRAM normally (SRAM mode) and as MRAM in the power gating mode. The 6T- and 8T-2MTJ are the same speed as SRAM mode because data is not stored to two MTJs. The 4T-2MTJ NV-SRAM is the same speed as SRAM, but its power is determined by performances of the MTJ device. However, these NV-SRAMs have leakage current paths that are an origin of static power, since these circuits are fundamentally SRAM circuits, as shown in Fig. 19.

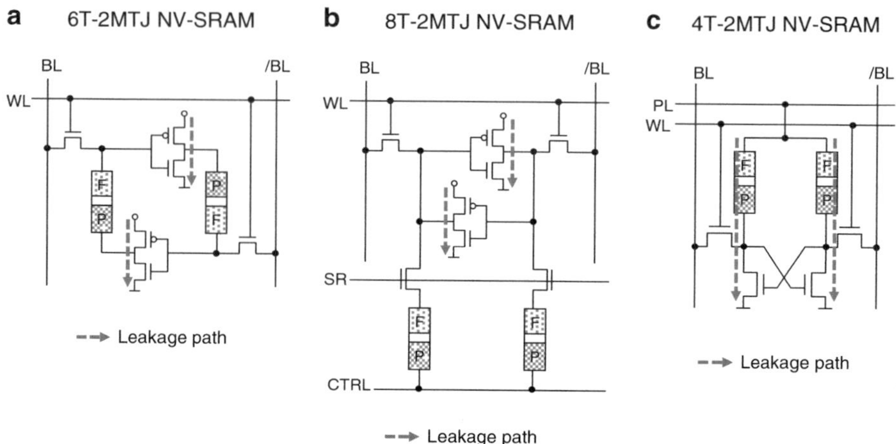

Fig. 19 Nonvolatile SRAM with 2 MTJs based on 6 transistors (6T) (**a**), 8T (**b**) and 4T (**c**)

Table 1 Features of nonvolatile RAM candidates for replacing LLC

Cell Type	SRAM (reference)	1T-1MTJ [1]	6T-2MTJ [2]	6T-2MTJ [3]	4T-2MTJ [4]
Non-volatile	**No**	Yes	Yes	Yes	Yes
Area	1	0.2–0.6	~1.3	**~2.5**	0.8–1.2
Speed	1	**Slower**	SRAM compatible	SRAM compatible	SRAM compatible
Active power	1	**Higher**	SRAM compatible	SRAM compatible	Much higher
Leakage power	**Large**	Small	**Large**	**Large**	**Much larger**
Leakage power@power gating	Small	Small	Small	Small	Small
Robustness to read/write error	(reference)	Comparable	**Worse due to MTJ variation**	Comparable	**Worse due to MTJ variation**

The 4T-2MTJ-NV-SRAM also has leakage current paths. Furthermore, it has much larger active current than SRAM. The power gating is used to avoid the power consumption of these leakage paths. However, it cannot be used while applications are running, which is the same situation as for SRAM, as described in the previous session. In addition, power gating imposes a circuit area overhead due to extra power lines and switching transistors.

The following sections describe a DRAM-MRAM hybrid memory design which enables the reduction of energy consumption without the power gating and a 2T-2MTJ dual cell based memory design which decreases both the write-latency and read-latency overhead.

3.2 *DRAM-MRAM Hybrid Memory Design*

The authors propose a novel hybrid memory cell design involving cell-level hybridization of DRAM and STT-MRAM having no static leakage current path [32]. We have shown effective power reduction by applying the hybrid memory to LLC for mobile CPU. Furthermore, it has been revealed that the power reduction is enhanced by using advanced perpendicular magnetic tunnel junctions (p-MTJ) with low power and high speed for write operation. The proposed DRAM/MRAM hybrid memory named "D-MRAM" with 3T-1MTJ memory cell design is shown in Fig. 20. The capacitance of RT_D's gate and RT_M's source-drain capacitance (off-state) are equivalent to those of a capacitor in a DRAM cell, as shown in Figs. 20 and 21. The D-MRAM operates as DRAM (DRAM-mode) when the nonvolatility is not required and operates as MRAM (MRAM-mode) when the nonvolatility is required. The DRAM-mode write operation drives WWL high to store 1/0 in the cell capacitor. The DRAM-Mode read operation drives RWL high, then RBL current rises or remains because it is charged up through RT_D depending on cell capacitance. The difference compared to the reference for DRAM-mode (D_{ref}) is read in the sense amp (SA). The retention time calculated is about 10 µs. For the MRAM-mode, write operation is based on current injection into the MTJ through two transistors by the spin torque transfer, as shown in Fig. 21c. The MRAM-mode write operation drives WWL and SWL high to switch MTJ state (AP/P). The DRAM-mode read operation drives WWL and SWL below the voltage of MRAM-mode write operation. The difference of SL signal caused by the difference of MTJ resistance state is read in SA with reference voltage for MRAM-mode (M_{ref}). In read operation, the read signal lines and reference signal lines are switched to connect the sense amplifier according to the read mode (DRAM or MRAM), as shown in Fig. 21b, d.

We designed the 1 Mb D-MRAM. The subarray configuration of 1 Mb D-MRAM was designed for cache memory, as shown in Fig. 22a. Using this memory macro, performance and power are simulated based on low-power

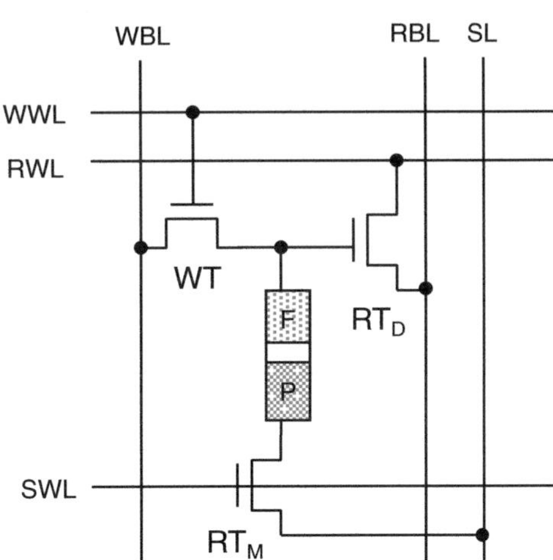

Fig. 20 Circuit diagram of Dynamic-MRAM (D-MRAM)

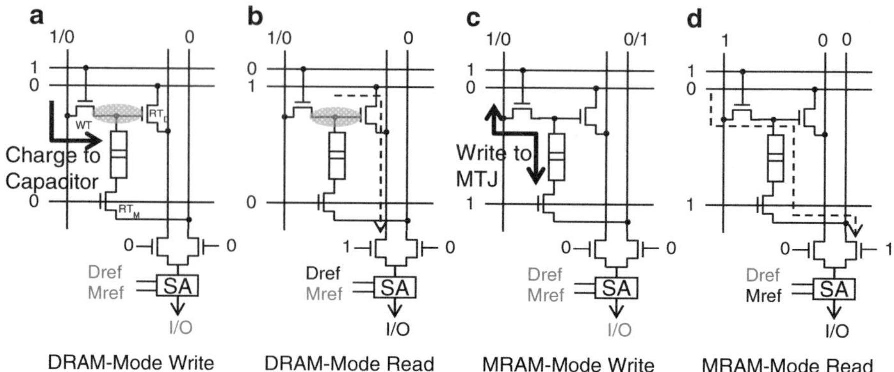

Fig. 21 Operation of D-MRAM

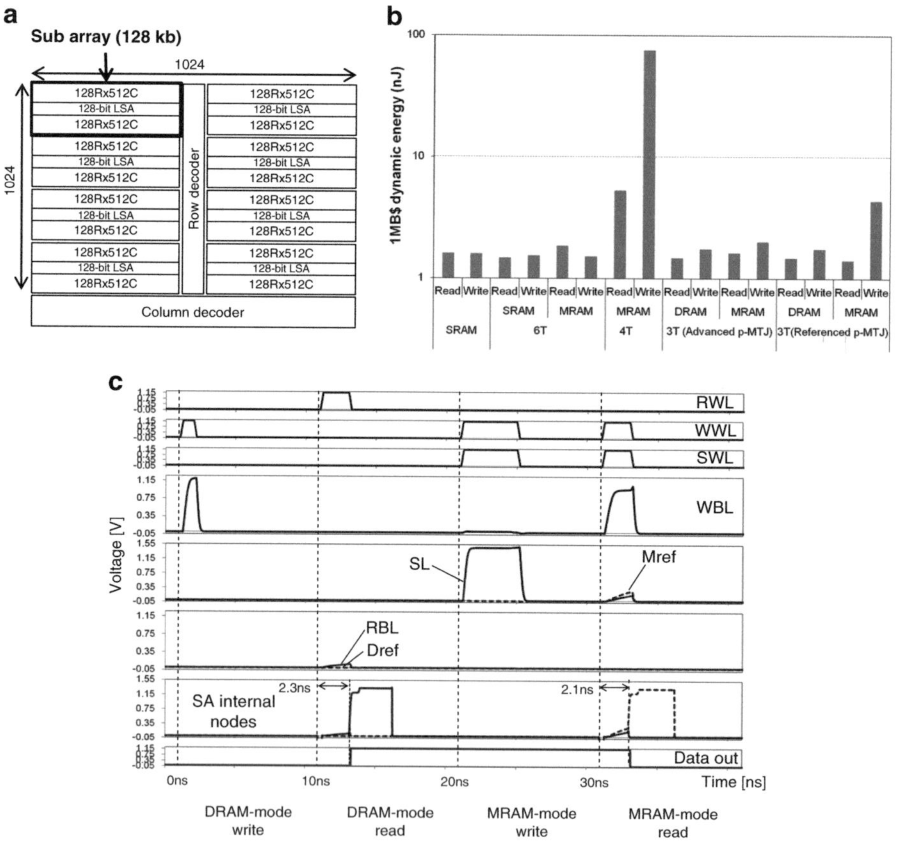

Fig. 22 (**a**) Subarray configuration of 1 Mb D-MRAM. (**b**) Dynamic energy of 1 Mb RAM.
(**c**) Simulated waveforms of D-MRAM operation

65 nm CMOS technology. As for the MTJs, we used an advanced perpendicular (p-) STT-MRAM having high speed and low current (write time: 3 ns, write current: 50 μA) that Toshiba has recently developed [21]. A conventional STT-MRAM model (write time: 25 ns, write current: 120 μA) was used as a reference, whose performance and power are close to those of the conventional p-MTJs reported by other groups [31], and its programming energy is about ×20 larger than that of the advanced p-MTJ. Simulated results are shown in Fig. 22b, c. The write access times of DRAM- and MRAM-mode are 1.5 and 4.5 ns, respectively, and the read access times are both about 2.2 ns, indicating that DRAM-mode has higher memory access speed than MRAM-mode. DRAM-mode also has lower memory access energy than MRAM-mode, as shown in Fig. 22b.

Figure 23 shows the scheme for memory-mode (DRAM/MRAM) selection. When a D-MRAM-cache line is initially accessed, it works as DRAM and the data are retained with refresh operation every retention time, ~10 μs. However, as the DRAM-mode is longer, the refresh power becomes larger. To optimize the overall power of LLC, it is suitable that the data are transferred to the MRAM-mode after the time to change from DRAM-mode to MRAM-mode, Tc. Tc also depends on the cache access pattern determined by the application running on the CPU. In our case study considering cache access analysis, 50 μs was selected for Tc. The memory-mode (DRAM-mode/MRAM-mode) is selected for each cache line. The "flag" data in the added bit on each line is changed from "0" to "1", and then D-MRAM cells start to operate as MRAM. When the cache lines are overwritten by other higher-priority data, the flag data is changed back to "0". This scheme is very effective for reducing energy for cache memory.

To evaluate the power of D-MRAM-based cache memory, we modified a CPU emulator, GEM5 [33] (ARM-mobile CPU architecture in Table 2) for nonvolatile cache memories, with SPEC-CPU2006 [34] benchmark suites. We calculated the cache operation power for comparison with 6T-2MTJ, 4T-2MTJ (with fine-grain power gating) and 3T-1MTJ as listed in Table 3, where 8T-2MTJ was not selected since its large memory cell area overhead (~×2.5 SRAM area) is unacceptable for mobile CPUs. The R-V curves of the advanced p-MTJ used for the simulations are shown in Fig. 24.

Calculated results in Fig. 25a show that the power reduction of D-MRAM-cache with advanced p-MTJ is about 40 % of that of conventional SRAM-based cache memory and much lower than that of other NV-SRAMs. The operation power of 6T-2MTJ NV-SRAM is almost the same as that of SRAM, since the power gating is rarely used. The operation power of 4T-2MTJ NV-SRAM is much higher than that of SRAM even though the very fast power gating scheme [31] is adopted, since the active power of 4T-2MTJ cell is very high because of large write/read current and high MTJ programming energy, as shown in Fig. 22b. 4T-2MTJ cell is suitable for low-frequency operation for low-end CPU (~several tens of MHz), as the average active power is proportionally reduced and standby time becomes longer with decreasing operation frequency. Figure 25b shows the difference of operation power of D-MRAM-based LLC with advanced / referenced p-MTJs. The power of D-MRAM-based LLC with advanced p-MTJ is decreased for all workloads,

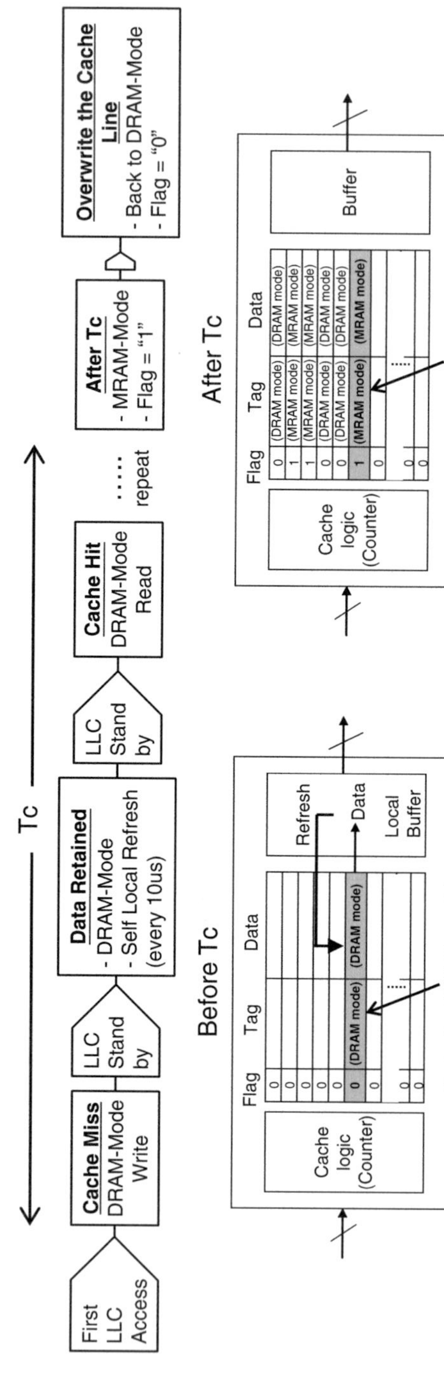

Fig. 23 Scheme for memory-mode (DRAM/MRAM) selection

Table 2 Architecture of mobile CPU used for calculations

Processors		Execution	
# of cores	1	Warm up	1 M inst.
Frequency	1GHz	Execution	10 M inst.
Issue width	1 (out of order)		
ISA	ARMv7		
Memory			
L1 cache	32 + 32kB, 4-way, 64B line, Write-through, 1 read/write port, 1 ns latency		
L2 cache	1 MB, 8-way, 64B line, Write-back, 1 read/write port		

Table 3 Lists of MTJ performances in L2 cache memory cell used for calculation

	MTJ device
Cell type	Write time/Current
SRAM (Reference)	–
2MTJ-6Tr [2]	3 ns/50 µA (Advanced p-MTJ)
2MTJ-4Tr [4]	25 ns/120 µA (Reference p-MTJ)
D-MRAM (1MTJ-3Tr, This work)	3 ns/50 µA (Advanced p-MTJ)

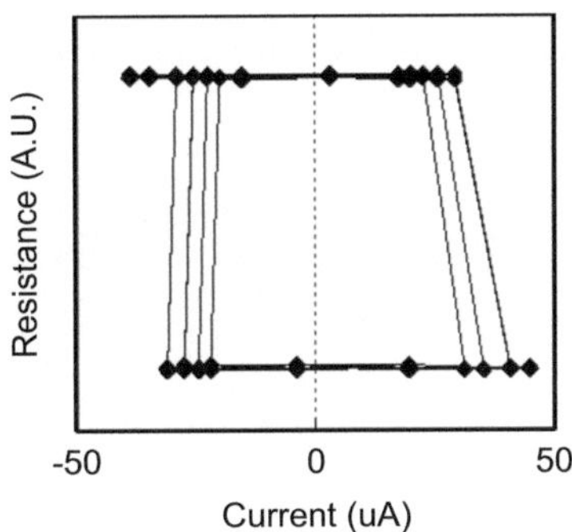

Fig. 24 I-V curve and resistance change of advanced p-MTJ

whereas the power with conventional (reference) p-MTJ is increased compared with that of SRAM-based cache for some workloads. This result indicates that high-speed and low-write-current p-MTJ is strongly required for low-power LLC in mobile CPU. It has been confirmed that reduction in write charge of MTJ is the most important factor to decrease power for the cache memory using MTJs, which is consistent with the expectation explained using Fig. 12. This is the first result to directly show that the low write charge of MTJs contributes to power reduction of cache memory for advanced CPU.

Figure 26 shows CPU performance, instruction per clock (IPC) calculated by GEM5. IPC of SRAM-based CPU is 1.0. There is little performance degradation of CPU with D-MRAM-based LLC. This result suggests the operation speed of

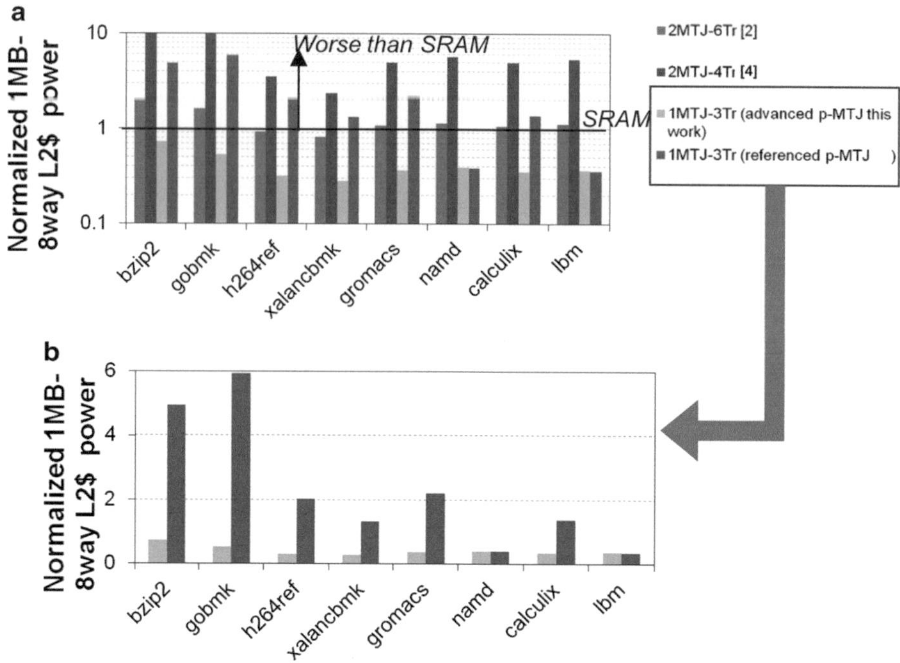

Fig. 25 (**a**) Normalized power of cache memory (SRAM is 1.0) calculated with a CPU simulator GEM5

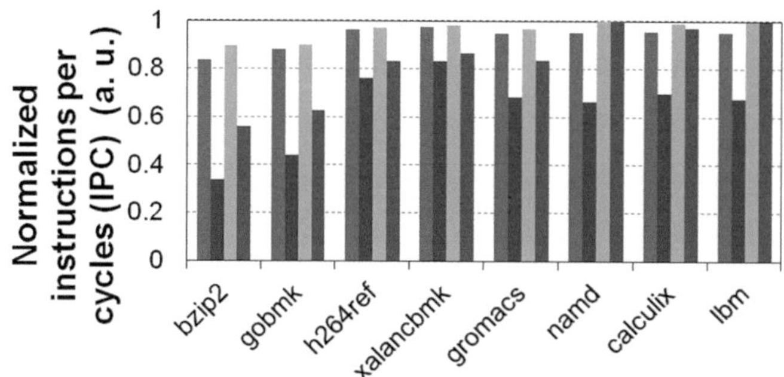

Fig. 26 CPU performance, instruction per clock, (IPC) calculated by GEM5

D-MRAM is fast enough for LLC operation. Since LLC has memory latency of 10 clock cycles or more, and less frequent access compared with memories in the CPU core, D-MRAM is applicable to LLC even though the access speed is slower than that of SRAM.

The cell layout of 3T-1MTJ D-MRAM is shown in Fig. 27. The area estimated from the layout based on 65 nm CMOS logic design rule is almost the same as that

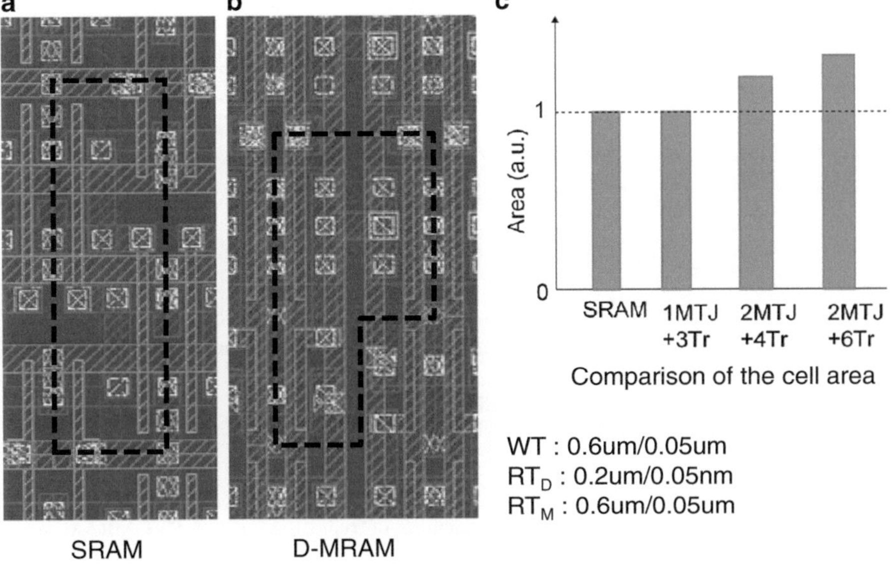

Fig. 27 Memory cell layout for (**a**) conventional 6T-SRAM cell and (**b**) proposed D-MRAM cell, and (**c**) comparison of area between SRAM and MTJ-based NVRAM cells

of conventional 6T-SRAM. Since the area can be reduced to less than half of this area by customizing the design rule, D-MRAM also has an advantage over SRAM in terms of area reduction.

Thus, it has been shown that the DRAM/MRAM hybrid memory design enables effective power reduction for high-performance mobile CPU. With D-MRAM-based cache memory in the CPU, power consumption can be reduced by about 60 % compared with that of SRAM-based cache while an application is running. This result is achieved by the novel D-MRAM memory design and low-write charge, 150 fC, of advanced p-MTJ with ultra-high-speed write and low-current write (3 ns, 50 μA).

3.3   2T-2MTJ Dual Cell Based Memory Design

Instead of DRAM hybridization, dual cell using two sets of single 1T-1MTJ cell is a novel memory circuit design to increase performance. To decrease the write-latency overhead, we have developed the high-speed-switching p-MTJ device described in the previous sections, because the write latency of STT-MRAM-based cache is dominated by the switching pulse width of the MTJ. On the other hand, to decrease the read-latency overhead, we have to develop not only high-MR-ratio p-MTJ, but also high-speed read circuits and memory architecture. The read latency is dominated by the propagation delay in asserting the accessed wordline

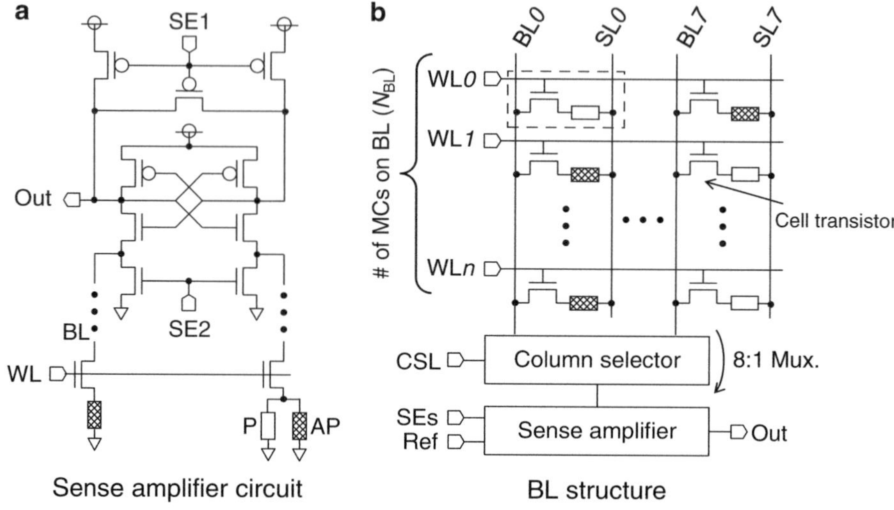

Fig. 28 Sense amplifier circuit and BL structure of typical 1T-1MTJ STT-MRAM using reference cell

Fig. 29 1T-1MTJ cell topology

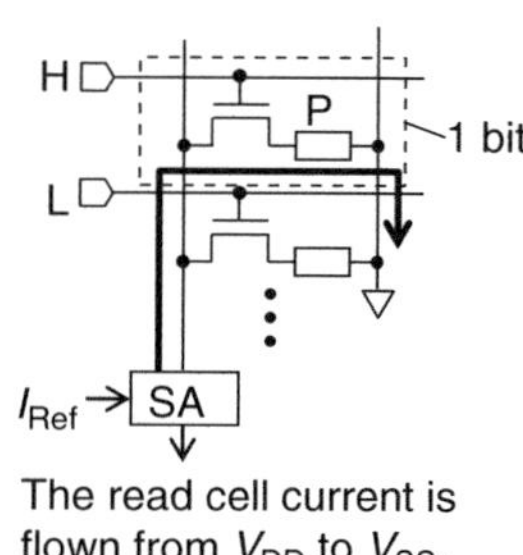

The read cell current is flown from V_{DD} to V_{SS}.

and targeted bitlines (including source lines) and evaluating the value of MTJ resistance. Hence, to minimize the propagation delay, the memory array architecture is required for high-performance cache. In addition, to decrease the evaluation time, high-MR-ratio p-MTJ device, cell topology, and high-speed sensing circuit (sense amplifier) are required.

The general application target of the STT-MRAM is large-capacity memory, such as DRAM, that utilizes a current-based sensing scheme. Figure 28 shows the sense amplifier circuit and BL structure of the prior current-based sensing scheme. For large capacity, greater priority is accorded to the total memory area than to the access latency in view of cost reduction. Thus, 1T-1MTJ cell topology is usually applied, as shown in Fig. 29, and reference cells are appended for read sense amplifiers, as shown in Fig. 28b. In this prior read sensing circuit, the read cell currents flow from a sense amplifier circuit to the accessed p-MTJ device and the reference cell, then the current difference caused by the resistance of the MTJ

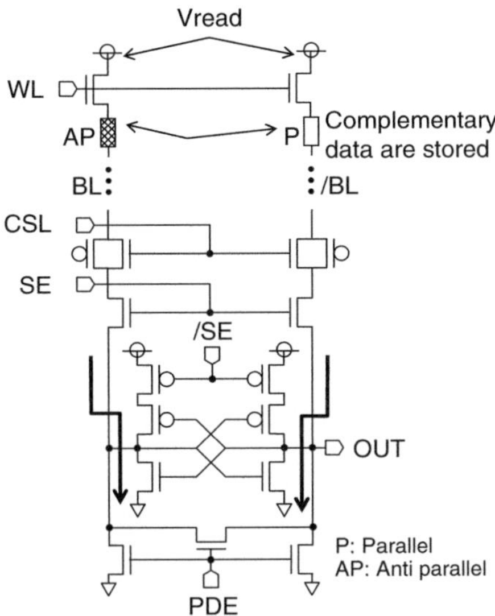

Fig. 30 Sense amplifier circuit and BL structure of proposed 2T-2MTJ STT-MRAM using current-integral scheme [36]

device is converted into a difference of the voltage drop of a transistor in the sense amplifier, and finally it is amplified.

According to this reading circuit, it is necessary to continue to pass the cell current to the MTJ device until the current value stabilizes during reading, posing problems of longer read operation and higher power consumption. This is a critical issue when STT-MRAM is used as a high-speed memory such as a cache memory of high-performance processors.

Also according to this circuit, a difference between a reference cell current and the accessed cell current dependent on the resistance (high resistance or low resistance) of the MTJ device becomes a read margin. Thus, noise produced at the time (instantaneous value) of converting a current value into a voltage value greatly affects the read operation. If it is assumed that, for example, the current value is of the order of 10 μA and the on-resistance of a transistor is a few kΩ, the read margin becomes about 5 mV and an extremely precise sense amplifier having large power and large size is needed to sense such a small read margin.

3.3.1 Current-Integral Sensing Circuit Design of STT-MRAM

Figure 30 shows a newly proposed sense amplifier circuit with current-integral scheme for cache memory. The sense amplifier adopts the differential voltage sensing method that compares potential V_{BL} of bitline (BL) and the reference potential $V_{/BL}$ of bitline bar (/BL). In Fig. 30, the sense amplifier is activated by asserting signal sense enable bar (/SE). A pre-discharge circuit is connected between two bitlines, BL and /BL, to equalize the potential of BL and /BL, with the ground potential Vss and also pre-discharge BL and /BL. These bitlines, BL and /BL, are

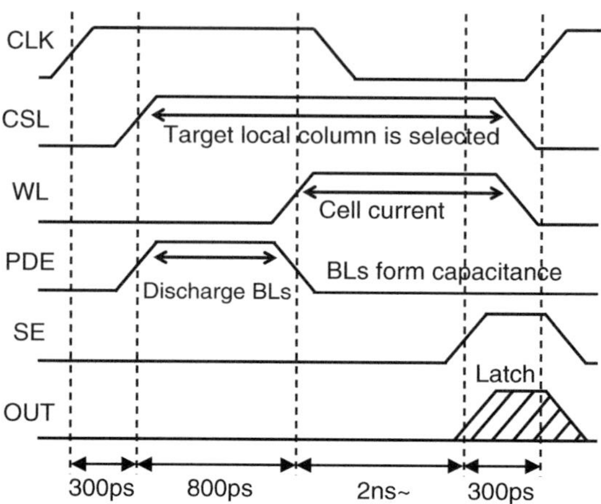

Fig. 31 Timing chart of proposed current-integral sensing scheme

pre-discharged to set them to a ground potential Vss before being changed to a floating state. By setting source lines, SL and /SL, to a read voltage Vread, which is slightly higher than the Vss, after asserting the accessed wordline WL, the potentials of BL and /BL, in a floating state, are changed to values in accordance with the resistance value of the MTJ.

Then, a difference between the read voltage of bitlines, V_{BL} and $V_{/BL}$, in a floating state is detected by using sense amplifier, and the difference is amplified to output read data, OUT.

Figure 31 shows a timing chart when reading is carried out by the current-integral scheme using the sense amplifier in Fig. 30. First, bitlines, BL and /BL, are discharged to set them to the ground potential Vss by setting activation signal PDE to "H". Row decoder circuit asserts the accessed wordline WL according to the row address signal during the discharge period. Further, the accessed column selector signal CSL is set to "H" to electrically connect bitlines, BL and /BL, to sense amplifier in this pre-discharge state, according to the column address signal. Then, by setting wordline WL to "H", bitlines, BL and /BL, are electrically connected to source line SL, which is charged to the read potential Vread, via cell transistor. Further, by setting activation signal PDE to "L", bitlines, BL and /BL, are set to a floating state. As a result, differential currents flow from source line SL toward bitlines, BL and /BL, to charge parasitic capacitance of each bitline and these cell currents are accumulated as the amount of charge of bitlines, BL and /BL. At this point, the rate of charging the parasitic capacitance of bitlines, BL and /BL, changes in accordance with the resistance value of the MTJ device in each cell. Therefore, if, for example, the MTJ device in a cell unit stores "1" and is in a high-resistance state, the rate of charging the parasitic capacitance of bitline becomes slower and if the MTJ device in a cell unit stores "0" and is in a low-resistance state, the rate of charging the parasitic capacitance of bitline becomes faster. That is, by setting activation signal /SE to "L", data of the MTJ device in a cell unit can be read out by sensing a difference between the charging potentials of bitlines, V_{BL} and $V_{/BL}$.

The period between the start of charging bitlines, BL and /BL, and the activation of sense amplifier can be made sufficiently shorter than that by the current sensing method. For example, the period can be made less than 1 ns by the voltage sensing method. In the voltage sensing method, by contrast, charges are accumulated and thus, a potential difference of 50 mV or more can be created and the design of the sense amplifier can be significantly simplified. In addition, there is no need to wait for saturation of the current value because the accumulation of charge starts immediately after the current starts to flow into the MTJ device. Thus, when compared with the current sensing method in which a sensing period delay is caused by an RC delay, the voltage sensing method can shorten the period between the start of passing a current to the MTJ device and the start of a sensing operation. Further, the voltage sensing method is a method by which data is read based on a potential difference, which is the accumulated value of charge, and thus, both the read margin and speed can be improved when compared with the current sensing method in which instantaneous values of current are compared.

3.3.2 Dual Cell Topology for Fast Read Operation

In order to achieve faster read operation, we apply the new cell topology, 2-cell/1-bit structure (2T-2MTJ cell) [36]. Figure 32 shows a memory cell topology. In this cell structure, 1-bit data is stored by using two cell units. Complementary data is stored in two 1T-1MTJ cell units. Each cell unit in the memory cell array includes an MTJ device and a selection transistor connected in series, as shown in Fig. 32. A gate terminal of the selection transistor is connected to wordline WL. One end of cell unit is connected to bitline BL*0* and the other end is connected to source line SL*0*. One end of the other cell unit is connected to bitline /BL*0* and the other end is connected to source line /SL*0*. Both bitlines and source lines extend in the column direction and one end thereof is connected to write circuit and read circuit via column selector. ON/OFF of column selector is controlled by column selection signal CSL from column decoder circuit.

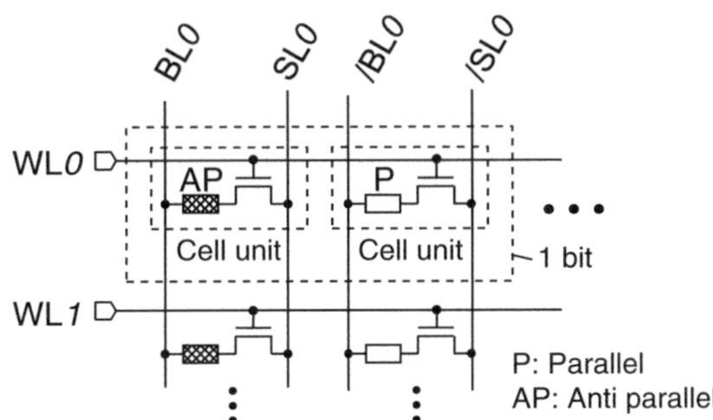

Fig. 32 2T-2MTJ cell structure

Complementary data are stored in each 1T-1MTJ cell unit.

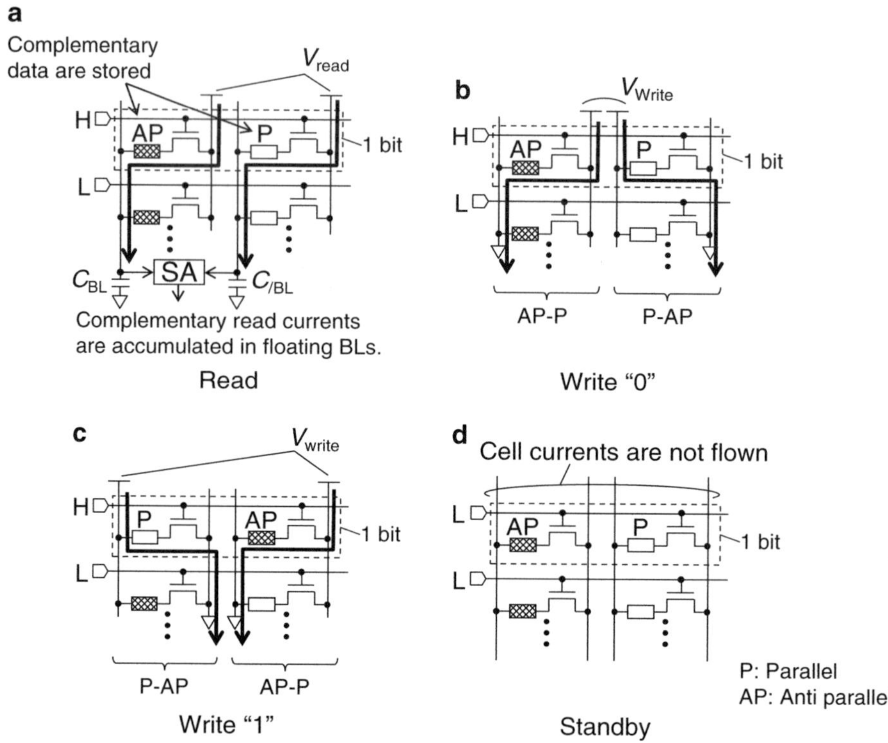

Fig. 33 Operation mechanism of proposed 2T-2MTJ cell

Figure 33 shows operation mechanisms of the 2T-2MTJ dual cell structure. Unlike in the 1T-1MTJ cell structure, since the sense amplifier circuits amplify the difference in cell current between a pair of MTJs, the reference cells are not needed. The cell size of the dual cell estimated in 65 nm CMOS technology is $0.45\ \mu m^2$, which is much smaller than that of SRAM. Read latency can be greatly reduced and the sensing margin is doubled because reference cells are eliminated. Figure 33a shows the readout cell currents. In the prior current-based read circuits, the read cell currents flow from Vdd to Vss continuously during read operation, as shown in Fig. 29. On the other hand, in the proposed sensing scheme, since while the accessed wordline, WL, is activated, the differential bitlines, BL and /BL, are left floating and work as "capacitance" as shown in Fig. 30, the read cell currents are stored in bitlines, BL and /BL, and no currents flow from Vdd to Vss directly. The sensing margin can be expanded by this "current-integral" scheme that accumulates the small difference between complementary cell currents through "L" resistance MTJ and "H" resistance MTJ. The area for the proposed sense amplifier circuit is 27 % less than that for the conventional current sensing circuit. Figure 33b, c show the write cell currents in the case of writing "0" datum and "1" datum, respectively. To write the complementary resistance states in a pair of MTJ devices in a 1-bit cell, symmetrical write currents continuously flow according to the write pulse width of the p-MTJ device, from a source line to a bitline and from a bitline to a source line. After read and write operation, the bitline equalizer circuits are asserted and cell leakages in standby states are suppressed, as shown in Fig. 33d.

3.3.3 Memory Array Architecture with Two-Level Hierarchical Bitline Structure

In order to achieve faster operation and reduce active energy due to bitline resistance (R) and capacitance (C), the authors adopt a two-level hierarchical bitline (BL) structure with local bitlines and global bitlines, as shown in Fig. 34. As total memory array size has been enlarged, the RC delay in bitlines and source lines can no longer be ignored. The RC delay is mainly caused by the junction capacitance of cell transistors in bitlines and source lines. Two-level hierarchical bitline structure can reduce the number of cells connecting to each bitline, and can eliminate the RC delay in bitline. In Fig. 34, 256-bit cells are connected to each local bitline and only column selector switches are connected to each global bitline. By using this hierarchical structure, when the number of bit cells in each local bitline is fixed, the memory capacity can be increased without large increase in RC delay. This is

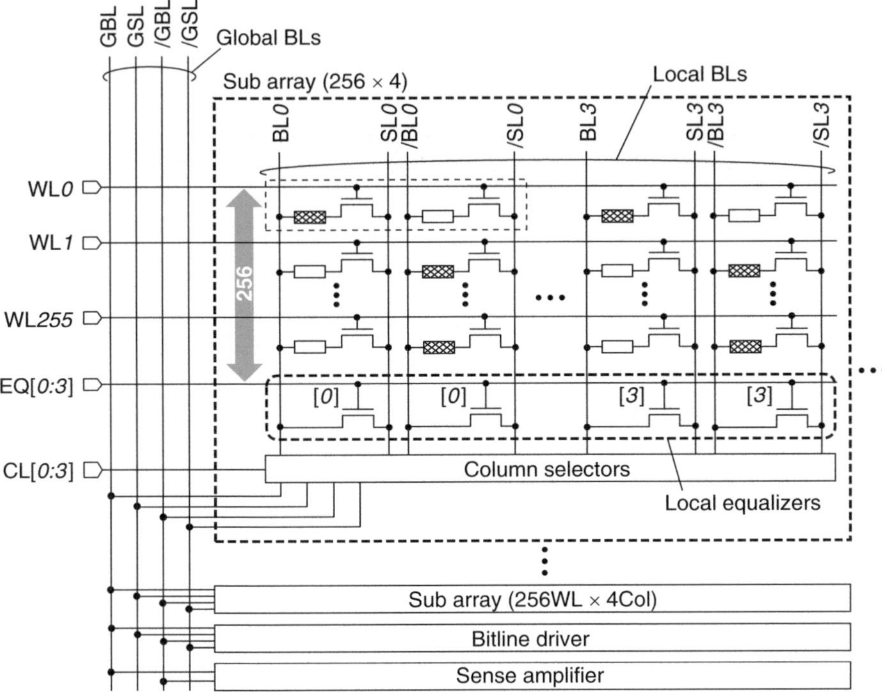

Fig. 34 Array structure of hierarchical BL with column selector and local-bitline equalizer

because the RC delay caused by global bitlines is relatively small compared to that caused by local bitlines.

3.3.4 1 Mb MRAM Design and Measurement Results of Test Chip

We designed 1-Mb STT-MRAM macro and fabricated the test chip in 65 nm CMOS for p-MTJ integration, as shown in Fig. 35. The total macro area and total cell area are 0.628 mm^2 (1,059.32 μm × 592.81 μm) and 0.472 mm^2 (0.5 μm × 0.9 μm × 1 Mb) respectively, thus the cell efficiency of over 75 % is achieved. In this chip, we apply a hierarchical bitline structure (Fig. 34) for fast read operation, and equalize the circuit between each local bitline (BL) and local source line (SL) to enhance the storage stability of half-selected cells. The global bitline length was set to 1,024 b and the wordline length was set to 512 b. The word length is 256 b. We measured the read operation speed with resistors in 65 nm CMOS. The measurement results indicate 4 ns cycle time operation at the core voltage of 1.05 V. Read and write power consumption with 256-b I/O width is 17.8 mW and 46.5 mW, respectively, by estimation from measured data. Although nonvolatile cache memories with STT-MRAM element had been investigated with a view to decreasing the leakage power of the cache, in the case of all conventional STT-MRAMs, total power is "increased" using nonvolatile caches, since active power (write and read power (Fig. 36)) is greatly increased compared with SRAM, although static power is decreased using STT-MRAM cell. On the other hand, our proposed STT-MRAM cell can decrease active cache power with high-speed operation, because of the advanced p-MTJ with low-power and high-speed write operation and high-MR, low-power and high-speed read circuit design, and hierarchical memory architecture. Table 4 summarizes a comparison of RAM for cache memory, where memory parameters are converted to cache line (=512 bits) access for fair comparison. These data have confirmed that 2T-2MTJ with the advanced perpendicular STT-MRAM is the best solution for nonvolatile cache memory applications on processors.

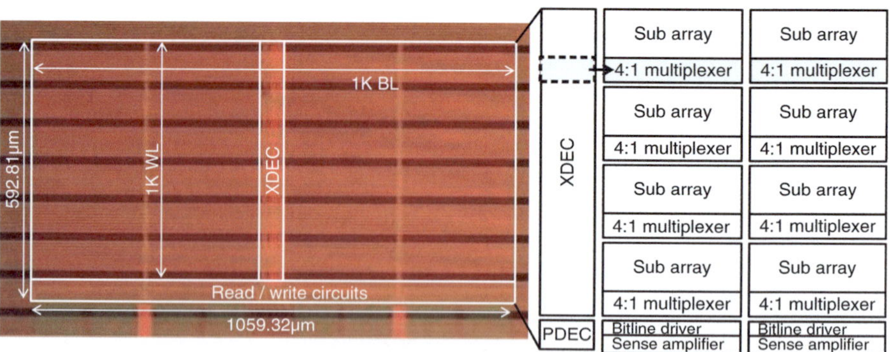

Fig. 35 Chip micrograph fabricated in 65 nm process

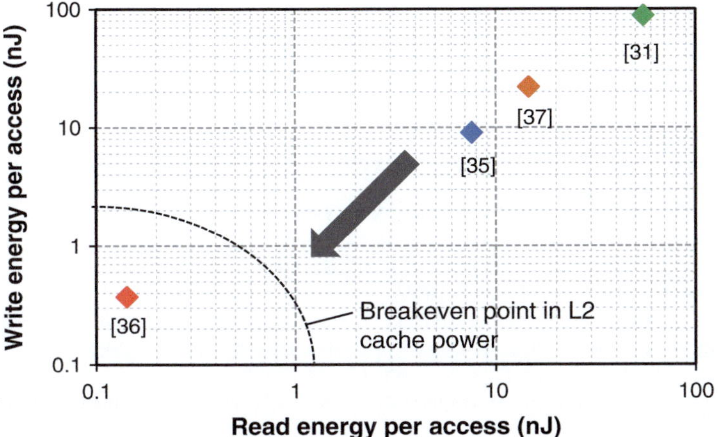

Fig. 36 Energy per access comparison of STT-MRAMs

Table 4 Comparison of STT-MRAMs

	Efficiency (%)	Cell size (F^2)	Latency (ns)	IO (bits)	Iw (μA)	Dynamic power (mW)	Energy per access (nJ)[a]
[37]	63	169.1	Read: 12, write: 12	32	600	Read: 60 (66 MHz), write: 91 (66 MHz)	Read: 14.5, write: 22.0
[35]	51	42.6	Read: 11, write: 30	16	N/A	Read: 7.8 (33 MHz), write: 9.3 (33 MHz)	Read: 7.56, write: 9.02
[38]	38	50.7	Read: 8, write: 10	32/64	N/A	N/A	N/A
[31]	64.8	270.3	Read: 8, write: 40	1	100	Read: 10.7 (100 MHz), write: 4.3 (25 MHz)	Read: 54.8, write: 88.1
[36]	75.1	106.5	Read: 4, write: 4	256	50	Read: 17.8 (250 MHz), write: 46.5 (250 MHz)	Read: 0.142, write: 0.372

[a] Converted to a cache line (=512 bits) access.

Acknowledgement A part of this work was supported by the New Energy and Industrial Technology Development Organization (NEDO) of Japan.

References

1. K. Ando, Roles of non-volatile devices in future computer system: normally-off computer, in *Energy-aware systems and networking for sustainable initiatives*, ed. by N. Kaabouch, W.-C. Hu (IGI Global, Hershey, 2012)
2. H. Yoda, S. Fujita, N. Shimomura, E. Kitagawa, K. Abe, K. Nomura, H. Noguchi, J. Ito, *IEDM Tech. Dig.* (2012), p. 259
3. T. Miyazaki, N. Tezuka, J. Magn, Magn. Mat. **139**, L231–L234 (1995)
4. J.S. Moodera, L.R. Kinder, T.M. Wong, R. Meservey, Phys. Rev. Lett. **74**, 3273 (1995)
5. J.C. Slonczewski, J. Magn. Magn. Mat. **159**, L1 (1996)
6. L. Berger, Phys. Rev. B **54**, 9353 (1996)
7. Y. Huai, F. Albert, P. Nguyen, M. Pakala, T. Valet, Appl. Phys. Lett. **84**, 3118 (2004)
8. Y. Huai, M. Pakala, Z. Diao, Y. Ding, Appl. Phys. Lett. **87**, 222510 (2005)
9. H. Meng, J.P. Wang, Appl. Phys. Lett. **88**, 172506 (2006)
10. S. Mangin, D. Ravelosona, J.A. Katine, M.J. Carey, B.D. Terris, E.E. Fullerton, Nat. Mater. **5**, 210 (2006)
11. T. Seki, S. Mitani, K. Yakushiji, K. Takanashi, Appl. Phys. Lett. **88**, 172504 (2006)
12. H. Yoda et al., *7th IWFIPT*, Session IIIc
13. M. Nakayama, T. Kai, N. Shimomura, M. Amano, E. Kitagawa, T. Nagase, M. Yoshikawa, T. Kishi, S. Ikegawa, H. Yoda, J. Appl. Phys. **103**, 07A710 (2008)
14. T. Nagase, K. Nishiyama, M. Nakayama, N. Shimomura, M. Amano,T. Kishi, H. Yoda, Presented at American Physical Society March meeting 2008, New Orleans (2008)
15. T. Kishi et al., *IEDM 2008 digest*, 12-6
16. T. Daibou et al., 11th Joint MMM-Intermag Conference 2010 Digest, DA-08
17. S. Ikeda, K. Miura, H. Yamamoto, K. Mizunuma, H.D. Gan, M. Endo, S. Kanai, J. Hayakawa, F. Matsukura, H. Ohno, Nat. Mater. **9**, 721 (2010)
18. D.C. Worledge, G. Hu, D.W. Abraham, J.Z. Sun, P.L. Trouilloud, J. Nowak, S. Brown, M.C. Gaidis, E.J. O'Sullivan, R.P. Robertazzi, Appl. Phys. Lett. **98**, 022501 (2011)
19. D. Saida, N. Shimomura, E. Kitagawa, C. Kamata, M. Yakabe, Y. Osawa, S. Fujita, J. Ito, IEEE Trans. Magn. Accepted (2014)
20. N. Shimomura, S. Fujita, J. Ito, E. Kitagawa, D. Saida, T. Daibou, Y. Kato, C. Kamata, S. Kashiwada, Y. Osawa, H. Noguchi, H. Yoda, The 6th IEEE International Nanoelectronics Conference 2014 (INEC 2014), Sapporo
21. E. Kitagawa, S. Fujita, K. Nomura, H. Noguchi, K. Abe, K. Ikegami, T. Daibou, Y. Kato, C. Kamata, S. Kashiwada, N. Shimomura, J. Ito, H. Yoda, *IEDM Tech. Dig.* (2012), p. 677
22. G.E. Rowlands et al., Appl. Phys. Lett. **98**, 102509 (2011)
23. H. Liu, D. Bedau, D. Backes, J.A. Katine, J. Langer, A.D. Kent, Appl. Phys. Lett. **97**, 242510 (2010)
24. C. Papusoi, B. Delaët, B. Rodmacq, D. Houssameddine, J.P. Michel, U. Ebels, R.C. Sousa, L. Buda-Prejbeanu, B. Dieny, Appl. Phys. Lett. **95**, 072506 (2009)
25. D.C. Worledge, G. Hu, D.W. Abraham, J.Z. Sun, P.L. Trouilloud, J. Nowak, S. Brown, M.C. Gaidis, E.J. O'Sullivan, R.P. Robertazzi, Appl. Phys. Lett. **98**, 022501 (2011)
26. K. Miura, M. Yamanouchi, H. Sato, S. Ikeda, F. Matsukura, H. Ohno, *MMM-INTERMAG 2013*. FT-05, Jan. 2013
27. L. Thomas et al., J. Appl. Phys. **115**, 172615 (2014)
28. G. Jan, Y.J. Wang, T. Moriyama, Y.J. Lee, M. Lin, T. Zhong, R.Y. Tong, T. Torng, P.K. Wang, Appl. Phys. Exp. **5**, 093008 (2012)
29. K. Abe, S. Fujita, T.H. Lee, Architecture of three-dimensional circuit using nanoscale memory devices. *European Micro and Nano Systems* (EMN04) (2004), pp. 225–229. K. Abe, S. Fujita, T.H. Lee, Novel nonvolatile logic circuits with three-dimensionally stacked nanoscale memory device. Proceedings of the 2005 NSTI Nanotechnology Conference, vol. 3 (2005), pp. 203–206. K. Abe, K. Nomura, S. Ikegawa, T. Kishi, H. Yoda, S. Fujita, Hierarchical nonvolatile memory with perpendicular magnetic tunnel junctions for normally-off

computing. The 2010 International Conference on Solid State Devices and Materials (SSDM), Tokyo (2010), pp. 1144–1145
30. S. Yamamoto, S. Sugahara, Jpn. J. Appl. Phys. **48**, 043001 (2009)
31. T. Ohsawa, H. Koike, S. Miura et al., *Symposium on VLSI Circuits* (2012), pp. 46–47
32. K. Abe, H. Noguchi, E. Kitagawa, N. Shimomura, J. Ito, S. Fujita, *IEDM Tech. Dig.* (2012), p. 243
33. http://gem5.org/
34. http://www.spec.org/
35. K. Tsuchida, T. Inaba, K. Fujita, Y. Ueda, T. Shimizu, Y. Asao, T. Kajiyama, M. Iwayama, K. Sugiura, S. Ikegawa, T. Kishi, T. Kai, M. Amano, N. Shimomura, H. Yoda, Y. Watanabe, *ISSCC Dig. Tech. Papers* (2010), pp. 258–260
36. H. Noguchi K. Kushida, K. Ikegami, K. Abe, E. Kitagawa, S. Kashiwada, C. Kamata, A. Kawasumi, H. Hara, S. Fujita, Symp. *VLSI Circuits Dig. Tech. Papers* (2013), pp. 108–109
37. R. Nebashi, N. Sakimura, H. Honjo, S. Saito, Y. Ito, S. Miura, Y. Kato, K. Mori, Y. Ozaki, Y. Kobayashi, N. Ohshima, K. Kinoshita, T. Suzuki, K. Nagahara, N. Ishiwata, K. Suemitsu, S. Fukami, H. Hada, T. Sugibayashi, N. Kasai, *ISSCC Dig. Tech. Papers* (2009), pp. 462–463
38. J.P. Kim, T. Kim, W. Hao, H.M. Rao, K. Lee, X. Zhu, X. Li, W. Hsu, S.H. Kang, N. Matt, N. Yu, *Symposium on VLSI Circuits Dig. Tech. Papers* (2011), pp. 296–297

Beyond STT-MRAM, Spin Orbit Torque RAM SOT-MRAM for High Speed and High Reliability Applications

Guillaume Prenat, Kotb Jabeur, Gregory Di Pendina, Olivier Boulle, and Gilles Gaudin

1 Introduction

For 40 years, microelectronics has been following the Moore's law, stating that the density and speed of integrated circuits would double every 18 months. However, this trend is presently getting out of breath, because of incoming insurmountable physical limits. Due to decreasing devices size, leakage current is becoming the main contributor to power dissipation of CMOS. Indeed, the increased density and reduction in die size lead to heat dissipation and reliability issues. Moreover, the dynamic power keeps on growing up with both clock frequency and global capacitance while the power supply is not scaled down accordingly. Several solutions are investigated to try to push forward these limits at technology, circuit or architecture levels. The "more than Moore" concept consists in using new devices beside or in replacement of standard CMOS transistors. For instance, the use of non-volatile devices is seen as a promising solution to reduce power consumption, improve reliability and offer new functionalities. Several technologies are intensively investigated like Phase Change Random Access Memory (PCRAM), Ferroelectric RAM (FeRAM), RedoxRAM (ReRAM) and Magnetic RAM (MRAM). In its 2010 report, ITRS identified RedoxRAM and STT-MRAM as the two most promising technologies for embedded memories at technology nodes below 16 nm. The combination of non-volatility, fast access time and endurance in MRAM technology paves the path toward a universal memory. Although an expanding attention is given to two-terminal Magnetic Tunnel Junctions (MTJ) with writing based on Spin-Transfer Torque (STT) switching as the potential candidate for future memories, it suffers

G. Prenat (✉) • K. Jabeur • G. Di Pendina • O. Boulle • G. Gaudin
University of Grenoble Alpes, SPINTEC, Grenoble 38000, France

CEA, INAC-SPINTEC, Grenoble 38000, France

CNRS, SPINTEC, Grenoble 38000, France
e-mail: guillaume.prenat@cea.fr

© Springer International Publishing Switzerland 2015
W. Zhao, G. Prenat (eds.), *Spintronics-based Computing*,
DOI 10.1007/978-3-319-15180-9_4

from weaknesses. Indeed, two main shortcomings are still limiting the reliability and endurance of STT-MRAMs: i) The high current density required for writing can occasionally damage the MTJ barrier, specially for switching on the nanosecond time scale ii) It remains a challenge to fulfill a reliable reading without ever causing switching for very advanced technology nodes, since writing and reading operations share the same path, through the junction. Indeed, the smaller the MTJ the lower the writing current without having the possibility to reduce the reading current to maintain a reliable sensing. Three-terminal MTJ with writing based on Spin-Orbit Torque (SOT) approach revitalizes the hope of an ultimate RAM. It represents a pioneering way to triumph over current two-terminal MTJ's limitations by separating the reading and the writing paths, completely avoiding tunnel barrier damaging and read disturb issues.

2 MTJs Written by Spin Orbit Torque

Recently, a new effect was discovered, leading to original devices with three terminals. This effect is called Spin Orbit Torque (SOT), a generic term encompassing in reality different possible mechanisms originating from the bulk of the material and the interfaces between the different layers. These mechanisms issued from the spin orbit interaction are often named the Rashba effect and the Spin Hall Effect (SHE) [1–4] for the contribution of respectively the interfaces and the bulk. However this denomination is often a shortcut since the contribution from interfaces and bulk are intricated. SOT allows switching the magnetization of a magnetic Storage Layer (SL) by passing a current in the plane of the conductive contact line underneath the MTS, instead of through the MTJ itself (Fig. 1). This allows separating the reading and writing paths with great benefits: firstly, since no large writing current is passing through the tunnel barrier damaging it, the

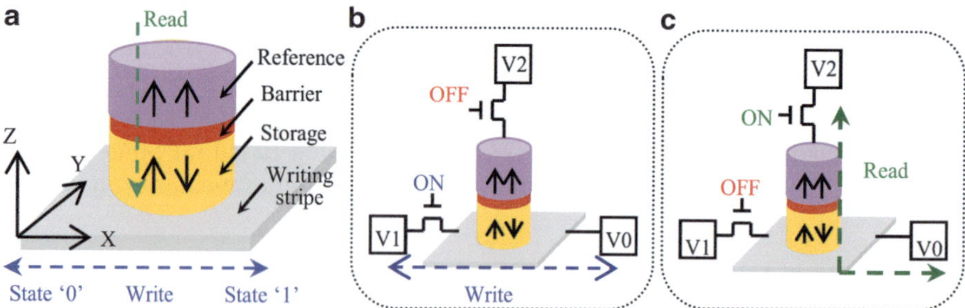

Fig. 1 (**a**) Schematic of the three-terminal SOT device and the two independent paths for (**b**) write and (**c**) read operations. In-plane current injection through the write line (writing stripe) induces the perpendicular switching of the storage layer

endurance can be truly infinite. Secondly, risks of accidental writing during reading are drastically reduced. Moreover, the switching duration using this scheme can be very fast, typically some hundreds of picoseconds experimentally demonstrated, providing the injection of relatively large current densities. Compared to STT [5, 6] the fast SOT-switching is deterministic and does not show any precessional behavior [7]. Today, the current density is higher than for STT (typically $5 \cdot 10^7$ A/cm^2), but since it flows through a very thin line (typically 5 nm), the resulting writing current can be very small. Moreover, since the writing time can be very short, the resulting energy becomes very interesting. The main limitation of this technology compared to STT is apparently the area density because of the additional terminal. However, this picture is not so simple considering a complete memory since i) transistors can be shared ii) the area of MRAM is often limited by the size of the addressing transistors more than that of the single magnetic cell. However, since the writing current decreases proportionally to the lateral dimension of the device and not with its surface, the scalability is weaker than for STT writing, but remains much better than for magnetic field writing. An important feature of the device is the thermal stability of the device which is ensured down to 20 nm.

Another specificity of this writing scheme is that a small static longitudinal magnetic field is required to avoid stochastic switching: without it, for a given direction of the writing current, the magnetization can switch randomly in either P (Parallel) or AP (Anti Parallel) configuration. Adding this longitudinal magnetic field makes the switching deterministic, with a direction of the current writing the P state and the opposite direction the AP state. This field can be generated by the stray field of an in-plane magnetized layer, integrated to the pilar stack, acting like a permanent magnet, with no additional power consumption. This particularity can be even turned into an advantage. Indeed, if the direction of the magnetic field is switched, the operation of the device is inverted: the direction of the current previously used to write a logic "0" will write a logic "1" and vice-versa (Fig. 2). It can be useful for reconfigurable computing. In this case, the magnetization of the layer generating the magnetic field could be chosen using a current line to control it like in original Field Induced Magnetic Switching (FIMS) writing schemes. The corresponding power consumption would then depend on the reconfiguration frequency.

Although this technology is at very early stage development, these properties make it particularly interesting for high speed applications that do not require very high densities but high endurance. Typically, it can be suitable for low-cache levels replacement in processors, which is not possible with classical STT. Since this technology solves some of the major issues of STT, a lot of academic organisms and industrial companies are intensively working on it and we have good confidence that it can have a predominant place in the future of spintronics, for a lot of applications.

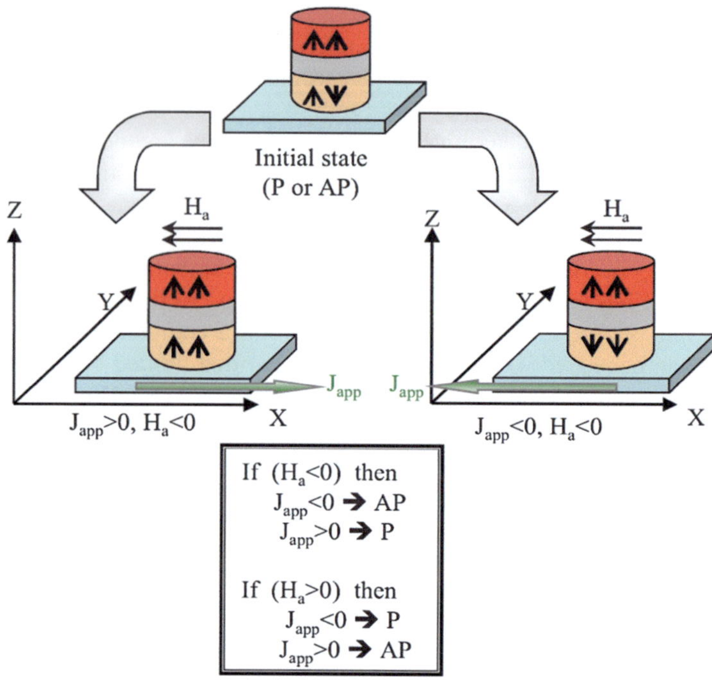

Fig. 2 Switching dependence on the external longitudinal applied field Ha and current direction

3   Applications of SOT-RAM

3.1   Introduction in the Memory Hierarchy of Processors

As seen in the previous chapters, a lot of studies have been recently carried out to introduce STT-RAM in the memory hierarchy of processors. Indeed, it is the only technology that offers an intrinsic non-volatility together with a speed suitable for use as a working memory and a quasi-infinite endurance. Adding non-volatility in the memory hierarchy has a lot of advantages, in particular the ability to ease the power gating techniques, which consist in cutting off the power supply of unused blocks to drastically reduce the leakage. Indeed, having a local non-volatility allows directly cutting off the power supply without having to save the content of the memory in distant non-volatile or very low-leakage storage devices. The first and more obvious approach consists in directly replacing the parts of the memory hierarchy whose operating speed is compatible with those of STT-RAM (DRAM main memory, or L2 cache level). Going deeper into the memory hierarchy seems difficult since STT-RAM cannot compete with fast SRAM in terms of writing speed. However, some works have shown that the better density of MRAM allows having a larger L1 cache capacity, which can compensate the loss of speed for some applications [8]. This approach requires rethinking the architecture of processors and memory hierarchy and is adapted only to some specific applications. Other works propose the use of a hybrid SRAM/MRAM memory. Several architectures have been proposed, allowing using the memory as a standard SRAM in operation, but with the capability of backing-up the content in MTJs and restore it at any moment. This allows bringing non-volatility in low levels of cache, but with a loss in terms of density.

Since SOT-MRAM have a writing speed approaching that of SRAM, it is possible to consider directly replacing L1 cache by SOT-RAM. The result would be the introduction of non-volatility in L1 cache, without penalty in terms of speed. Moreover, thanks to its density which remains much better than the one of SRAM, larger caches could be used, resulting in probable performance improvement.

3.2 Non-volatile Flip-Flops, Normally-Off/Instant-On Computing and Memory-in-Logic

The architecture of latches and Flip-Flops (FF), distributed memory containing the active data of the logic part of the circuit, is very similar to that of SRAM. It is thus possible to introduce non-volatility in the logic itself, in the same way that SRAM can be made non-volatile [9–11]. It is particularly interesting since it allows freezing the state of the processor or any digital/mixed signal IC, and restart in the same state. It allows developing new paradigms like normally-off/instant-on computing [12], in which the circuit is off by default and power supplied only when required. It can be switched-off instantaneously and restarted in the same state. This still reduces the power consumption but also improves the reliability. It is possible for instance to implement a checkpoint for rollback operation in processors, to restore the system state with a snapshot taken in the past. Thanks to its high speed, low-energy writing and infinite endurance, SOT-MRAM could allow saving the context more often, even possibly at each clock edge, without significant energy overhead or performance loss and thus still improving the reliability.

Although the intrinsic non-volatility of MTJs naturally encourages using them as memory elements, it also offers the possibility to intrinsically mix logic and memory to shorten and multiply the communication between them, leading to a new paradigm in logic architectures. This concept, called "logic-in-memory" has been proposed in [13]. Examples of circuits based on Current Mode Logic (CML) have been proposed and implemented in silicon demonstrators giving encouraging results [14]. For this purpose, the SOT-MRAM has another advantage thanks to its third terminal offering an additional degree of freedom and the possibility to reverse its behavior by changing the direction of the longitudinal magnetic field.

4 Design Tools and Environment for the Design of Hybrid CMOS/SOT-RAM Components

4.1 Compact Electrical Model of SOT-MTJ

As the semiconductor industry is progressively going toward "hybrid CMOS" Integrated Circuits (ICs), compact model development has become a cornerstone in the circuit/system design and verification tool flow.

Recently, a compact model of the SOT-MRAM has been proposed by Spintec [15]. The most significant physical phenomena involved in the writing (SOT) and the reading (TMR) are taken into account and realistic parameters, obtained from characterization of single SOT-MRAM cells [16] were used to calibrate the model. This model is written in Verilog-A language and based on physical equations modeling the behavior of the device. The dynamics of the magnetization is described, based on the Landau-Lifshitz-Gilbert (LLG) equation [17], including additional SOT terms [1–4, 16, 18]. The dependence of the resistance upon the magnetic configuration (Tunnel Magneto-Resistance or TMR, relative resistance variation of the stack between Parallel and Anti-Parallel resistance states) and the applied voltage is described using Julière's model [19] and Simmons' model [20]. Moreover, for an improved accuracy, we have integrated the dynamic conductance given by the Brinkman model [21] and we take into consideration the dependence of TMR upon bias voltage. In this model, all the physical quantities, such as the magnetization components (m_x, m_y, m_z) are represented by voltages. Figure 3 shows a physical view and a block diagram of the model, with two modules, one dedicated to the computation of the dynamics of the magnetization and the other dedicated to the TMR.

Figure 4 shows a simulation result using SPECTRE electrical simulator from Cadence. It illustrates the operation of the device, by showing the dynamics of the m_z component of the magnetization for different values of the writing current density J_{app}. In this particular case, the magnetization of the two magnetic layers constituting the MTJ is oriented perpendicularly to the layers. The value of m_z is then 1 for P state and -1 for AP state. Intermediate values correspond to transient behavior during the switching process. The duration of the writing pulse is 3 ns. We see that during the current pulse application the magnetization aligns in the plane of the layers ($m_z = 0$). Once the magnetization is stabilized, the current is removed and the magnetization switches in its final state according to the direction of the current. For $J_{app} = 10^{12}$ A/m^2, the SOT is not strong enough to initiate the switching and the MTJ remains in its P state. For $J_{app} = 2 \cdot 10^{12}$ A/m^2, the 3 ns pulse is not long enough for the magnetization to stabilize and the MTJ switches back to its initial P state. For higher values of J_{app}, switching occurs, with a duration decreasing with the increase of the current density, allowing a trade-off between speed and power consumption, depending on the targeted application. In the same way, increasing the pulse duration allows improving the reliability of the switching. We see that for high current density values ($4 \cdot 10^{12}$ A/m^2 for instance), the switching time can be less than 1 ns. In Fig. 5, the influence of the longitudinal magnetic field on the switching is illustrated. Without it, the switching is stochastic. When the field is applied, the

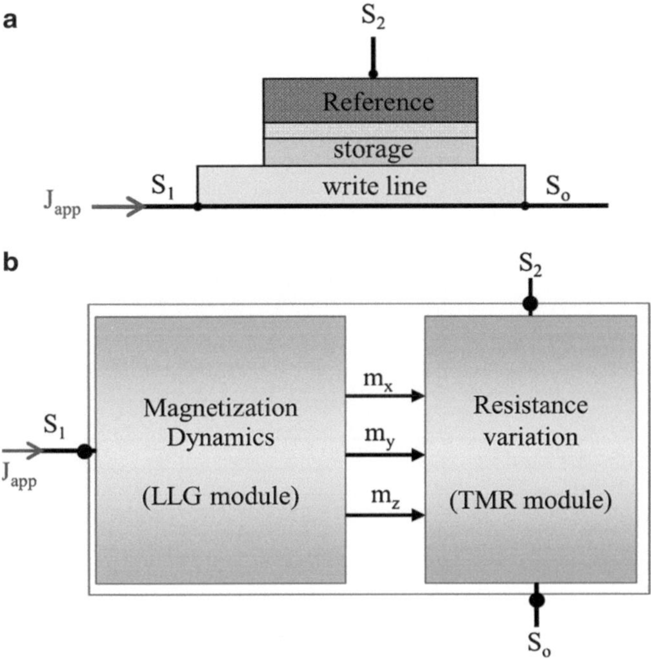

Fig. 3 Modeling strategy of the three-terminal SOT-MRAM. (**a**) Physical view of the SOT-MRAM (cross section). (**b**) Block diagram of the compact model

switching becomes deterministic, with a switching time duration decreasing with the increase of the magnitude of the field. This model was successfully validated by comparison of the results with micromagnetic simulations.

4.2   *Ultra-fast SOT-MRAM Based Non-volatile Flip-Flop*

In order to illustrate the potential of SOT-MRAM, we describe here a Non-Volatile Flip-Flop (NVFF) architecture based on this technology (Fig. 6), developed at Spintec [22]. It is composed of a master latch made non-volatile using a pair of SOT-MRAMs (Fig. 6b) connected to a standard slave latch (Fig. 6c). The operation is managed by a non-overlapping two-phase clock signal generator (ck1, ck2). Four transistors are used to generate a bidirectional current flowing in the writing lines of the SOT-MRAMs devices in series. The direction of the current is determined by the input data D. The current is generated during the first phase clock ck1. During the second phase clock ck2, the master latch reads the data of the SOT-MRAMs and transfers it to the output Q. This operation is illustrated in Fig. 7, showing a simulation of the NVFF using the compact model presented in the previous section. In this work, the technology node is 40 nm, for both CMOS and SOT-MRAM. In these conditions, the writing current is 60 μA, comparable to perpendicular STT-MRAM. However as

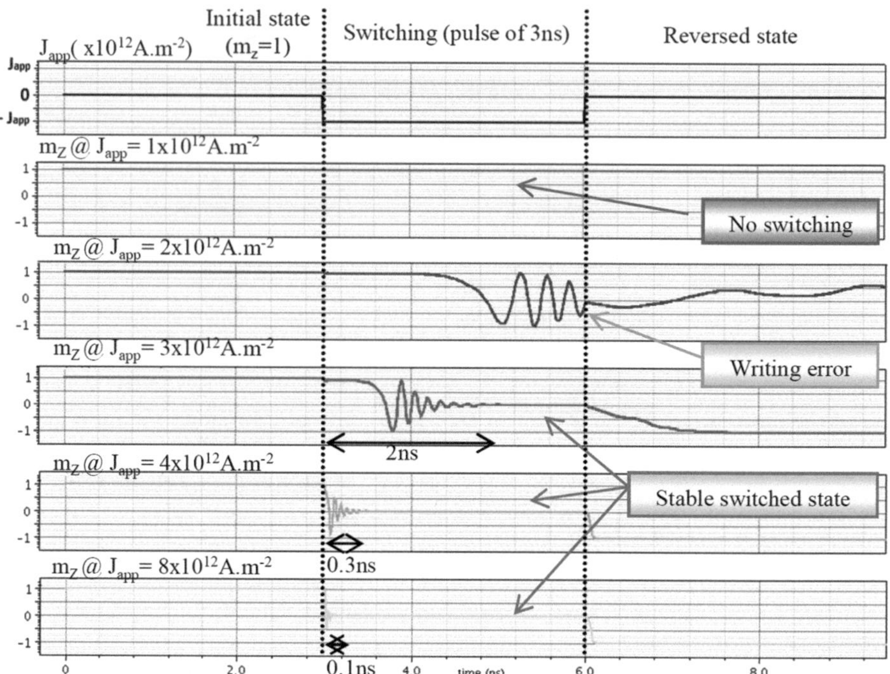

Fig. 4 Simulation of the dynamics of the perpendicular magnetization m_z of a SOT-MRAM as a function of the injected current

seen in inserts, the switching time is only 250 ps, four times faster than STT, dividing the energy by four.

The operation of this NVFF fundamentally differs from those presented so far [9–11]: the magnetic part is written at every clock cycle contrary to the previous cases, where the data was backed up in the magnetic part only when required, on demand. This is made possible because of the high writing speed and low-power consumption of the SOT technology. It means that with such a device, the power supply can be cut-off at any moment, without further action and restarted in the same state. This opens the door to the concept of "normally-off" or "instant-on" circuits [12]. This Flip-Flop can be used as a primitive cell into the ASIC design library allowing the design of non-volatile, high speed and low-power digital circuits [23, 24].

4.3   System Level Integration

To evaluate the benefits that can be expected from integrating MRAM and in particular SOT-MRAM in the different levels of processors hierarchy, it is necessary to carry-out system level simulations, using processor simulators, like Gem5 or

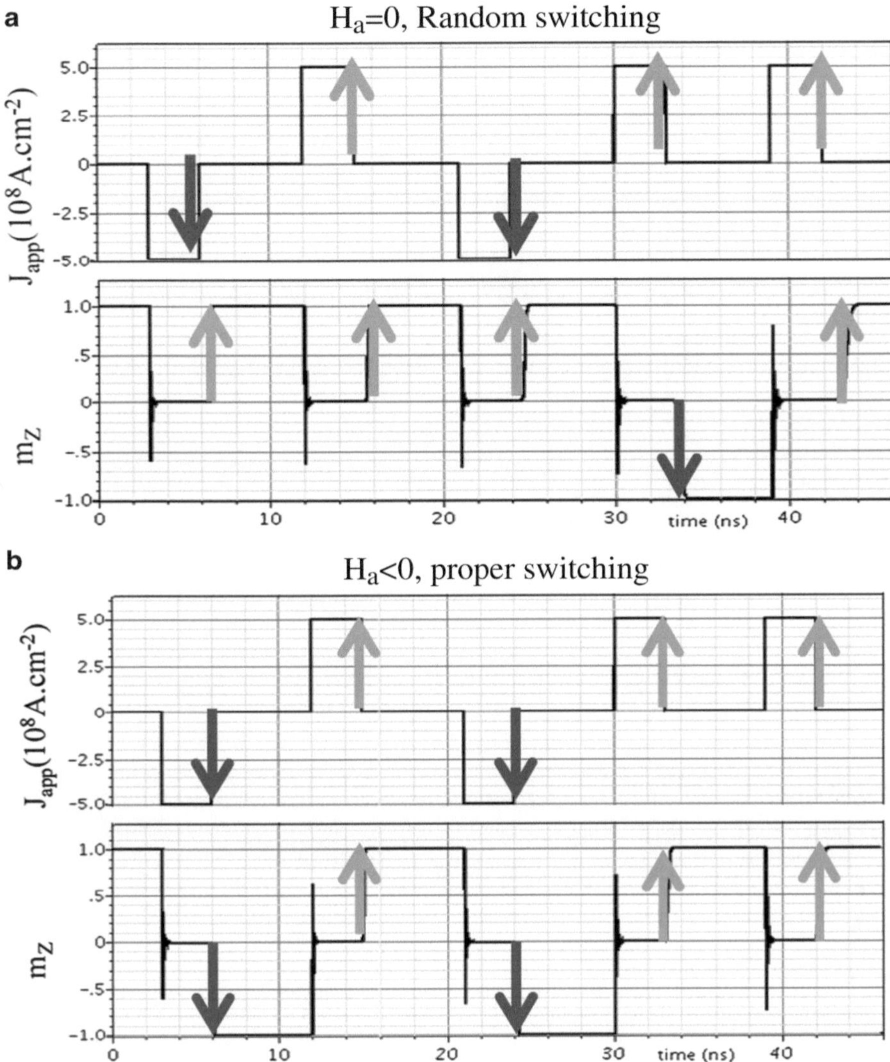

Fig. 5 Study of the dynamic behavior of the perpendicular magnetization m_z according to the value of the external applied field Ha: (**a**) $H_a = 0$, random switching and (**b**) $H_a < 0$, proper switching

simple scalar (Fig. 8). These tools take as inputs the architecture of the processor, including the memory hierarchy and the models of the memory blocks. It allows running an application and evaluating the performance in terms of CPU time, number of cycles, memory transactions (read/write, cache hit/miss...). The models of the memory can be provided by characterization of an existing memory (by test or simulation) or using a memory simulator, like NVSIM tool (specific for non-volatile resistive memories). This work is currently carried out for MRAM in general [8] and for SOT-MRAM in particular, giving encouraging preliminary results [25].

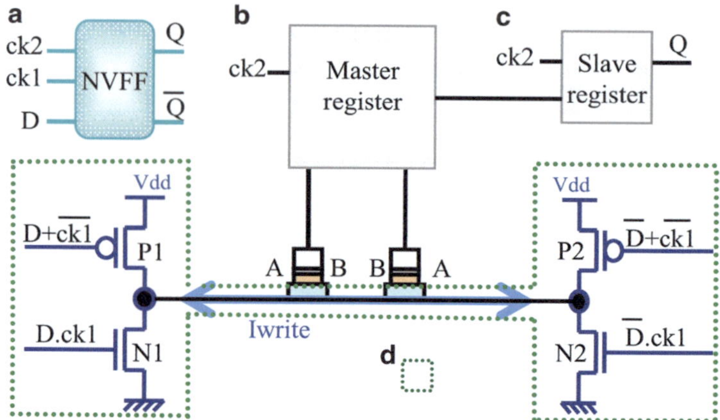

Fig. 6 SOT-MRAM based NVMFF (**a**) symbol, (**b**) master register architecture, (**c**) slave register, (**d**) writing circuit

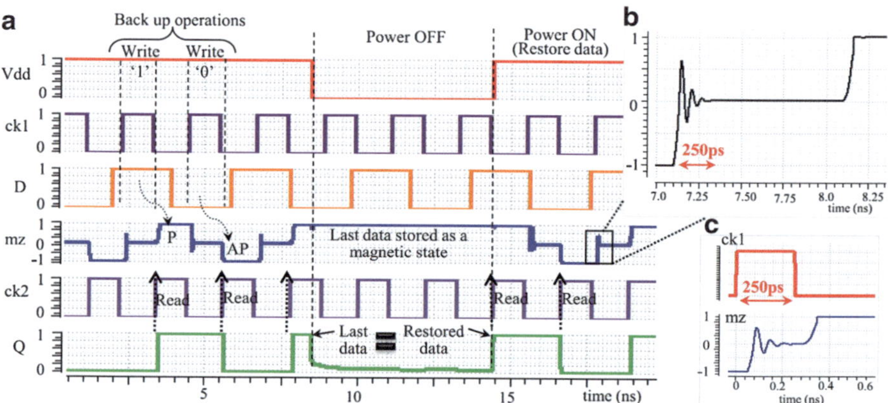

Fig. 7 SOT-MRAM based NVMFF simulation results (**a**) write and read operations (**b**) perpendicular magnetization m_z behavior of the SOT-MRAM during a 1 ns current pulse to switch from AP ($m_z = -1$) to P ($m_z = 1$) (**c**) SOT-MRAM witching with only a 250 ps pulse (exactly after oscillations stabilization)

5 Conclusion

The advent of new MRAM generations has led to a keen interest in the microelectronics world for this technology as a possible solution to push forward the limits of microelectronics. Several tools are developed to allow exploring new circuits and system architectures using MRAM devices. As seen in this chapter, both STT-MRAM and SOT-MRAM technologies have their own strengths and limits and is thus suitable for a given set of applications. A wide field of investigation is open covering technology developments to improve the existing devices, discover

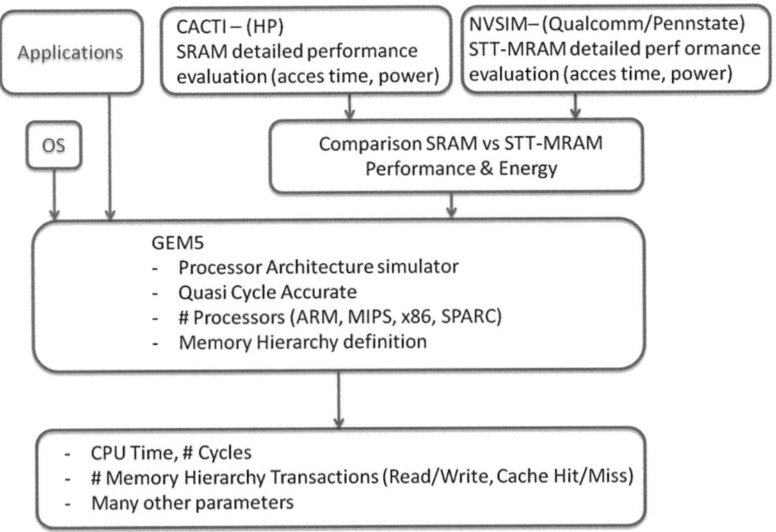

Fig. 8 System level design flow to evaluate the performance of processors embedding MRAM

new ones, and design of innovative circuits and systems. Two paths must be followed in parallel: identifying relevant target applications for each technology and adapting the system architectures to gain maximum benefit.

The performance of the most recent technologies and in particular SOT-MRAM, in terms of speed and power consumption which could compete with SRAM, paves the way towards a true revolution in the memory hierarchy. It allows dreaming of a new paradigm consisting in introducing non-volatility at all levels of this memory hierarchy and possibly writing the data in the MTJs at every clock cycle, opening the door to real normally-off/instant-on circuits.

Acknowledgement This work has been partially funded by the European Commission under the spOt project (grant agreement n318144) in the framework of the Seventh Framework Program.

References

1. M. Miron, K. Garello, G. Gaudin, P.-J. Zermatten, M.V. Costache, S. Auffret, S. Bandiera, B. Rodmacq, A. Schuhl, P. Gambardella, Perpendicular switching of a single ferromagnetic layer induced by in-plane current injection. Nature **476**, 89–193 (2011)
2. M. Miron, G. Gaudin, S. Auffret, B. Rodmacq, A. Schuhl, S. Pizzini, J. Vogel, P. Gambardella, Current-driven spin torque induced by the Rashba effect in a ferromagnetic metal layer. Nat. Mater. **9**, 230–234 (2010)
3. P. Gambardella, I.M. Miron, Current-induced spin-orbit torques. Philos. Trans. A Math. Phys. Eng. Sci. **369**, 3175–3197 (2011). doi:10.1098/rsta.2010.0336

4. L. Liu, C.-F. Pai, Y. Li, H.W. Tseng, D.C. Ralph, R.A. Buhrman, Spin-torque switching with the giant spin hall effect of tantalum. Science **336**, 555–558 (2012)
5. C. Papusoi, B. Delaët, B. Rodmacq, D. Houssameddine, J.-P. Michel, U. Ebels, R.C. Sousa, L. Buda-Prejbeanu, B. Dieny, 100 ps precessional spin-transfer switching of a planar magnetic random access memory cell with perpendicular spin polarizer. Appl. Phys. Lett. **95**, 072506 (2009)
6. M. Marins de Castro, R.C. Sousa, S. Bandiera, C. Ducruet, A. Chavent, S. Auffret, C. Papusoi, I.L. Prejbeanu, C. Portemont, L. Vila, U. Ebels, B. Rodmacq, B. Dieny, Precessional spin-transfer switching in a magnetic tunnel junction with a synthetic antiferromagnetic perpendicular polarizer. J. Appl. Phys. **111**, 07C912 (2012)
7. K. Garello, C.O. Avci, I.M. Miron, M. Baumgartner, A. Ghosh, S. Auffret, O. Boulle, G. Gaudin, P. Gambardella, Ultrafast magnetization switching by spin-orbit torques. Appl. Phys. Lett. **105**, 212402 (2014)
8. L. Torres, R. Brum, L. Cargnini, G. Sassatelli, in *Trends on the Application of Emerging Nonvolatile Memory to Processors and Programmable Devices.* Circuits and Systems (ISCAS), 2013 I.E. International Symposium (2013), pp. 101–104
9. W. Zhao, E. Belhaire, V. Javerliac, C. Chappert, B. Dieny, in *A Non-volatile Flip-Flop in Magnetic Fpga Chip.* Design and Test of Integrated Systems in Nanoscale Technology, 2006 DTIS. International Conference (2006), pp. 323–326
10. N. Sakimura, T. Sugibayashi, R. Nebashi, N. Kasai, in *Nonvolatile Magnetic Flip-Flop for Standby-Power-Free Socs.* Custom Integrated Circuits Conference, CICC 2008. IEEE (2008), pp. 355–358
11. Y. Guillemenet, L. Torres, G. Sassatelli, N. Bruchon, On the use of magnetic rams in field programmable gate arrays. Int. J. Reconfig. Comput. **2008**, 1:1–1:9 (2008). http://dx.doi.org/10.1155/2008/723950
12. T. Kawahara, Scalable spin-transfer torque ram technology for normally-off computing. IEEE Des. Test Comput. **28**(1), 52–63 (2011)
13. A. Mochizuki, H. Kimura, M. Ibuki, T. Hanyu, Tmr based logic-in-memory circuit for low-power vlsi*. IEICE Trans. Fundam. Electron. Commun. Comput. Sci. **E88-A**(6), 1408–1415 (2005). http://dx.doi.org/10.1093/ietfec/e88-a.6.1408
14. S. Matsunaga, J. Hayakawa, S. Ikeda, K. Miura, H. Hasegawa, T. Endoh, H. Ohno, T. Hanyu, Fabrication of a nonvolatile full adder based on logic-in-memory architecture using magnetic tunnel junctions. Appl. Phys. Express **1**(9), 091301 (2008). http://apex.jsap.jp/link?APEX/1/091301/
15. K. Jabeur, G. Prenat, G. Di Pendina, L. Buda-Prejbeanu, I. Prejbeanu, B. Dieny, in *Compact Model of a Three Terminal Mram Device Based on Spin Orbit Torque Switching.* Semiconductor Conference Dresden-Grenoble (ISCDG) 2013 International (2013), pp. 1–4.
16. K. Garello, I.M. Miron, C.O. Avci, F. Freimuth, Y. Mokrousov, S. Blügel, S. Auffret, O. Boulle, G. Gaudin, P. Gambardella, Symmetry and magnitude of spin–orbit torques in ferromagnetic heterostructures. Nat. Nanotechnol. **8**, 587 (2013)
17. L.D. Landau, E. Lifshitz, On the theory of the dispersion of magnetic permeability in ferromagnetic bodies. Phys. Z. Sowjetunion **8**(153), 101–114 (1935)
18. S. Manipatruni, D.E. Nikonov, I.A. Young, Voltage and energy-delay performance of giant spin hall effect switching for magnetic memory and logic. **arXiv**:1301.5374v1 [cond-mat.mes-hall]) (Submitted on 23 Jan 2013)
19. M. Julliere, Tunneling between ferromagnetic films. Phys. Lett. A **54**, 225–226 (1975)
20. J.G. Simmons, Generalized formula for the electric tunnel effect between similar electrodes separated by a thin insulating film. J. Appl. Phys. **34**(6), 1793–1803 (1963)
21. W.F. Brinkman, R.C. Dynes, J.M. Rowell, Tunnelingconductance of asymmetrical barriers. J. Appl. Phys. **41**, 1915–1921 (1970)
22. K. Jabeur, G. Di Pendina, F. Bernard-Granger, G. Prenat, Spin orbit torque non-volatile flip-flop for high speed and low energy applications. Electron Device Lett. IEEE **35**(3), 408–410 (2014). doi:10.1109/LED.2013.2297397. http://ieeexplore.ieee.org/stamp/stamp.jsp?tp=&arnumber=6714403&isnumber=6746036

23. G. Di Pendina, K. Torki, G. Prenat, Y. Guillemenet, L. Torres, in *Integrated Circuit and System Design. Power and Timing Modeling, Optimization, and Simulation*. Ultra Compact Non-volatile Flip-Flop for Low Power Digital Circuits Based on Hybrid Cmos/Magnetic Technology (Springer, Berlin Heidelberg, 2011), pp. 83–91
24. G. Di Pendina, G. Prenat, B. Dieny, K. Torki, A hybrid magnetic/complementary metal oxide semiconductor process design kit for the design of low-power nonvolatile logic circuits. J. Appl. Phys. **111**(7), 07E350–07E350-3 (2012)
25. R. Bishnoi, M. Ebrahimi, F. Oboril, M.B. Tahoori, in *Architectural Aspects in Design and Analysis of SOT-based Memories*. Design Automation Conference (ASP-DAC), 20–23 Jan 2014 19th Asia and South Pacific, (2014), pp. 700, 707. doi:0.1109/ASPDAC.2014.6742972

Challenge of Nonvolatile Logic LSI Using MTJ-Based Logic-in-Memory Architecture

Takahiro Hanyu

1 Introduction

In this chapter, a new architecture, called "nonvolatile logic-in-memory (NV-LIM) architecture," is presented, where the NV-LIM architecture could overcome performance wall and power wall due to the present CMOS-only-based logic-LSI processors [1–3]. Figure 1a shows a conventional logic-LSI architecture, where logic and memory modules are separately implemented together and these modules are connected each other through global interconnections. Even if the device feature size is scaled down in accordance with the semiconductor technology roadmap, the global interconnections are not shorten, rather than are getting longer, which resulting in longer delay and higher power dissipation due to inside wires. In addition, since on-chip memory modules are "volatile", they always consume the static power to maintain the stored data.

On the other hand, several emerging storage devices are getting developed to overcome the weak points of conventional semiconductor memories; dynamic random-access memory (DRAM) and static random-access memory (SRAM). Especially, magnetoresistive random-access memory (MRAM) that has already undergone a few incarnations, is now converging on a scheme for upending the memory business. Spin-transfer torque (STT) MRAM promises speed and reliability comparable to that of SRAM, where SRAM is the quick-access memory embedded inside microprocessors, along with the "nonvolatility" of flash, the storage of smartphones and other portables. Since magnetic tunnel junction (MTJ) device, the key element of MRAM, is easily distributed over a logic-circuit plane by using a three-dimensional (3D) stack structure as shown in Fig. 1b, performance degradation due to intra-chip global wires could be drastically

T. Hanyu (✉)
Research Institute of Electrical Communication, Tohoku University, Sendai, Japan
e-mail: hanyu@irec.tohoku.ac.jp

© Springer International Publishing Switzerland 2015
W. Zhao, G. Prenat (eds.), *Spintronics-based Computing*,
DOI 10.1007/978-3-319-15180-9_5

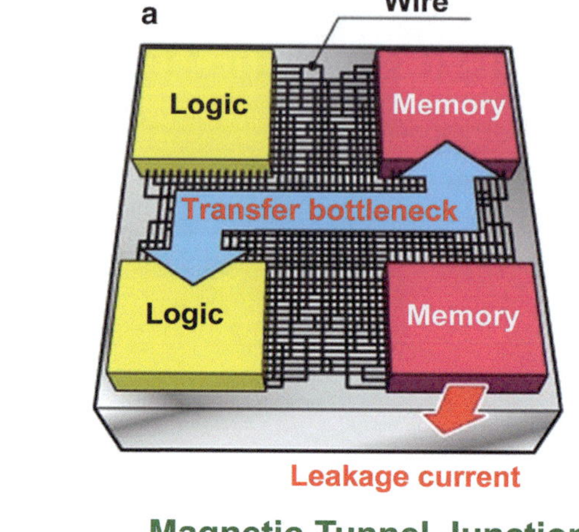

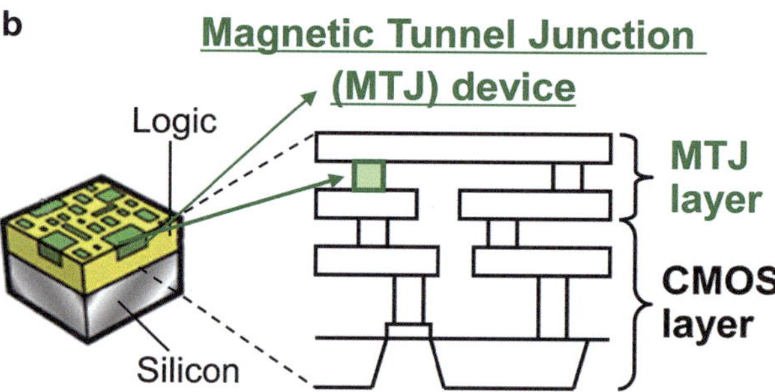

Fig. 1 Comparison of logic-LSI architectures; (**a**) conventional, (**b**) nonvolatile logic-in-memory

mitigated, which leads to a high- performance, ultra-low-power and highly reliable (or highly resilient) logic LSI.

One of the most useful methods to cut off the leakage power is to use power gating. Figure 2a shows a time chart of power dissipation in conventional logic LSI without power gating. If you apply the power gating in the conventional logic LSI, a part of standby power can be eliminated, but two additional operations, "back-up" and "boost-up" procedures, must be performed before and after applying the power gating, which may discourage to apply the power-gating technique as shown in Fig. 2b. In contrast, non-volatility is a good combination of applying the power gating, which ideally eliminates the wasted power dissipation as shown in Fig. 2c.

Figure 3a shows nonvolatile VLSI processor architecture, where a high-density and high-speed MRAMs and nonvolatile flip-flops are used to simply realize a nonvolatile logic LSI [4, 5]. When you could merge a part of nonvolatile on-chip memory into logic-circuit modules as shown in Fig. 3b, it would be possible to improve the performance of the nonvolatile logic LSI. In the following description, some concrete design examples using MTJ-based nonvolatile NV-LIM architecture such as nonvolatile field programmable gate array (FPGA) [6–14], nonvolatile ternary content-addressable memory (TCAM) [15–22], and nonvolatile random-access logic-LSI unit (MCU) [23, 24] are demonstrated and their usefulness is discussed.

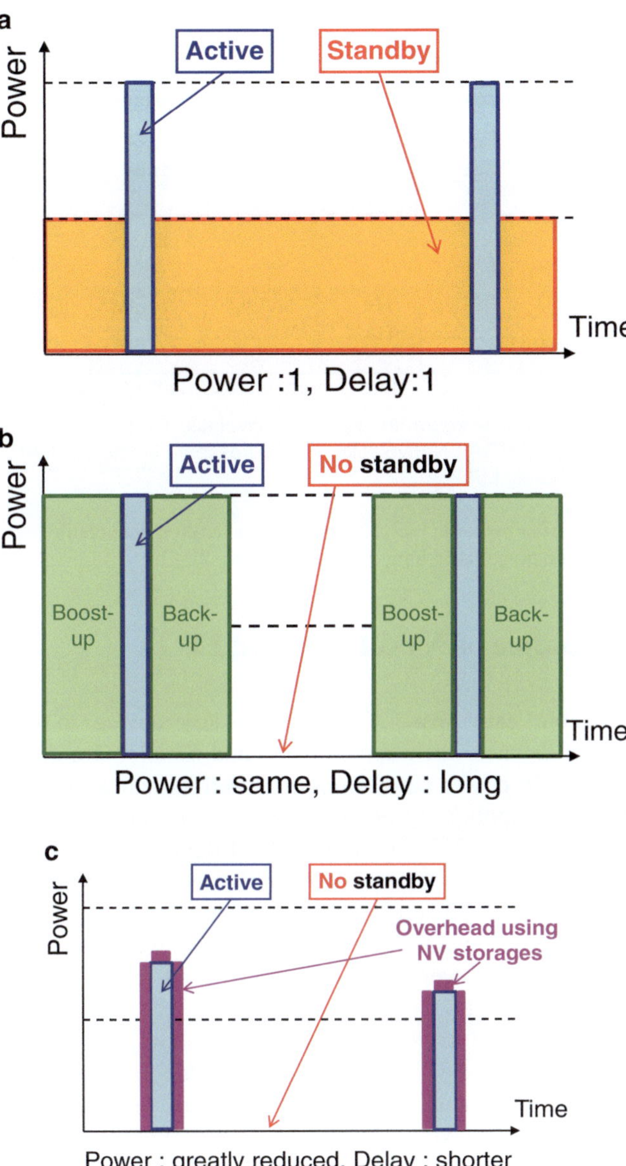

Fig. 2 Combination of power-gating and nonvolatile logic techniques; (**a**) Conventional CPU without power gating, (**b**) Conventional CPU with power gating, (**c**) NV-LIM CPU with power gating

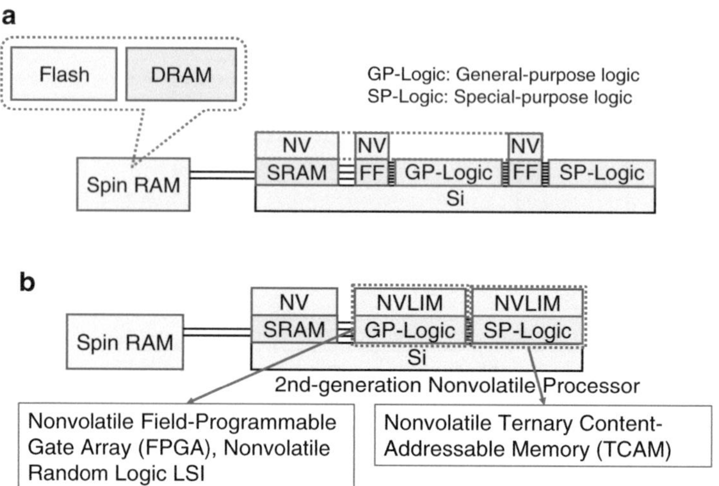

Fig. 3 Configuration of nonvolatile logic LSIs; (**a**) 1st-generation nonvolatile logic-LSI architecture, (**b**) 2nd-generation nonvolatile logic-LSI architecture

2 Design Example of NV-LIM-Based FPGA

Field programmable gate array (FPGA) is a key device to quickly realize prototyping systems, where their specification and function are directly programmable by users, while power consumption as well as hardware cost is a serious problem in expanding application fields of FPGAs, especially in the field of mobile and portable applications [25]. The use of MTJ devices could solve the power-dissipation problem. Figure 4 shows the overall structure of a nonvolatile FPGA, where each lookup table (LUT) circuit in the configuration logic block (CLB) stores logical configuration data into MTJ devices. Therefore, whenever an LUT circuit is in a standby mode, its power supply could be shut down, which completely eliminates the wasted standby power dissipation.

Although the use of MTJ devices makes the LUT circuit nonvolatile, its hardware cost is rather than increased when you simply replaces conventional SRAM-cell-based volatile storage elements with nonvolatile ones, because every MTJ-based nonvolatile storage element generally requires sense amplifier (SA) as shown in Fig. 5a. In order to reduce the hardware overhead, MTJ devices are merged into the combinational logic circuit as shown in Fig. 5b, where the technique is the circuit-level NV-LIM architecture [6]. As a result, the LUT circuit becomes compact because only a single SA is required in the proposed LUT circuit. Figure 6a shows the circuit diagram of the MTJ-based two-input nonvolatile LUT circuit, and Fig. 6b shows a fabricated two-input nonvolatile LUT-circuit test chip and its features. The immediate wakeup behavior of the nonvolatile LUT circuit has been confirmed by the measured waveforms as shown in Fig. 7.

Fig. 4 Overall structure
of a nonvolatile FPGA

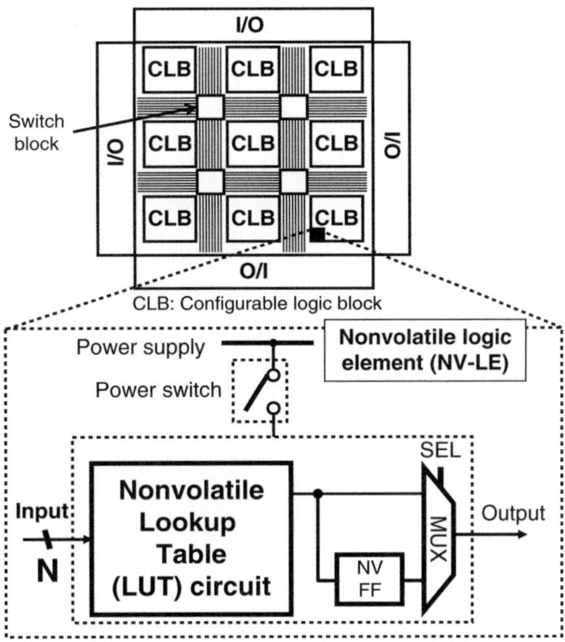

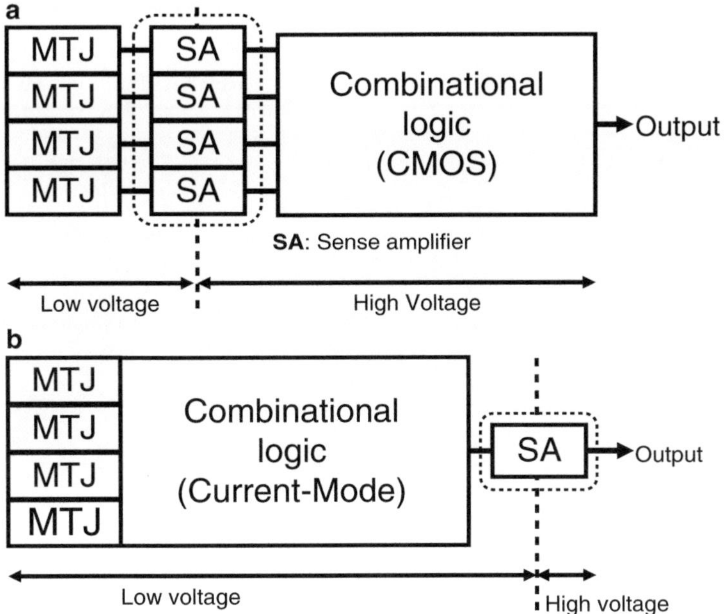

Fig. 5 Design philosophy of a compact nonvolatile LUT circuit; (**a**) Conventional approach, (**b**) proposed NV-LIM architecture-based approach

Fig. 6 MTJ-based 2-input nonvolatile LUT circuit with NV-LIM architecture; (**a**) circuit diagram, (**b**) fabricated chip photomicrograph and its features

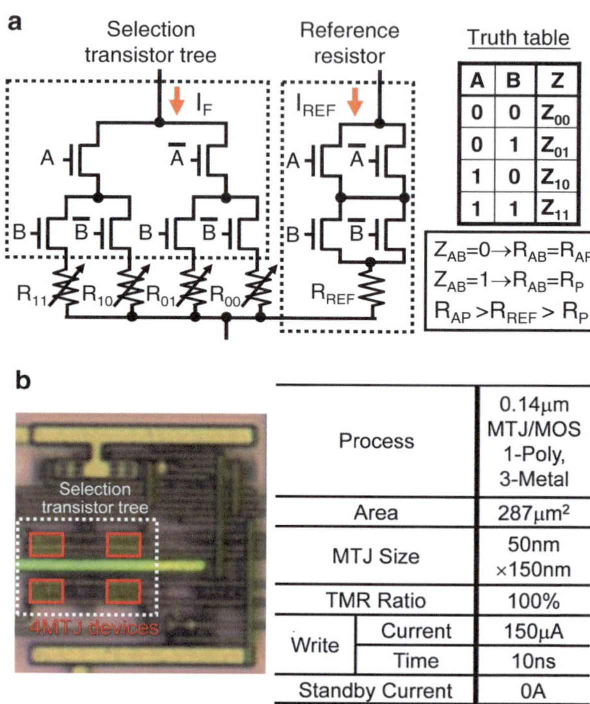

A	B	Z
0	0	Z_{00}
0	1	Z_{01}
1	0	Z_{10}
1	1	Z_{11}

$$Z_{AB}=0 \rightarrow R_{AB}=R_{AP}$$
$$Z_{AB}=1 \rightarrow R_{AB}=R_{P}$$
$$R_{AP} > R_{REF} > R_{P}$$

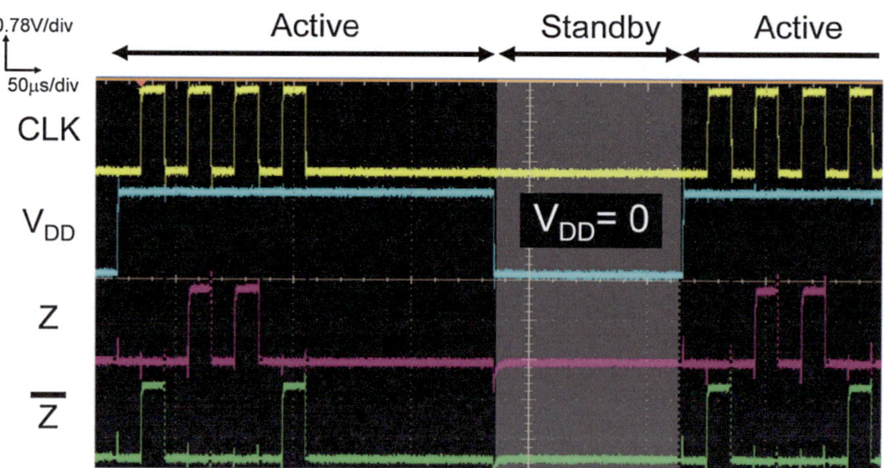

Process		0.14μm MTJ/MOS 1-Poly, 3-Metal
Area		287μm²
MTJ Size		50nm ×150nm
TMR Ratio		100%
Write	Current	150μA
	Time	10ns
Standby Current		0A

Fig. 7 Immediate wakeup behavior of the 2-input nonvolatile LUT circuit

In the practical FPGA, the number of input variables in the LUT function must be four or more, while the variation of the resistance values of MTJs devices becomes critical, because multi-input LUT circuit requires many MTJ devices and MOS transistors, where they are connected serially. For the stable and reliable

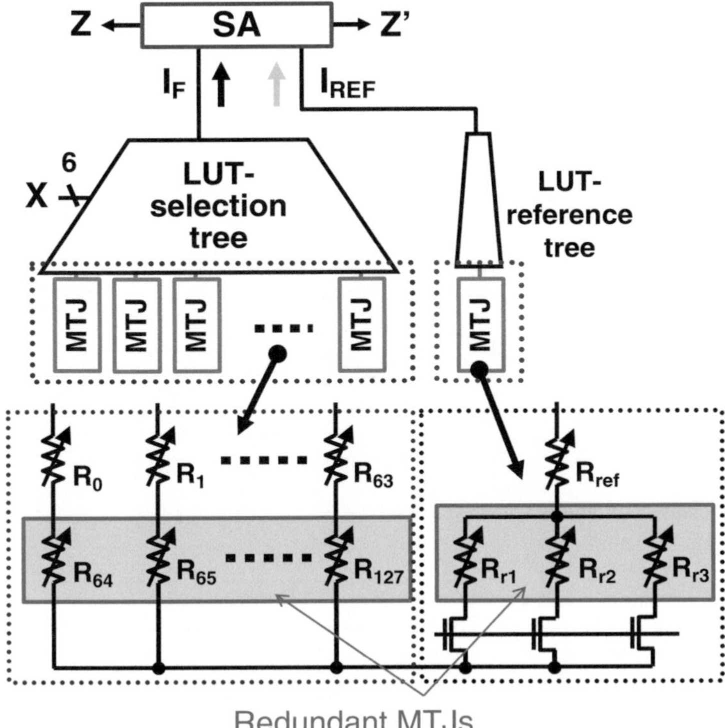

Fig. 8 Resistance-variation compensation technique using redundant MTJ devices

operation of the multi-input LUT circuit, we insert "redundant" MTJ devices to adjust the operating point of the LUT circuit. Figure 8 shows a design example of the multi-input nonvolatile LUT circuit, where both twice number of MTJs and three additional MTJs are inserted into the LUT-selection tree and the LUT-reference tree, respectively [7, 8]. Figure 9 summarizes the comparison of multi-input LUT circuits. It is clearly demonstrated that the proposed the NV-LIM-based NV-LUT circuit is implemented.

Not only LUT circuit but also other components, switch block (SB) and connection block (CB), in FPGA chip are efficiently implemented by using the circuit-level NV-LIM architecture. Since the write-current characteristic of MTJ device is left-right asymmetric as shown in Fig. 10a, only a single MOS transistor with a large width is shared by every NV latch, while each NV latch has only MOS transistors with a small width as shown in Fig. 10b, which greatly reduces the effective chip area of routers [9]. Figure 11 shows a fabricated nonvolatile FPGA chip, where almost 1,000 tiles (each tile consists of LE, CB and SB, that is, the minimum set of basic components in FPGA chip) are integrated in the area of 3.4×2.0 mm^2 under 90 nm CMOS/perpendicular-MTJ technologies. This high-density integration of nonvolatile FPGA chip has firstly succeeded by using the

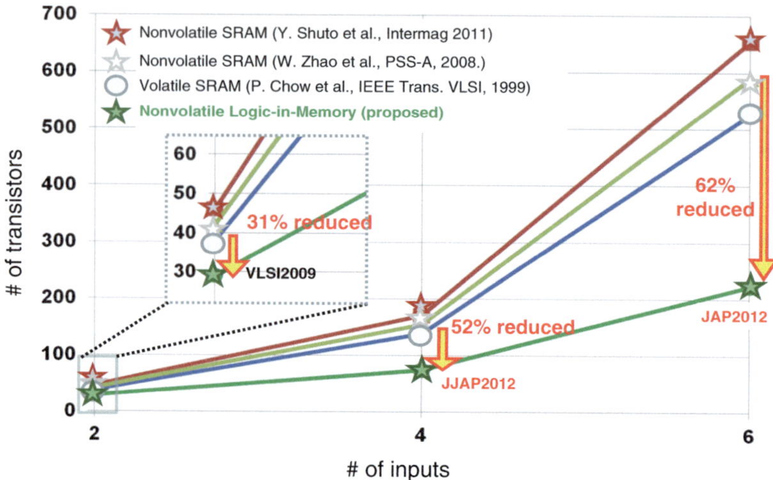

Fig. 9 Comparison of multi-input LUT circuits

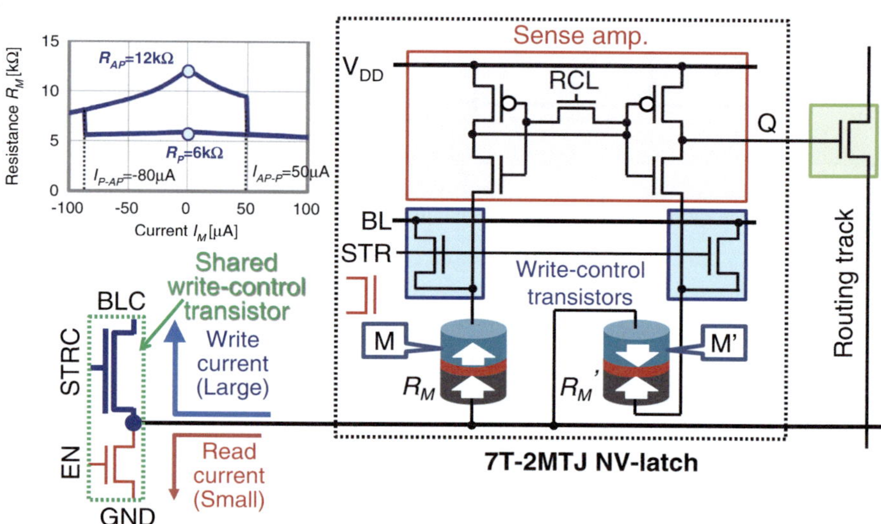

Fig. 10 Compact nonvolatile routing switch sharing write driver

NV-LIM architecture [10, 11]. As a future prospect, it is also important to design nonvolatile logic LSI using three-terminal MTJ (3T-MTJ) device [12–14], because write current path is separated from read current path in the 3T-MTJ device [26, 27], which greatly mitigates the circuit-design restricts of nonvolatile logic LSIs. Figure 12 shows a 3-T MTJ-based nonvolatile LE. The use of 3-T MTJ devices makes the LE compact and improved the switching speed, because read-current level could be appropriately determined independent of write-current level.

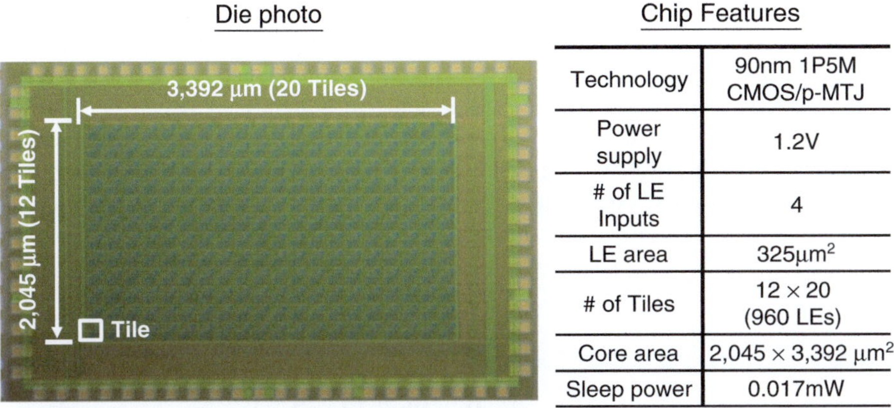

Technology	90nm 1P5M CMOS/p-MTJ
Power supply	1.2V
# of LE Inputs	4
LE area	325μm^2
# of Tiles	12 × 20 (960 LEs)
Core area	2,045 × 3,392 μm^2
Sleep power	0.017mW

Fig. 11 Resistance-variation compensation technique using redundant MTJ devices

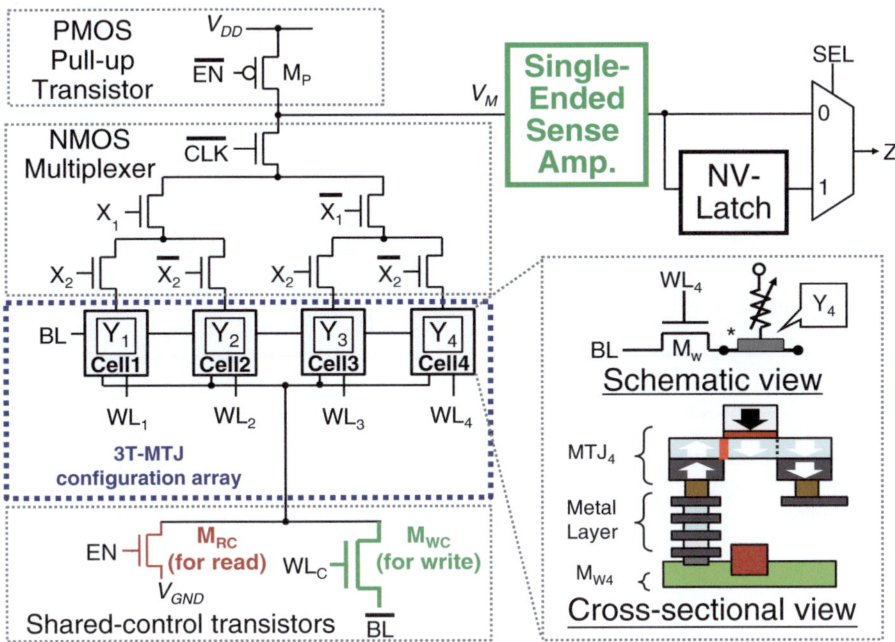

Fig. 12 Design example of a 3-T MTJ-based logic element using NV-LIM architecture

3 Design Example of NV-LIM-Based TCAM

As a typical nonvolatile special-purpose logic-LSI example using NV-LIM structure, MTJ-based non-volatile ternary content-addressable memories (TCAM) have been designed and fabricated [15–22]. TCAM is a functional memory for high-speed data retrieval that performs a fully parallel search and fully parallel comparison between an input key and stored words. Currently, its high bit cost and high power dissipation, higher than those of standard semiconductor memories such as static random access memory limits the fields to which TCAM can be applied. Figure 13 shows the truth table of a TCAM cell function. Its rich functionality makes data search powerful and flexible but with conventional CMOS realization there is an associated cost of a complicated logic circuit with two-bit storage elements. Figure 14 shows the design philosophy of realizing the TCAM cell circuit

Stored data		Search input	Current comparison	Match result
B	**(b$_1$,b$_2$)**	**S**		**ML**
0	(0,1)	0	$I_Z < I_Z'$	**1** (*Match*)
		1	$I_Z > I_Z'$	**0** (*Mismatch*)
1	(1,0)	0	$I_Z > I_Z'$	**0** (*Mismatch*)
		1	$I_Z < I_Z'$	**1** (*Match*)
X (don't care)	(0,0)	0	$I_Z < I_Z'$	**1** (*Match*)
		1	$I_Z < I_Z'$	**1** (*Match*)

Fig. 13 Truth table of the TCAM cell function

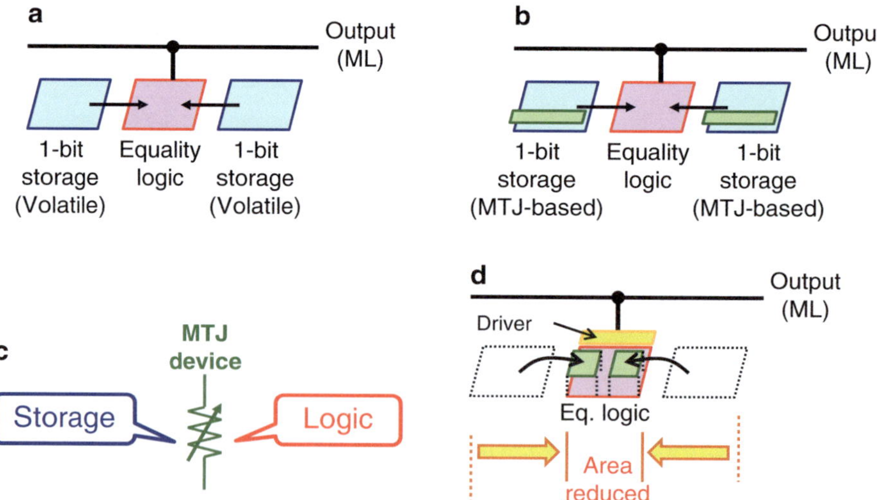

Fig. 14 Design philosophy of making a compact TCAM cell circuit; (**a**) conventional TCAM cell structure, (**b**) conventional NV-TCAM cell structure, (**c**) MTJ device merging storage and logic functions, (**d**) proposed NV-TCAM cell structure

compactly with a non-volatile storage capability. In the case of both conventional volatile TCAM cell structure and conventional non-volatile TCAM cell structure without using LIM architecture shown in Fig. 14a, b, respectively, the bit cost is high. In contrast, when two-bit storage elements are merged into a logic-circuit part by using the LIM architecture as shown in Fig. 14c, the proposed TCAM cell structure becomes compact and non-volatile as in Fig. 14d. Figure 15a, b compare a conventional volatile TCAM cell circuit and the proposed non-volatile one, respectively. The conventional CMOS-based volatile TCAM cell circuit consumes 12 MOS transistors (12T-TCAM circuit structure) while the proposed one takes just 4 MOS transistors with two MTJ devices (4T-2MTJ circuit structure) [15–20]. Note that MTJs do not affect the total TCAM cell-circuit one, because MTJs are fabricated onto the CMOS plane. Compact realization due to NV-LIM architecture has the advantage of improving the performance of the circuit by inserting a driver as shown in Fig. 15c. Figure 16 summarizes the comparison of TCAM word circuits with 144 cells. By the appropriate division of the TCAM word circuit, the activation ratio of the TCAM can be minimized. Figure 17 shows the variety of the segment-based TCAM word-circuit structures. In the case of the 3-segment-based NV-TCAM word-circuit structure, where the first segment, the second-segment and the rest consist of 3-bit, 7-bit and 134-bit cells respectively, its average activation ratio becomes as low as 2.8 %, which indicates that 97 % or more TCAM cells can be in standby mode on average by the fine-grained power gating

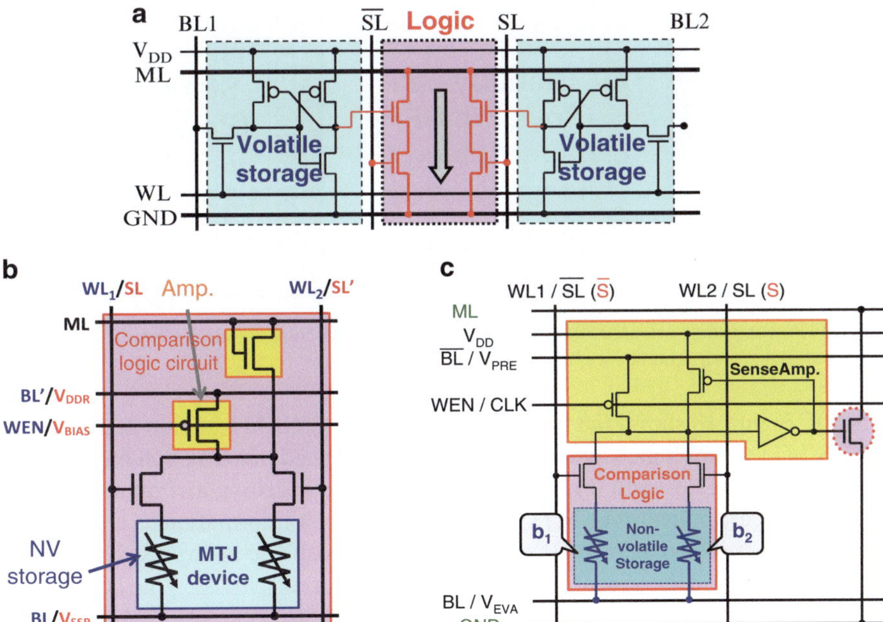

Fig. 15 TCAM cell-circuit design; (a) conventional volatile TCAM cell circuit, (b) proposed 4T-2MTJ NV-TCAM cell circuit, (c) proposed 7T-2MTJ NV-TCAM cell circuit

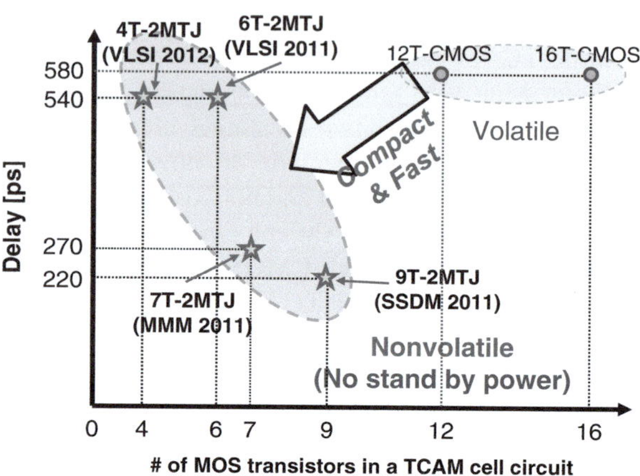

Fig. 16 Comparison of delays and cell-transistor counts in TCAMs

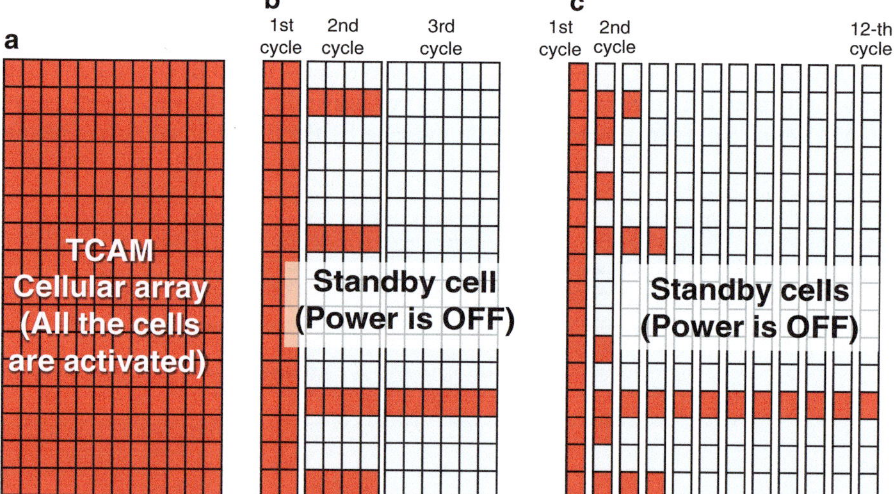

Fig. 17 Power-gating-oriented TCAM search schemes; (**a**) fully parallel search scheme, (**b**) series-parallel search scheme, (**c**) bit-serial search scheme

[15–20]. Figure 18 shows the fabricated non-volatile TCAM test chip under 90-nm CMOS/MTJ technologies, which is used as a high-speed index search engine [17].

Robustness against soft error due to particle strike is getting an important factor in the practical applications. MTJ device stores one-bit information as a resistance whose value is robust against alpha particle and atmosphere neutron strikes, which significantly lower the probability of single-event upsets (SEUs). The TCAM also becomes robust against delay variations caused by single event transients (SETs) as it is designed based on four-phase dual-rail encoding realized using complementary NAND and NOR-type word circuits as shown in Fig. 19 [21, 22]. The dual-rail

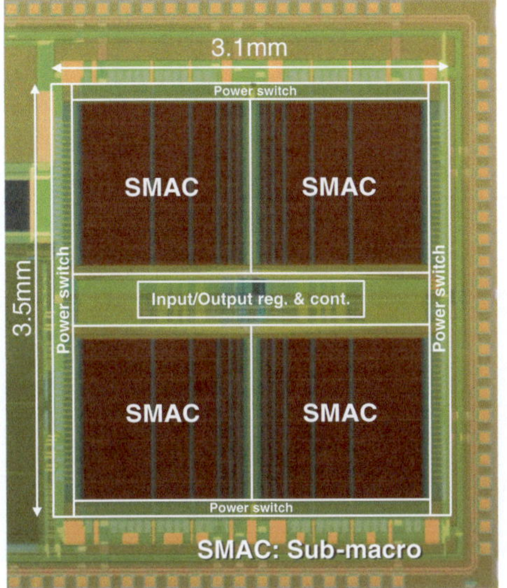

Process	90nm CMOS/ Perpendicular MTJ
Capacity	1 Mb
Cell size	4.5 μm^2
MF-CAM Macro	256b x 1024w x 4SMAC
Max. Freq.	200 MHz

Power	Dynamic	25.7 mW (Typical case)
	Static	0.4968 mW (Standby)
		0.00486 mW (Sleep)
Power supply		1.2 V

Fig. 18 Fabricated nonvolatile TCAM test chip and its features

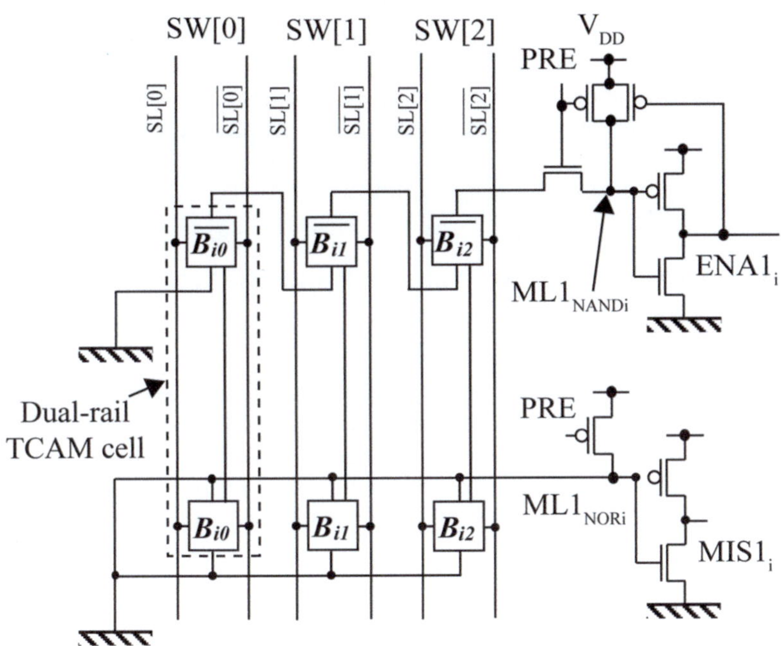

Fig. 19 Design of Asynchronous dual-rail nonvolatile TCAM word circuit with soft-error tolerance

	Synchronous (CMOS)	Extension of ASYNC'13 (CMOS)	Proposed (CMOS/MTJ)
Cycle time [ns]	3.398	N/A	3.410
(search delay [ns])	1.699	N/A	2.330 (data)
(precharge delay [ns])	1.699	N/A	1.060 (spacer)
Eneegy metric [fJ/bit/search]	0.580	N/A	0.686
TCAM cell	24T	48T (2x 24T)	20T-4MTJ
SEU tolerant in cell	**Yes**	**Yes**	**Yes**
SEU free in cell	No	No	**Yes ($<10^{-50}$)**
Delay-variation tolerant	No	**Yes**	**Yes**
Soft-error detection	No	No	**Yes**

Fig. 20 Comparison of TCAMs with soft-error tolerance

TCAM cell is compactly designed using 20 transistors (20T) and 4 MTJ devices stacked on a CMOS layer as opposed to a single-rail 24T TCAM cell that consists of soft-error tolerant storage elements. In addition, soft errors can be detected using the dual-rail signals. As a design example, a 256-word × 64-bit TCAM is designed under a 90-nm CMOS/MTJ technology and is evaluated with a collected charge caused by a particle strike, which induces the SET and hence the delay variation. Figure 20 summarizes the performance comparison of TCAMs. The proposed TCAM properly operates under the delay variation, while achieving comparable performance to a synchronous single-rail TCAM in which an up to 25 % timing error occurs.

4 Design Example of Nonvolatile Random-Access Logic LSI

In order to design and implement MTJ-based nonvolatile random-access logic circuit, we must make a basic gate family using NV-LIM architecture. We have employed a nonvolatile full adder circuit to demonstrate a circuit based on logic-in-memory architecture [28]. Figure 21 shows the circuit diagram of the full adder. It consists of SUM-circuit and CARRY-circuit parts, where the symbols A (A′; the complement of A) and Ci (Ci′) are the external inputs and the symbol B (B′) is a stored input. The use of a dynamic logic style [29] (where pre-charged sense amplifier [30] has been also presented as a high-speed, highly stable and low-power logic style of nonvolatile logic) controlled by clock signals, CLK and CLK′, cuts off the steady current flow from the supply voltage VDD to GND, which reduces the dynamic power dissipation of the circuit. The stored data is programmed by controlling external signals. Complementary stored inputs, B and B′, are programmed by using individual current-flow path, which is selectable by

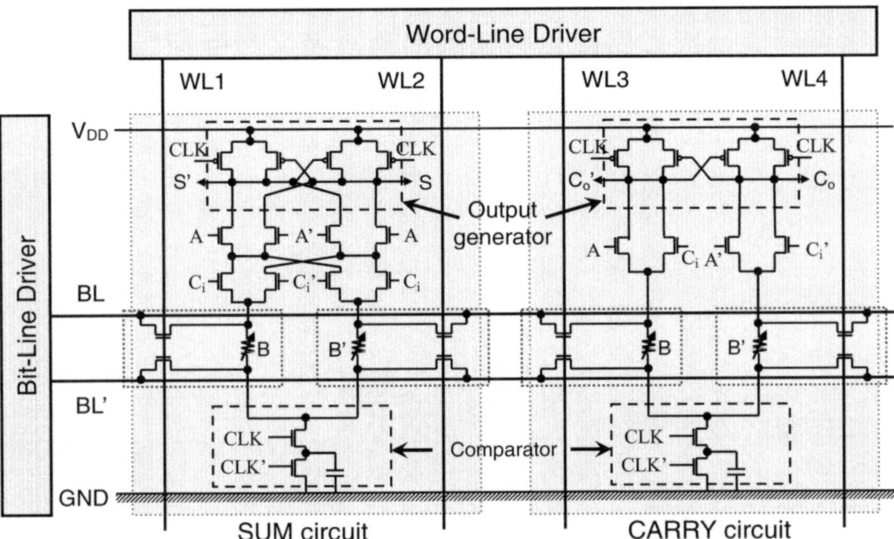

Fig. 21 Circuit diagram of a nonvolatile full adder with MTJ-based logic-in-memory architecture

the word lines, WL1, WL2, WL3, and WL4, and the bit lines, BL and BL$'$. For example, in the case of storing B = 0 into the corresponding MTJ in the SUM circuit, the word line WL1 is set to the supply voltage VDD, and BL and BL$'$ are set to GND and VDD, respectively, which makes the current-flow path through the MTJ set up as shown in Fig. 21. All the external inputs and the complementary clock signals are turned off during the above write operation.

Figure 22 shows the measured waveforms of the SUM circuit chip, where the stored inputs, B and B$'$, are fixed to "0" and "1", respectively and periodic 1.0-V-peak-to-peak voltage signals are applied to CLK, CLK$'$, A, A$'$, Ci, and Ci$'$, respectively, under periodic turn on and off of VDD = 1.0 V. It can be clearly seen in the traces of Fig. 22 that the output S_{after} (S right after power-on) is the same as S_{before} (S just before power-off), which means that stored data remain intact even if VDD is shut down and is turned on again. It should be noted that nonvolatile storage function of the present circuit is realized without employing complex reload/write-back from/into an off-chip nonvolatile storage device.

In order to design practical-scale MTJ-based NV-LIM LSIs, it is important to establish an (semi-)automated design flow. We have developed this flow by combining de facto standard engineering design automation (EDA) tools and new supplementary design tools for precise simulation of MTJ device characteristics as shown in Fig. 23 [31]. By using the proposed flow, various MTJ-based NV-LIM circuits can be designed by using HDL, and the corresponding layout including MOS and MTJ/MOS-hybrid cells can be automatically synthesized, as shown in Fig. 24, where its layout validity can be completely verified through DRC and LVS.

As a typical example of nonvolatile random-access logic LSI, we have developed the motion-vector prediction unit [23, 24]. Figure 25 shows a test-chip

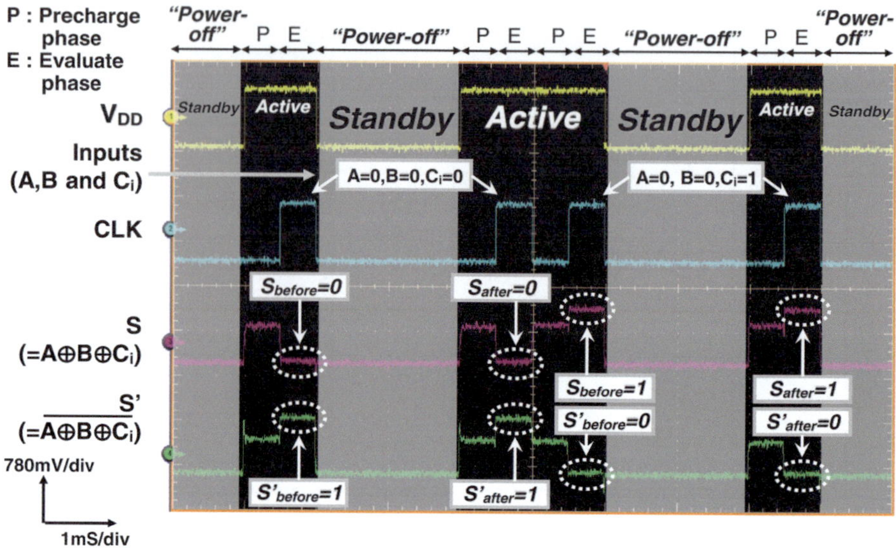

Fig. 22 Measured waveforms of the SUM circuit chip with the proposed NV-LIM architecture

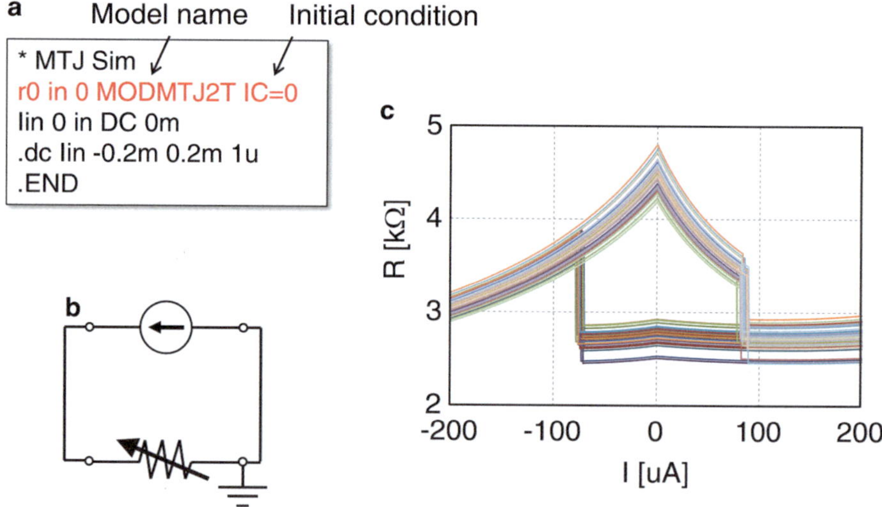

Fig. 23 STT-MTJ device model built in SPICE simulator; (**a**) example of a netlist, (**b**) corresponding equivalent circuit, (**c**) simulated waveforms

photomicrograph of the motion-vector prediction unit using 90 nm MTJ/MOS process made on a 300 mm wafer fabrication line. Twenty-five processing elements (Pes) are arranged in a 5×5 grid, which reduces the dissipation to one-fourth. The number of MOS transistors is about 0.5 million and that of MTJ devices is about 13,000.

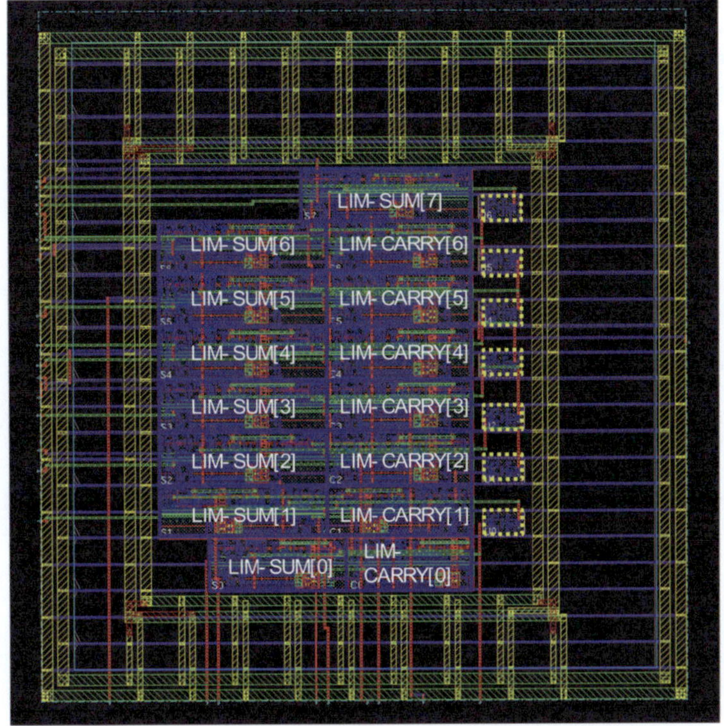

Fig. 24 Layout-design example of an NV-LIM-based random logic circuit

Fig. 25 Fabricated NV-LIM-based hardware-accelerator LSI chip for motion-vector extraction

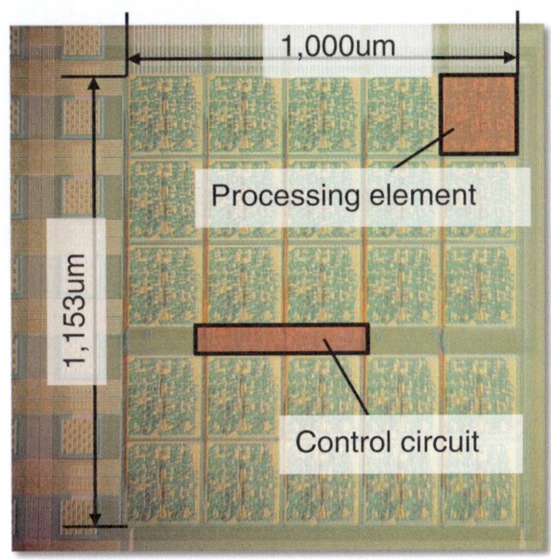

Acknowledgement The author thanks M. Natsui, A. Mochizuki, S. Matsunaga, D. Suzuki, N. Onizawa, S. Ikeda, T. Endoh, and H. Ohno for their great contribution concerning technical support and chip-fabrication support. A part of this research was supported by JSPS FIRST Program.

References

1. H. Ohno, T. Endoh, T. Hanyu, N. Kasai, S. Ikeda, in *International Electron Device Meeting (IEDM)*, 2010, p. 9.4
2. T. Hanyu, SPIN **3**(4), 1340014 (2013)
3. T. Hanyu, D. Suzuki, A. Mochizuki, M. Natsui, N. Onizawa, T. Sugibayashi, S. Ikeda, T. Endoh, H.I. Ohno, in *International Electron Device Meeting (IEDM)*, 2014, in press
4. N. Sakimura, Y. Tsuji, R. Nebashi, H. Honjo, A. Morioka, K. Ishihara, K. Kinoshita, S. Fukami, S. Miura, N. Kasai, T. Endoh, H. Ohno, T. Hanyu, T. Sugibayashi, in *IEEE International Solid-State Circuits Conf. (ISSCC)*, 2014, pp. 184–185
5. H. Koike, T. Ohsawa, N. Sakimura, R. Nebashi, Y. Tsuji, A. Morioka, K. Miura, H. Honjo, T. Sugibayashi, S. Ikeda, T. Hanyu, H. Ohno, T. Endoh, in *IEEE Asian Solid-State Circuits Conference (ASSCC 2013)*, 2013, pp. 317–320
6. D. Suzuki, M. Natsui, S. Ikeda, H. Hasegawa, K. Miura, J. Hayakawa, T. Endoh, H. Ohno, T. Hanyu, in *IEEE Symposium on VLSI Circuits, Dig. Tech. Papers*, 2009, p. 80
7. D. Suzuki, M. Natsui, T. Hanyu, Trans IEICE-C, **J92-C**(7), 233–240, (2009) (Japanese).
8. D. Suzuki, M. Natsui, T. Endoh, H. Ohno, T. Hanyu, J. Appl. Phys. **111**, 07E318 (2012)
9. D. Suzuki, D. Suzuki et al., J. Appl. Phys. **115**, 17B742 (2014)
10. D. Suzuki et al., IEICE ELEX **10**, 20130772 (2013)
11. D. Suzuki, et al., IEEE Trans Magn (2014, in press).
12. D. Suzuki, Y. Lin, M. Natsui, T. Hanyu, Jpn. J. Appl. Phys. **52**(4), 04CM04 (2013)
13. D. Suzuki, M. Natsui, A. Mochizuki, T. Hanyu, Jpn. J. Appl. Phys. **53**(4S), 04EM03 (2014)
14. D. Suzuki, N. Sakimura, M. Natsui, A. Mochizuki, T. Sugibayashi, T. Endoh, H. Ohno, T. Hanyu, IEICE Electronics Express (2014, in press).
15. S. Matsunaga, A. Katsumata, M. Natsui, S. Fukami, T. Endoh, H. Ohno, T. Hanyu, in *IEEE Symp VLSI Circuits*, 2011, pp. 298–299
16. S. Matsunaga, S. Miura, H. Honjou, K. Kinoshita, S. Ikeda, T. Endoh, H. Ohno, T. Hanyu, in *IEEE Symp VLSI Circuits*, 2012, pp. 44–45
17. S. Matsunaga, N. Sakimura, R. Nebashi, Y. Tsuji, A. Morioka, T. Sugibayashi, S. Miura, H. Honjo, K. Kinoshita, H. Sato, S. Fukami, M. Natsui, A. Mochizuki, S. Ikeda, T. Endoh, H. Ohno, T. Hanyu, in *IEEE Symp VLSI Circuits*, C106–C107, 2013, pp. 106–107
18. S. Matsunaga, A. Katsumata, M. Natsui, T. Endoh, H. Ohno, T. Hanyu, Jpn. J. Appl. Phys. **51**(2), 02BM06 (2012)
19. S. Matsunaga, A. Katsumata, M. Natsui, T. Endoh, H. Ohno, T. Hanyu, J. Appl. Phys. **111**(7), 07E336 (2012)
20. S. Matsunaga, M. Natsui, S. Ikeda, K. Miura, T. Endoh, H. Ohno, T. Hanyu, in *Proc. Asia and South Pacific Design Automation Conf. (ASP-DAC)*, 2012, pp. 475–476
21. S. Matsunaga, A. Mochizuki, T. Endoh, H. Ohno, T. Hanyu, IEICE Electron. Express **11**(3), 20131006 (2014)
22. N. Onizawa, S. Matsunaga, and T. Hanyu, in *20th IEEE International Symposium on Asynchronous Circuits and Systems (ASYNC 2014)*, pp. 1–8, 2014
23. M. Natsui, D. Suzuki, N. Sakimura, R. Nebashi, Y. Tsuji, A. Morioka, T. Sugibayashi, S. Miura, H. Honjo, K. Kinoshita, S. Ikeda, T. Endoh, H. Ohno, T. Hanyu, in *IEEE International Solid-State Circuits Conf. (ISSCC), Dig. Tech. Papers*, pp. 194–195, 2013

24. M. Natsui, D. Suzuki, N. Sakimura, R. Nebashi, Y. Tsuji, A. Morioka, T. Sugibayashi, S. Miura, H. Honjo, K. Kinoshita, S. Ikeda, T. Endoh, H. Ohno, T. Hanyu, in *IEEE Journal of Solid-State Circuits*, 2015, in press
25. S. Brown, J. Rose, IEEE Des. Test Comput. **13**, 42 (1996)
26. S. Fukami et al., IEEE TMAG **48**, 2152 (2012)
27. S. Fukami et al., IEDM Tech. Dig. 72 (2013)
28. S. Matsunaga, J. Hayakawa, S. Ikeda, K. Miura, H. Hasegawa, T. Endoh, H. Ohno, T. Hanyu, Appl. Phys. Express **1**, 091301 (2008)
29. A. Mochizuki, H. Kimura, M. Ibuki, T. Hanyu, IEICE Trans. Fundam. **E88-A**, 1408–1415 (2005)
30. W. Zhao, C. Chappert, V. Javerliac, J.-P. Noziere, IEEE Trans. Magn. **45**, 3784–3787 (2009)
31. N. Sakimura, R. Nebashi, Y. Tsuji, H. Honjo, T. Sugibayashi, H. Koike, T. Ohsawa, S. Fukami, T. Hanyu, H. Ohno, T. Endoh, in *IEEE International Symposium on Circuits and Systems (ISCAS)*, 2012, pp. 1971–1974.

Logic Circuits Design Based on MRAM: From Single to Multi-States Cells Storage

Bojan Jovanović, Raphael Martins Brum, and Lionel Torres

1 Introduction

Decades of effective miniaturization of microelectronic switches (metal-oxide-semiconductor transistors) have brought us tremendous benefits. Every new generation of the electronic devices was smaller in area, faster in speed and more power efficient. However, as we entered a deep-submicron era, some side effects emerged. Due to short transistor channels and thin oxide gates, transistor leakage currents are dramatically increased. Moreover, it is becoming harder and harder to control the threshold voltage of a transistor (to make efficient transitions from ON to OFF state and vice-versa) and to reliably incorporate millions of them in an integrated circuit [1]. These obstacles threaten the continued scaling of CMOS devices and, if Moore's law is to hold, we certainly must to overcome them.

The roadblocks are present at system and architectural levels as well. The mainstream Von-Neumann computing architectures consist of a pure computational part (central processing unit—CPU) and a memory part in which the computing recipes (programs) and the input/output data of the calculations are stored [2]. Such complex systems have a memory hierarchy comprising different semiconductor memory types, as illustrated in Fig. 1. Dense, slow, and non-volatile storage memory with limited endurance (usually of NAND/NOR Flash type, located far from the CPU) is combined with fast, volatile, power and area consuming SRAM/DRAM working memory (close to the CPU) in order to ensure both

B. Jovanović
Faculty of Electronic Engineering, University of Niš, A. Medvedeva 14, Niš 18000, Serbia
e-mail: bojan.jovanovic@elfak.ni.ac.rs

R.M. Brum • L. Torres (✉)
LIRMM—University of Montpellier2/UMR CNRS 5506, 161 Rue Ada,
Montpellier 34095, France
e-mail: raphael.brum@lirmm.fr; lionel.torres@lirmm.fr

© Springer International Publishing Switzerland 2015
W. Zhao, G. Prenat (eds.), *Spintronics-based Computing*,
DOI 10.1007/978-3-319-15180-9_6

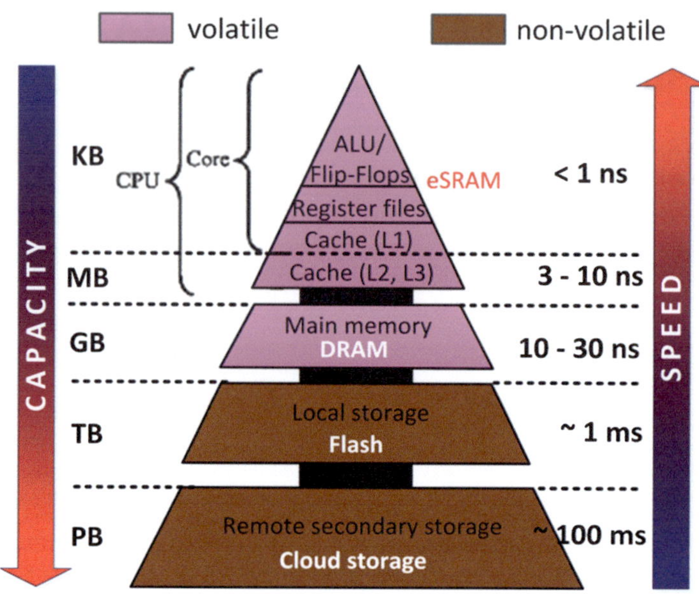

Fig. 1 Typical structure of a computer memory hierarchy

Table 1 Main characteristics of today's semiconductor memories [6] (shortcomings given in bold)

	SRAM	DRAM	NOR Flash	NAND Flash
Cell latency	$>$**60 F^2**	4–6 F^2	10 F^2	4–5 F^2
Read latency	$<$3 ns	7–20 ns	**5 µs**	**25 µs**
Write latency	$<$3 ns	7–20 ns	**1 ms**	**200 µs**
Static power	**Yes**	**Yes**	No	No
Endurance	$>10^{15}$	$>10^{15}$	**10^4**	**10^4**
Non-volatility	**No**	**No**	Yes	Yes

rapid accessibility and data non-volatility [3]. The main characteristics of today's semiconductor memories are summarized in Table 1.

However, this sort of design hierarchy requires complex control. During the system start-up process, data located in storage memory is copied to the working memories in order to set up memory systems for use. This booting procedure usually takes a long time and wastes a significant amount of power. This is illustrated in Fig. 2a that shows the energy profile of the computing system based on the Von-Neumann architecture. The same happens when such a system is shutting down. It requires time and power to store the status data back in the non-volatile storage memory. When such a system is in use, the working SRAM and DRAM memories consume a lot of energy to keep the data because they are both volatile. To additionally save the energy from the power supply, there is a

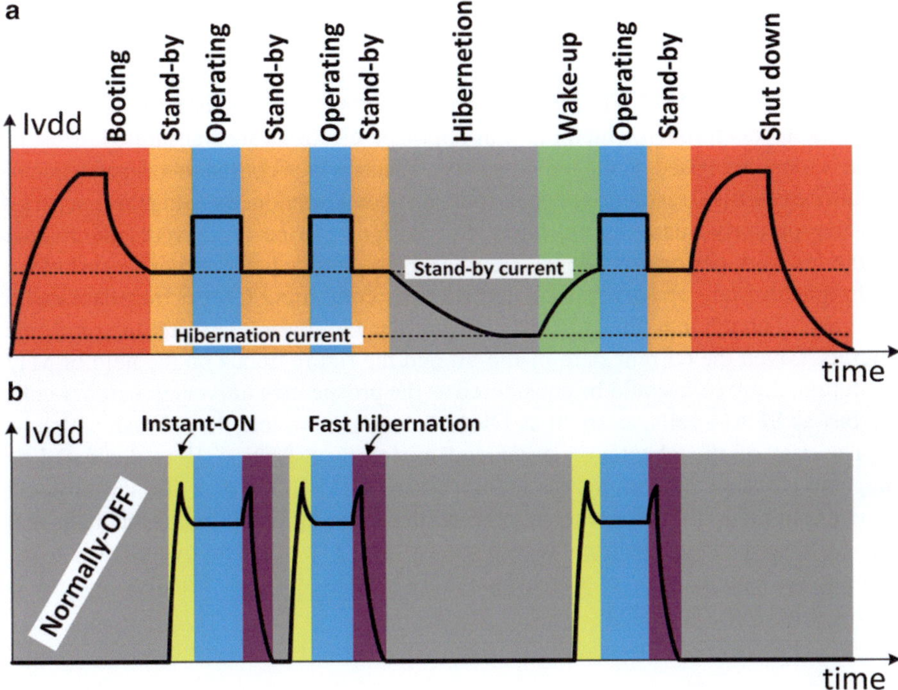

Fig. 2 Energy profiles of a: (**a**) mainstream Von-Neumann computing architecture; (**b**) normally-off and instant-on green computing approach

possibility for a system to be hibernated. In this case, the power is cut off from every computing block except from some CPU cache volatile memories that are still supplied with power in order to remember the context of a hibernated system.

Furthermore, there is a growing disparity of speed between CPU and storage memory outside the CPU chip. In the last 10 years, CPU speeds have improved at a rate of approximately 25 % per year, while the memory speed experienced only a 10 % increase [4]. This significantly limits the efficiency of data processing (clock frequency). To mitigate the limitations posed by off-chip storage memory bandwidth, large size working cache memories need to be used. If such caches are designed with SRAM, they may occupy 90 % of the chip area in future CMOS generations [5]. Besides, SRAM-based cache memory is becoming the major source of the leakage power consumption. The power gating technique applied on a SRAM cache is not efficient since it is paid by data loss and by the significant time and the energy required to retrieve the lost data.

Finally, having in mind that with the technology scaling, existing semiconductor memories get increasingly less reliable and more expensive to fabricate, it is clear that the conventional memories are approaching "the memory wall" becoming a bottleneck in computer performance [6].

Spintronic devices that exploit both the intrinsic spin of the electron and its associated magnetic moment, in addition to its fundamental electronic charge, are promising solutions to co-integrate with the CMOS and to circumvent the aforementioned scaling threats [7]. They generally comprise magnetic layers separated by non-magnetic layers through which the spin-polarized electrons are transmitted. Most of these devices are based on magneto-resistive effects which consist in a dependence of the device electrical resistance upon its magnetic configuration. By using this magneto-resistive property, many hybrid innovative designs can be conceived: non-volatile memories, high-performance logic circuits, RF oscillators, field/current sensors etc.

In this chapter, we survey some hybrid cells combining CMOS transistors with Magnetic Tunnel Junctions (MTJs) that are the key basic elements in spintronic circuits. Given the current state-of-the-art and the future trends on the field of MTJ evolution, these cells could be considered as the prospective universal memory cells (as fast as SRAM cells, as small as DRAMs and as non-volatile as Flash cells).

The rest of the chapter is organized as follows: Section II is dedicated to magneto-resistive random access memories (MRAMs). We describe the physical structure of the MTJ nanopillars together with the evolution of the mechanisms that are used to control its electrical resistance. In Sect. 3, we give an overview of some prospective applications of the hybrid MTJ/CMOS designs. Finally, Sect. 4 is reserved for the conclusions.

2 MRAM

Motivated by the robustness of magnetic devices against radiation, the development of the MRAMs is in the focus of the worldwide research teams for the last 30 years. Though early MRAM devices were unreliable, slow, and energy inefficient [8, 9], the discovery of MTJs that exhibit large enough margin at room temperature between two stable states (resistances) really opened new perspectives for MRAM.

In addition to radiation immunity and reliability, spin-based MTJs can be co-integrated with the conventional CMOS technology without imposing an area overhead, as illustrated in Fig. 3a. Furthermore, MTJs provide non-volatility, energy efficiency (little energy is needed to change the electron spin [10]), high speed data switching, improved density, infinite endurance, and the ability to continue shrinking in size [11], as will be demonstrated in the rest of the chapter.

An MTJ is a nanopillar composed of an ultra thin (0.8–1.5 nm) layer of insulator (oxide barrier) sandwiched between two ferromagnetic (FM) metals (Fig. 3a). The insulating layer is so thin that electrons can tunnel through the barrier if a bias voltage is applied between two FM electrodes. The resistance of MTJ depends on the relative orientation of the magnetization in the two FM layers.

In standard applications, the magnetization of one FM layer (the reference layer) is commonly pinned, whereas the other (storage) layer is free to take a parallel (P) or an anti-parallel (AP) orientation, thus determining parallel (R_p) or anti-parallel (R_{ap}) MTJ resistance and storing a binary state. The margin between

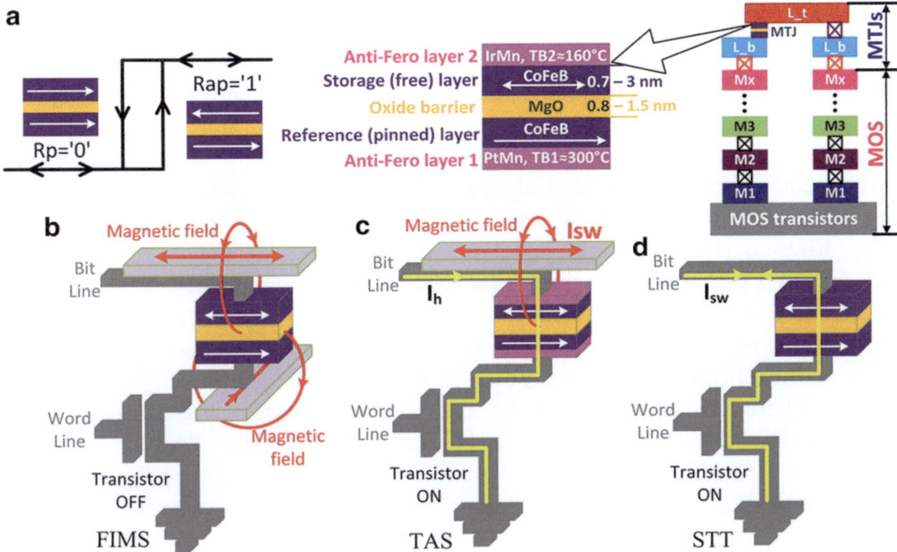

Fig. 3 (a) CMOS/MTJ co-integration; (b) FIMS MT switching; (c) TAS MTJ switching; (d) STT MTJ switching

these two MTJ resistances is defined by the tunnel magneto-resistance (TMR) ratio, $TMR\ [\%] = (Rap-Rp)/Rp$. There are two kinds of MTJs: in-plane MTJ (i-MTJ) where magnetization of ferromagnetic layers lies in the film plane and perpendicular MTJ (p-MTJ) where magnetization direction is perpendicular to the film plane.

The mechanism of switching between two MTJ states (i.e., writing non-volatile data bit) directly influences the area, speed, and power performance of hybrid (MTJ/CMOS) designs. It is, consequently, very important to have reliable MTJ switching. Using either an external magnetic field or a spin-polarized switching current can perform it.

2.1 Field-Induced Magnetic Switching (FIMS)

The magnetization of a storage layer in early MTJ devices was controlled through the external magnetic fields. To achieve write selectivity, two perpendicular current lines are used to generate the write field, as illustrated in Fig. 3b. Only the MTJ at the crossing of these two lines can be written, provided that the current densities on the lines are strong enough. Commercial availability of toggle MRAM proves that integration of the FIMS MTJ devices with the CMOS is feasible technology option [12]. However, as FIMS switching mechanism requires two high currents (about few mA), it suffers from the energy efficiency. Low storage density seems to be an additional obstacle. Moreover, process variations and thermal fluctuations always limit the density of this type of memories.

2.2   Thermally-Assisted Switching (TAS)

In order to improve a bit selectivity and to have more efficient MTJ switching, the second generation of MTJ devices was taking advantage of a general property of magnetic materials, which is that in most cases their magnetization is more easily switched at elevated temperature than at room temperature [13]. For this purpose, the MTJ nanopillar is slightly modified by adding two anti-ferromagnetic layers with different (low and high) blocking temperatures. To switch MTJ, heating current (I_h) is passed through the MTJ stack to heat the junction above the blocking temperature of the free layer. In this way, the free layer is unlocked and ready to store the non-volatile data bit determined by an applied external magnetic field (I_{sw}), as illustrated in Fig. 3c. In addition to improved writing efficiency (the heating current is between 100 μA and 1 mA), TAS also provides very high data reliability due to its exchange bias mechanism [14]. On the other hand, TAS is essentially limited by the heating and cooling dynamics [13]. Under typical operating conditions, the heating takes place in 3–7 ns whereas the cooling requires 15–25 ns.

2.3   Current-Induced Magnetic Switching (CIMS)

One of the main weaknesses of classic magnetic field writing MRAMs is that the required writing current rapidly increases with reduced MTJ size. That is why recent current induced magnetic switching (CIMS) methods use the spin-transfer torque (STT) effect [15, 16] in order to control the MTJ resistance. This enables the magnetization of a free layer to be switched with only one low (in the order of few tens of μA), spin-polarized bi-directional current (I_{sw}) passing through the MTJ stack. In this case, there is no need for high switching currents that induce a magnetic field (Fig. 3d). This leads to the reduction in cell size. If the density of the spin-polarized writing current pulse is greater than the critical current density (J_{C0}), MTJ resistance is determined only by its direction. STT writing approach is greatly simplified, faster and more energy efficient in comparison to the previous two. However, it suffers from reliability issues including data thermal stability, erroneous write by read current [17], and short retention times [18].

Mutual comparisons of these three different MRAM technologies are summarized in Table 2.

2.4   Trends in Evolution of MTJ Parameters

To be widely accepted by the IC industry, MTJ-based MRAM devices must outperform their counterparts (e.g., phase change memories—PCMs, or ferroelectric RAMs—FeRAMs). For this purpose, MRAM technologies have made important advance in the last decade allowing denser, faster and low power applications.

Table 2 Comparison of various MRAM technologies

	FIMS MRAM	TAS-MRAM	STT-MRAM
Scalability	Poor (>60 nm)	Good (>40 nm)	The best (<10 nm)
Min. cell size	Large (~30 F^2)	Small (~10 F^2)	Very small (~6 F^2)
Endurance	10^{16}	10^{12}	10^{16}
Writability	Poor	Good	The best
Power	Very high	High	The best
Latency	Long (>10 ns)	Very long (>20 ns)	The best (<5 ns)

As the TMR ratio that defines the margin between two MTJ stable states (MTJ resistances) is one of the most essential parameters for a reliable storage and restoration of non-volatile data bits, a lot of research effort has been invested in improving an MTJ resistance margin. The emergence of MTJ devices with the MgO oxide barriers significantly improved the TMR ratio. Today's commercial MTJs that have electrodes made of CoFeB and the tunnel barrier made of MgO achieved TMR of over 200 % [20]. Some laboratory prototypes reached 600 % [21] whereas, theoretically, TMR ratio can be as high as 1,000 % [22].

Regarding the MTJ writing efficiency, STT-MTJs with in-plane magnetization have very fast MTJ switching speed (up to 100 ps, according to [23]). However, with the writing currents of hundreds of microamperes, this switching approach is still not efficient since it consumes a lot of energy and requires large driving transistors. Furthermore, with the miniaturization of the MTJ fabrication node (below 40 nm for lateral sizes), data thermal stability for in-plane STT-MRAM becomes a critical issue due to its planar anisotropy storage principle, which leads to important random sensing errors [24].

Recently emerging STT-MRAM structures based on perpendicular MTJs (p-MTJs) proved to be the breakthrough technology enabling a significant reduction in the required switching current (several tens of micro amperes) [25]. Thus, p-MTJs have solved the biggest obstacle that had prevented MRAM densification for a long time. Because of small p-MTJ switching currents, heating by the switching current was greatly reduced which has highly positive impact on the MTJ endurance. Moreover, p-MTJs significantly improved thermal stability of non-volatile data bits [26]. The fact that p-MTJs can be obtained in a circular shape makes them less sensitive to process variations (easily to fabricate) unlike that of elliptical i-MTJs.

The speed of switching p-MTJs is inversely proportional to its writing current and can be adjustable to the particular application. Increasing the amount of the switching current entails decreasing the duration of the current pulse (i.e., speeding-up p-MTJ writing). For some low power applications, reported switching current reached 15 μA with 30 ns current pulse width [23]. On the other hand, in [27], 3 ns fast p-MTJ switching with 50 μA current was reported for some high frequency applications. IBM and Toshiba proved by using p-MTJs that writing with 1 ns current pulse was possible [23].

As a consequence of drastically reduced switching current, p-MTJs consume very small amount of energy (<1 pJ) during the switching phase. 90 fJ of programming energy per-bit was reported in [27].

3 Hybrid MTJ/CMOS Logic Circuits

All the aforementioned development progress of the MTJ devices certainly makes them suitable for CMOS co-integration in wide range of application fields. In the rest of this section, we will try to list some of them.

3.1 Applications in Processor Domain

Bringing the non-volatility directly into the working memory and, especially, into the CPU core would pave the way for new green computing paradigm based on "normally-off and instant-on" operation. Computing equipment could be quickly turned-off when not in use, keeping the off state with zero stand-by power as long as possible. On the computing request, the equipment could be turned on instantly, with the full performance capabilities. Such computing approach would be far more energy efficient compared with the current "normally-on" computing systems, as illustrated in Fig. 2b.

So far, a lot of research effort was invested towards a green computing. Hybrid memory cells that combine one MOS transistor with one MTJ stack (1 T-1MTJ) are particularly suitable for high density non-volatile mass storage or DRAM replacements [28]. As illustrated in Fig. 4, the process of reading (restoring) non-volatile data bit consists in biasing the MTJ stack with a given bias voltage and comparing resulting current that is proportional to the MTJ resistance with a reference current by using the sense amplifier design. Though area efficient, such hybrid cells cannot

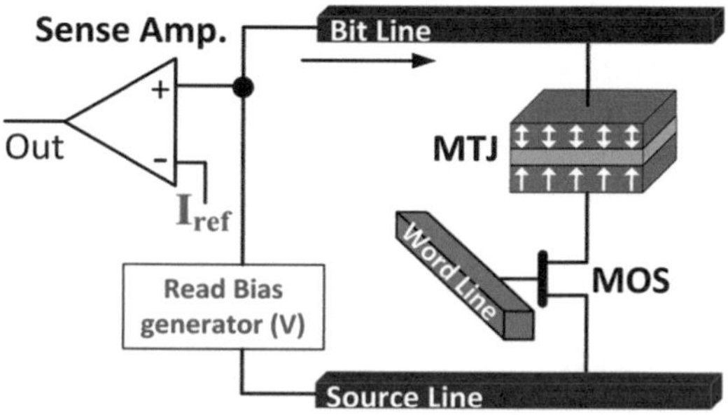

Fig. 4 Current sensing approach used to restore a non-volatile data bit from 1 T-1MTJ hybrid memory cell

read out information faster than 10 ns which makes them ineligible for the use in a CPU core (flip-flops, register files, cache memories) where the access to the information is very frequent (<1 ns).

To replace SRAM-based volatile memory cells or flip-flops located near the processor's arithmetic logic unit (ALU), many hybrid cells are proposed that are based on the hybrid (volatile/non-volatile) cross-coupled inverters [29–33]. The unique feature of these cells is that while CPU is in active state, they behave as a conventional CMOS-based flip-flop or SRAM memory cell with the very high speed of operation (>2 GHz). While CPU is in stand-by state, data are stored in MTJs and zero stand-by power is achieved by the power gating. After power supply returns, the cell itself operates as a sense amplifier automatically restoring the data saved in MTJs into the SRAM or flip-flop. This enables the processor core to quickly become ready to start arithmetic operation. Furthermore, such cells allow run-time saving of the processors' context (non-volatile check-pointing), thus significantly improving the reliability of data processing.

One hybrid cell of this kind [29] that has a structure similar to that of a six-transistor (6 T) SRAM cell is shown in Fig. 5. A volatile (SRAM) data context consists of the cross-coupled inverters (CMOS latch) used to store one data bit in its electrical, complementary form $(Q, !Q)$. In addition to the CMOS latch, the cell has two non-volatile (MRAM) contexts located in both pull-up and pull-down networks of the latch structure. Each MRAM context contains two perpendicular STT-MTJs that, for the correct operation of the cell, must be in mutually complementary states (Rp/Rap or vice versa).

The procedure of writing a volatile data bit is exactly the same as in the conventional SRAM memory cell. The volatile data bit to be written and its complementary value are connected to the BL and BLB lines, respectively. After

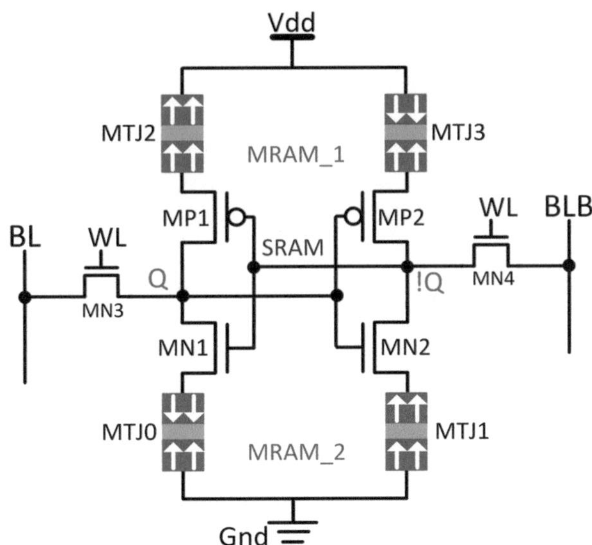

Fig. 5 Hybrid (volatile/ non-volatile) cross-coupled inverter design [29]

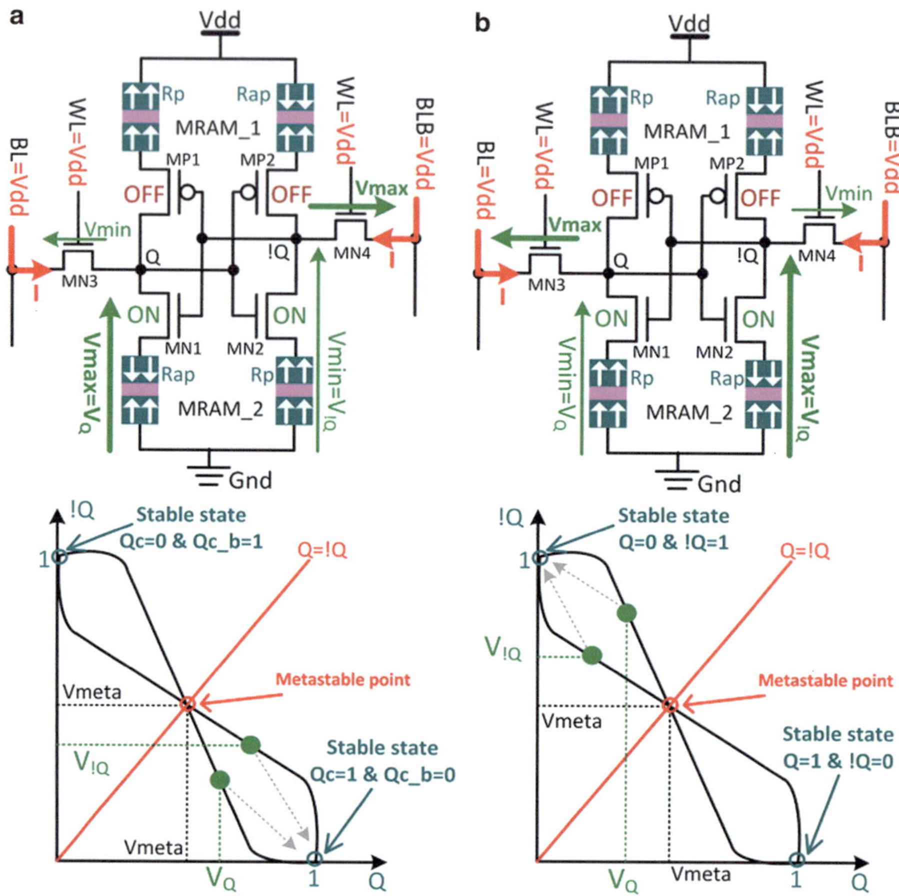

Fig. 6 The phase of reading MRAM_2 context that stores: (**a**) non-volatile data bit '1' (Rap/Rp); (**b**) non-volatile data bit '0' (Rp/Rap) [29]

activation of the access transistors (MN3 and MN4) with the WL signal pulse, the volatile data bit is stored in the CMOS latch.

Reading the non-volatile data bit (i.e., restoring the MRAM context to SRAM) consists of converting the physical value (resistance) stored in MTJs into its electrical equivalent which will be stored in the CMOS latch. Figure 6 illustrates the reading phase of MRAM_2 context (MTJs in the pull-down network). To read this MRAM context, BL and BLB lines need to be pre-charged to Vdd. The reading phase begins with activation of WL signal ($WL = Vdd$).

Consequently, pull-down transistors (MN1/MN2) of the CMOS latch are switched on, whereas the pull-up ones (MP1/MP2) are blocked (off). In both pull-down branches of the hybrid cell, there is a current flowing from the BL/BLB lines through the access transistors and NMOS pull-down transistors to the ground (Gnd). Provided that the cell is fully symmetrical (the transistors in both branches have equal on resistances since they have the same dimensions), the voltage drops on the Q and $!Q$ nodes entirely depends on the MTJ resistances in the MRAM_2 context that are in the path of the current. Furthermore, if both the transistors and the MTJs are carefully sized, the voltages on the latch nodes Q and $!Q$ can be adjusted to be one below and another above the meta-stable voltage (*Vmeta*), depending on the non-volatile data bit stored in MRAM_2 context. As illustrated on the transfer curve in Fig. 6a, non-volatile data bit '1' stored in MRAM_2 context (*Rap/Rp* configuration) will cause the voltage on the Q node to be greater than the meta-stable voltage ($VQ > Vmeta$). The opposite will occur if MRAM_2 context stores non-volatile data bit '0' (*Rp/Rap* configuration, Fig. 6b: Q and $!Q$ voltages will be below and above meta-stable voltage, respectively ($V_Q < Vmeta$; $V_{!Q} > Vmeta$).

In both scenarios, at the end of MRAM reading phase when the WL signal is deactivated and the access transistors are turned off, the CMOS latch converges from an unbalanced state to one of its stable states, which is strictly determined by the state (resistance) of MTJs in MRAM_2 context. The procedure of reading MRAM_1 context is the same. The only difference is that, in this case, BL and BLB lines need to be pre-charged to Gnd. Consequently, the pull-up network is activated, the current flows in both branches from the power supply (Vdd) to the BL/BLB nodes (which are now on the ground potential) putting the latch in a meta-stable state. Finally, when the access transistors are deactivated, the latch converges from an unbalanced state to a stable one determined by the non-volatile data bit stored in MRAM_1 context (MTJ2 and MTJ3). *Rp/Rap* configuration for MTJ2/MTJ3 stores non-volatile data bit '1' whereas the *Rap/Rp* combination is used to store non-volatile '0' bit.

As reported in [29], measured performance of this hybrid cell implemented in 28 nm FD-SOI technology combined with 45 nm round perpendicular STT-MTJs show that the cell is able to benefit from dual storage facility needing less than 1 ns to restore a data bit from one of the non-volatile contexts. Moreover, it is shown that the devices based on this hybrid cell would benefit from the run-time (on-the-fly) reconfiguration ability given that the cell is able to write non-volatile data bit without disturbing the volatile one (Q, $!Q$).

In order to additionally decrease required implementation area, there are some hybrid cells that have a structure similar to that of a 4 T-SRAM loadless volatile memory cells [34–36]. Two MTJs are "embedded" within a CMOS part making the cell inherently non-volatile. Unlike 1 T-1MTJ cells, hybrid cells of this kind have double-ended read system that allows achieving of the very fast access appropriate for embedded applications. As shown in Fig. 7, 4 T-NVSRAM hybrid cell presented in [34] consists of two PMOS access transistors (MP1 and MP2) with low threshold voltage (*Vth*) and two cross-coupled NMOS transistors (MN1 and MN2) used to store one volatile data bit. In addition, the cell has one non-volatile (MRAM) context located in the pull-down network.

The low threshold voltage of the PMOS access transistors implies increased sub-threshold leakage current compared to the leakage of the pull-down NMOS transistors (*Ioffp > Ioffn*). This, in turn, ensures volatile data retention when the cell is

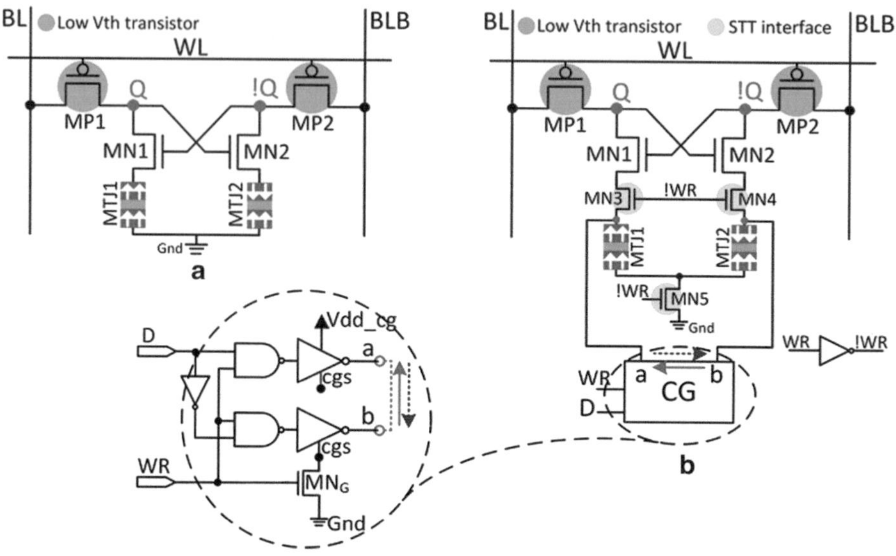

Fig. 7 (**a**) Hybrid (MTJ/CMOS) cell with a structure similar to that of 4 T-SRAM loadless cell; (**b**) STT writing interface together with the current generator (CG) design [34]

on stand-by (BL,BLB,WL = Vdd). The procedure of writing a volatile data bit is exactly the same as in conventional 4 T-SRAM loadless memory cells whereas the restoring phase is similar to that of a previously described hybrid cell. Figure 7b shows STT writing interface. In addition to the current generator (CG) that supplies the bi-directional, spin-polarized current needed to write a non-volatile data bit (D), it contains the footer transistor MN5 as well as the pass transistors MN3 and MN4. In normal cell operation, these three transistors are always switched on (*WR='0'*). Conversely, during the phase of writing a non-volatile data bit (*WR='1'*), they cut the MTJs off from the ground rails and cross-coupled NMOS transistors, ensuring that spin-polarized CG current passes through both MTJs in mutually opposite directions. The direction of the CG current is strictly determined by the non-volatile data bit to be written (D). Given that in the idle state CG inverters are with the active pull-down networks (logic zero at the inverters' outputs), the volatile data bit (electrical charge) stored in NMOS cross-coupled transistors could discharge through the CG. To prevent this happening, a power-gating transistor MNG is used to cut-off the CG from the ground rails during its idle state.

Reported cell's performance [34] show that a non-volatile data bit can be restored for about 100 ps. p-MTJs are reliably switched with 50–60 μA current pulse taking 4–5 ns of time and spending about 300 fJ of energy. It is also shown that body biasing technique offered by the FD-SOI [37] can reduce the size of the cell by up to 60 % while boosting its speed and dynamic energy performance by about 80 %.

Many research groups were exploring the feasibility of replacing a SRAM CPU cache memory with a non-volatile cache based on the hybrid cells [38–53]. So far, general conclusion is that the implementations of hybrid last level caches (LLC), L2, L3, and L4, can be superior over the conventional SRAM ones. Having in mind increasing LLC capacity, which is due to increased number of processor cores and limited clock frequency, it is shown that hybrid LLC replacements can lead to a significant reduction in both static and dynamic power while keeping the same performance. In [38], 1-Mb STT-MRAM L2 cache design is presented that is able to operate at the speed of 250 MHz. Compared with 0.8 V operating SRAM design, this hybrid L2 cache could reduce the energy per instruction (EPI) by 64 %, while maintaining instructions per cycle (IPC) performance degradation within 6 %. Similarly, authors in [39] report that the use of L2 and L3 STT-MRAM caches can eliminate performance loss while still reducing the energy-delay product by more than 70 %.

To boost the processor's performance with STT-MRAM LLC caches, researchers use different approaches. Some of them get benefit from the ability of p-STT-MTJs to trade-off the non-volatile data retention period with writing speed and required energy [39, 40]. They relax STT-MRAM retention time in order to reduce both write latencies and write energy. In [40], it is shown that an MTJ retention period of 10 ms is suitable for L2 cache. Some authors try to profit from the fact that STT-MRAM density is around four times higher than the SRAM density [41, 42]. They use hybrid LLC cache that has either the same capacity as the SRAM one (to reduce area and interconnect delays) or the same area as the SRAM cache (to increase capacity and reduce misses). Finally, there are approaches that try to minimize or avoid non-volatile writes [43–48], to address high write latency of STT-MRAM [49–51] or to address 0/1 write asymmetry [52, 53].

On the other hand, the implementation of STT-MRAMs into L1 caches, register files and CPU core logic circuits is still questionable. Though STT-MRAM can eradicate static power, it drastically increases the active power in the CPU core and completely negates the reduction in static power [54]. In other words, write latency and dynamic write energy of the state-of-the-art p-MTJs are still high compared to SRAM-based L1 cache performance. Consequently, the superiority border of STT-MRAMs against SRAMs lies between L1 and L2 caches. Further improvements of the p-MTJ performance and hybrid circuit design are certainly needed to push the border upward into the L1.

3.2 *Programmable Logic*

MTJs can also provide benefits in reconfigurable computing systems which offer major advantages such as reduced design cost, rapid prototyping and a generic hardware for mapping arbitrary applications [55].

Field-programmable gate arrays (FPGAs) are one of the most popular hardware reconfigurable platforms. They consist of elementary logic functions (lookup tables—LUTs) connected by a network of wire segments and programmable switches. Both the content of LUTs and the states of programmable switches are fully determined by the bits stored in a configuration memory.

Present FPGA platforms can be divided into SRAM-based ones which employ SRAM cells to store the configuration and flash or anti-fuse FPGAs which use non-volatile memory for storing configuration [56]. SRAM-based FPGAs are generally quite fast to program. However, volatile nature of SRAM requires an additional off-chip non-volatile memory to permanently store configuration bits. Consequently, the start-up configuration process is slow (~100 ms). Moreover, SRAM cells are significantly area consuming. On the other hand, anti-fuse or flash-based FPGAs save the silicon area and generally provide fast start-up. Nevertheless, since anti-fuse devices can be programmed only once, they are restricted to applications that do not require run-time reconfiguration. Flash-based FPGAs have slow writing time and consume a lot of writing energy.

MRAM-based FPGAs have the ability to combine the benefits from both worlds. Hybrid designs of the core computational and routing logic eliminate the need for an additional off-chip non-volatile storage. This removes the boot-up time and alleviates design complexity at the chip/board level while enabling comparable speeds to SRAM-based FPGAs and partial run-time reconfiguration ability. More-over, MRAM-based FPGAs are able to successively improve FPGA reliability regarding a single event upset (SEU) at ground level which becomes particularly expressed in a deep-submicron technology nodes [57].

First pioneering work on the use of emerging non-volatile memories for FPGAs was reported in 2006 [58]. The authors used 130 nm CMOS technology in combination with 100 nm TAS MTJs. Since then, many works on this subject appeared, ranging from proposed architecture-level enhancements to actual circuit implementations [59–64]. They all profit from the fact that MTJ resistance can be used to store a non-volatile configuration bit. As shown in Fig. 8b, by applying suitable bias voltage to the series of p and n transistors, the configuration stored in non-volatile MTJs is written into the cross-coupled SRAM inverters of an FPGA LUT.

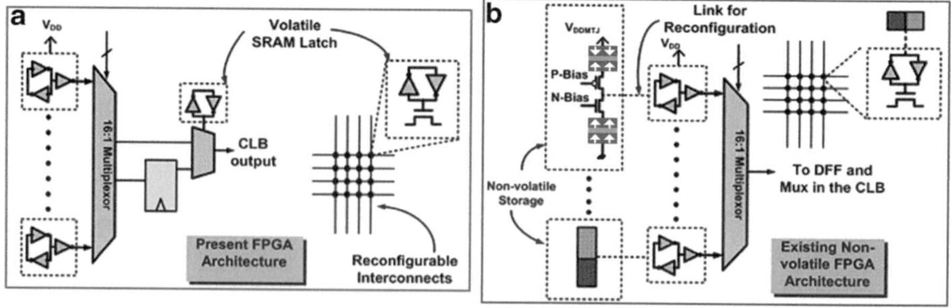

Fig. 8 Proposed scheme for integration of STT-MRAM in the configurable logic block and programmable interconnects of a CMOS/MRAM hybrid FPGA architecture [60]

In [60], authors propose new LUT architectural approach that improves integration density at the cost of increased power overhead. The proposed architecture leverages on the high integration density of emerging STT MRAM devices and minimizes the total logic area by reducing the contribution of CMOS cross-coupled inverters serving as a configuration bits. For a set of benchmark circuits, simulation results showed improvements of 44.39 % in logic area and 22.28 % in delay of a configurable logic block (CLB) and average improvement of 16.1 % in dynamic power over a conventional CMOS FPGA design.

In [62], 2-input radiation hardened LUT architecture based on TAS MTJs was physically implemented and tested. Such hybrid FPGA approach is appropriate for some space applications since MTJs are less sensitive to total ionizing dose (TID) in regard to the flash memories. MTJs can be used as a reference memory to perform "scrubbing" techniques, which consists in regularly reloading the configuration of the FPGA to fix the radiation induced errors that might have occurred.

However, despite many different hybrid FPGA architectural approaches, there is no commercially available MRAM-based FPGA chip. Some exciting combinations of Everspin's STT-MRAM memories with Altera's Stratix FPGA devices have been announced for the near future [65]. We believe that the current improvements on the field of MTJ performance will certainly encourage other players on the field to take a risk.

3.3 Logic-in-Memory

MTJ technology is equally attractive for building logic configurations which merge non-volatile memories and logic circuits (so called logic-in-memory architecture) to overcome the communication bottleneck between off-chip memories and on-chip logic as well as to reduce leakage and dynamic power. As illustrated in Fig. 9, non-volatile storage elements (MTJs) are distributed over a logic-circuit plane, right above the CMOS logic. Two functions are interconnected by vertical vias which can greatly reduce global routings (interconnection delay) in today's VLSI chips [19]. Consequently, wide memory bandwidth as well as instant power gating without escaping/reloading data can be realized.

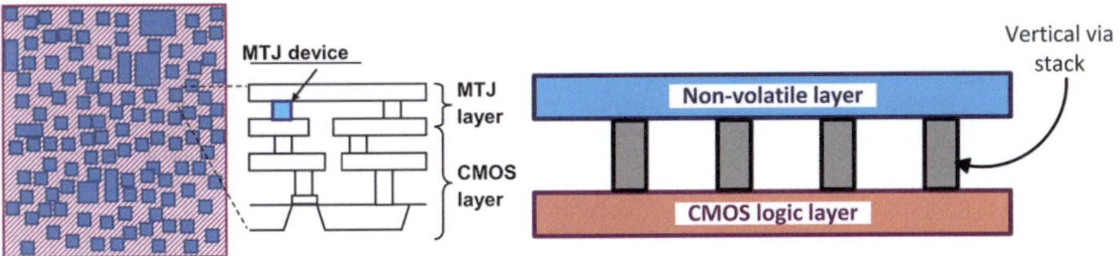

Fig. 9 Merging MTJ non-volatile storage elements with the CMOS logic to obtain hybrid Logic-in-Memory architecture [19]

To demonstrate the possibilities brought by the non-volatile back-end technology for logic-in-memory concepts, researchers from Tohoku University and Hitachi were fabricated a first test chip of non-volatile full adder circuit [66]. It was possible to control power consumption of a design by a unit of the command line as well as to reduce leakage current in the intermittent stand-by states. The experimental results showed that the dynamic power of a hybrid full-adder design was reduced to less than one-fourth while maintaining the same performance compared to nominal CMOS circuit.

In [67], logic-in-memory approach is applied to the motion-vector prediction algorithm that is important in video compression. A 75 % of the leakage power reduction was recorded with proposed hybrid hardware in comparison with that of a conventional CMOS-only-based hardware.

Author in [68] presents an MTJ-based bit-parallel/bit-serial non-volatile ternary content-addressable memory (TCAM) design demonstrating its superiority over a corresponding CMOS-only realization, whereas in [69] interested reader can found the designs of hybrid NAND, NOR, and XOR circuits.

3.4 Multi Bit MRAM Cells

Continuous increasing of MTJ's TMR ratio inspired the development of multi-level cell (MLC) STT-MRAM, which allows multiple data bits to be stored in a single memory cell thus additionally increasing the storage density. The concept is illustrated in Fig. 10. Two MTJs with different areas (TMR ratios) that are connected in series exhibit four-level resistance, depending on a combination of their magnetization directions. Having the efficient control circuits for reliable reading and writing the state of MTJs, it is possible to use the cell to store two non-volatile data bits. However, as a result of narrowed storage margin (distinction between two resistance values—ΔR), the performance and reliability of MLC STT-MRAM cells become more sensitive to both the MOS/MTJ device variations and the thermal-induced randomness of MTJ switching.

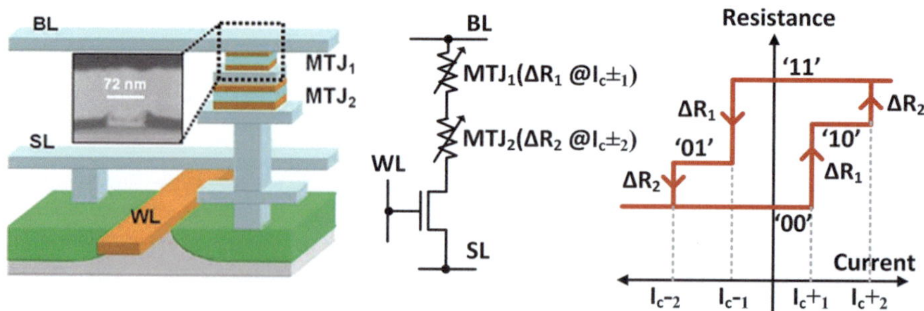

Fig. 10 The series-connected MTJs in multi-level hybrid memory cell that is able to store two data bits [71]

In [70], the variation sources of MLC STT-MRAM designs together with their impacts on the reliability of the read and write operations are systematically analyzed. Simulation results showed that although the stacking MLC STT-MRAM has not yet satisfy the requirements of commercial product under realistic fabrication conditions, it has shown the great potentials under careful design optimization.

Two-step techniques for reliable reading/writing stacked MTJs are proposed in [71]. The number, direction, and amount of the write current pulses are controlled according to the write data. For reading the cell data, a multi-reference sense amplifier for the two-step read operation is introduced.

Authors in [72] present a flexible hybrid cell based on stacked MTJs that can be dynamically configured between two possible functional modes (high-reliability mode and multi-level cell mode), based on the data requirements of the addressed application. The cell is ready to be used in future high-density non-volatile memories and neuromorphic computing circuits.

The physical arrangements and control circuitry for three bits-per-cell MRAM concept are presented in [73]. Particular focus is on the conversion of three conventional binary bits to octal encoded data and the required sequence for writing eight states per cell using current driven magnetic fields. Write sequences for controlling arrays of multi-bit memory cells are also analyzed.

The study of the use of MLC STT-MRAM in cache design of embedded systems and microprocessors can be found in [74]. The authors conclude that two-step read/ write accesses and inflexible data mapping strategies in the existing MLC STT-MRAM cache architecture may even result in system performance degradation. They propose cross-layer solution introducing reverse MTJ into MLC cell offering a more balanced device. At the architectural level, they propose a cell split mapping method to divide cache lines into fast and slow regions and data migration policies to allocate the frequently-used data to fast regions. Simulation results proved that the proposed techniques can improve the system performance by 10 % and reduce the energy consumption on cache by 26 % compared with conventional MLC STT-MRAM cache design.

3.5   Emerging MRAM Devices

The lack of the commercial availability of hybrid MTJ/CMOS devices is mainly due to the fact that current p-MTJs are not yet mature enough to be mass produced. The main roadblocks on the way of p-MTJ evolution are its writing time (which is still long for some CPU core applications), high current densities required for writing, and reliable reading without causing switching since reading and writing operation share the same path through the junction. Non-volatile spintronic devices whose active power is much lower than that of a state-of-the-art STT-MRAMs are urgently required to expand the boundary of applications and to evolve computer systems. Something similar to what happened in the history of CMOS evolution, when current-driven bipolar transistors were replaced by the voltage-driven field-effect transistors (this drastically improved power efficiency) is necessary to happen on the MTJ side as well. The change to voltage-induced magnetization switching is one of the most promising paths to lower power consumption.

Three-terminal MTJ devices that have separated reading and writing paths promise an increased reliability and both short and energy efficient switching time while keeping the non-volatility [75–77].

Although it is too early to estimate the feasibility of new spintronic devices to be integrated into CPU core, recent advances in spintronic technologies are very encouraging.

4 Conclusions

Spintronic devices that benefit from the electron's spin degree of freedom provide a promising synergy to the overmature CMOS technology. After decades of efforts in making MTJ-based MRAM devices more reliable, more efficient, and faster, they are ready to be co-integrated with the CMOS and used in wide variety of hybrid applications and innovative architectures. In this chapter, we listed some of them.

Hybrid STT-MRAM cells are particularly suitable to bring a non-volatility into some parts of the memory hierarchy of the mainstream computing architectures. So far, they are ready to replace working volatile DRAM memories as well as some levels of LLC SRAM-based cache memories, providing a green computing approach which is based on normally-off and instant-on operation. For breaking the border and enter into the CPU core (L1 cache, CPU registers and core logic circuits), MRAM performance (especially the speed and energy efficiency of non-volatile data writing) need to be further improved.

MTJs, used as programmable resistors, can be planted over the plane of CMOS logic circuits thus mixing the memory and logic aspects and improving the speed end the efficiency of data traffic. They can be equally used as configuration memories in reconfigurable computing systems, providing high-density, fast FPGAs with the dynamic reconfiguration ability.

Recent improvements in MTJ performance additionally indicate that MRAM hybrid devices are expecting a bright future.

Acknowledgments The authors wish to acknowledge support from French National Agency for Scientific Research (ANR), through the projects DIPMEM and MARS.

References

1. J. Rabaey, *Low Power Design Essentials* (Springer, New York, 2009)
2. J. Hennessy, D. Patterson, *Computer Architecture: A Quantitative Approach* (Morgan Kaufmann, San Francisco, 2011)

3. S. Przybylski, *Cache and Memory Hierarchy Design: A Performance-directed Approach* (Morgan Kaufmann, San Francisco, 1990)
4. C. Liu, A. Sivasubramaniam, M. Kandemir, Organizing the last line of defense before hitting the memory wall for CMPs. Software, IEE Proceedings, 14–18 February 2004
5. B.-M. Rogers, A. Krishna, G.-B. Bell et al., Scaling the bandwidth wall: challenges in and avenues for CMP scaling. *Proceedings of the 36th Annual International Symposium on Computer Architecture*, Austin, 20–24 June 2009
6. M. Indaco, P. Prinetto, E.-I. Vatajelu, On the impact of process variability and aging on the reliability of emerging memories. Paper presented at the 19th IEEE European Test Symposium, Paderborn, 26–30 May 2014
7. International technology roadmap for semiconductors. (Semiconductor Industry Association (SIA), San Jose, 2011)
8. A. Pohm, C. Comstock, A. Hurst, Quadrupled nondestructive outputs from magnetoresistive memory cells using reversed word field. J. Appl. Phys. **67**(9), 4881 (1990)
9. B. Dieny et al., Giant magnetoresistance in soft ferromagnetic multilayers. Phys. Rev. B **43**, 1297–1300 (1991)
10. C. Chappert, A. Fert, V. Dau, The emergence of spin electronics in data storage. Nat. Mater. **6**(11), 813–823 (2007)
11. W. Zhao, E. Belhaire, C. Chappert, P. Mazoyer, Spintronic device based non-volatile low standby power SRAM. *Paper presented at the IEEE Annual Symposium on VLSI*, Montpellier, 7–9 April 2008
12. J. Slaughter et al., Toggle and spin-torque MRAM: status and outlook. J. Magn. Soc. Jpn. **5**, 171 (2010)
13. I. Prejbeanu, W. Kula, K. Ounadjela et al., Thermally assisted switching in exchange-biased storage layer magnetic tunnel junctions. IEEE Trans. Magn. **40**(4), 2625–2627 (2004)
14. I. Prejbeanu et al., Thermally assisted MRAM. J. Phys. Condens. Matter **19**(16), 165218 (2007)
15. L. Berger, Emission of spin waves by a magnetic multilayer traversed by a current. Phys. Rev. B **54**(13), 9353–9358 (1996)
16. J.-C. Slonczewski, Current-driven excitation of magnetic multilayers. J. Magn. Magn. Mater. **1859**(1/2), L1–L7 (1996)
17. A. Singh, S. Schwarm, O. Mryasov, S. Gupta, Interlayer exchange coupled composite free layer for CoFeB/MgO based perpendicular magnetic tunnel junctions. J. Appl. Phys. **114**(20), 203901 (2013)
18. R. Takemura, T. Kawahara, K. Ono et al., Highly-scalable disruptive reading scheme for Gb-scale SPRAM and beyond. Proceedings of IEEE International Memory Workshop, Seoul, 16–19 May 2010
19. T. Kawahara, K. Ito, R. Takemara, H. Ohno, Spin-transfer torque RAM technology: review and prospect. Microelectron. Reliab. **52**(4), 613–627 (2012)
20. W. Kim et al., Extended scalability of perpendicular STT-MRAM towards sub-20 nm MTJ node. Paper presented at the IEEE International Electron Devices Meeting, Washington, 5–7 December 2011
21. S. Ikeda et al., Tunnel magnetoresistance of 604% at 300 K by suppression of Ta diffusion in CoFeB/MgO/CoFeB pseudo-spin-valves annealed at high temperature. Appl. Phys. Lett. **93**(8), 082508 (2008)
22. J. Mathon, A. Umerski, Theory of tunneling magnetoresistance of an epitaxial Fe/MgO/Fe (0 0 1) junction. Phys. Rev. B **63**(22), 220403 (2001)
23. H. Yoda, S. Fujita, N. Shimomura et al., Progress of stt-mram technology and the effect on normally-off computing systems. Paper presented at the IEEE International Electron Devices Meeting, San Francisco, 10–13 December 2012
24. W. Zhao, L. Torres, L. Cargnini et al., High performance SoC design using magnetic logic and memory, in *VLSI-SoC: Advanced Research for Systems on Chip*, ed. by S. Mir, C.-Y. Tsui, R. Reis, O. Choi (Springer, Berlin, 2012), pp. 10–33

25. M. Nakayama, T. Kai, N. Shimomura et al., Spin transfer switching in TbCoFe/ CoFeB/MgO/ CoFeB/TbCoFe magnetic tunnel junctions with perpendicular magnetic anisotropy. J. Appl. Phys. **103**(7), 07A710 (2008)
26. T. Kishi, H. Yoda, T. Kai et al., Lower-current and fast switching of a perpendicular TMR for high speed and high density spin-transfer-torque MRAM. Paper presented at the IEEE International Electron Devices Meeting, San Francisco, 15–17 December 2008
27. E. Kitagawa, S. Fujita, K. Nomura et al., Impact of ultra low power and fast write operation of advanced perpendicular MTJ on power reduction for high-performance mobile CPU. Paper presented at the IEEE International Electron Devices Meeting, San Francisco, 10–13 December 2012
28. R. Dorrance, F. Ren, Y. Toriyama, A. Hafez, C. Yang, D. Markovic, Scalability and design-space analysis of a 1T-1MTJ memory cell for STT-RAMs. IEEE Trans. Electron Devices **59**(4), 878–887 (2012)
29. B. Jovanović, R.-M. Brum, L. Torres, A hybrid magnetic/ complementary metal oxide semiconductor three-context memory bit cell for non-volatile circuit design. J. Appl. Phys. **115**(13), 134316 (2014)
30. W. Black, B. Das, Programmable logic using giant magnetoresistance and spin-dependent tunneling devices. J. Appl. Phys. **87**(9), 6674–6679 (2000)
31. K. Hass, G. Donohoe, Y. Hong, B. Choi, Magnetic flip flops for space applications. IEEE Trans. Magn. **42**(10), 2751–2753 (2006)
32. W. Zhao, C. Chappert, V. Javerliac, J. Nozière, High speed, high stability and low power sensing amplifier for MTJ/CMOS hybrid logic circuits. IEEE Trans. Magn. **45**(10), 3784–3787 (2009)
33. Y. Guillemenet, Magnetic logic for the exploration of reconfigurable architectures, Dissertation, University of Montpellier 2 (2011)
34. B. Jovanović, R.-M. Brum, L. Torres, Evaluation of hybrid MRAM/CMOS cells for "normally-off and instant-on" computing. Analog Integr. Circuits Signal Process. **81**, 607 (2014)
35. T. Ohsawa, H. Koike, S. Miura et al., A 1 Mb nonvolatile embedded memory using 4T2MTJ cell with 32 b fine-grained power gating scheme. IEEE J. Solid-State Circuits **48**(6), 1511–1520 (2013)
36. T. Ohsawa, S. Ikeda, T. Hanyu, H. Ohno, T. Endoh, 1Mb 4T-2MTJ nonvolatile STT-RAM for embedded memories using 32b fine-grained power gating technique with 1.0ns/200ps wake-up/power-off times. Paper presented at the IEEE Symposium on VLSI Circuits, Honolulu, 13–15 June 2012
37. F. Arnaud, N. Planes, O. Weber et al., Switching energy efficiency optimization for advanced CPU thanks to UTBB technology. Paper presented at the IEEE International Electron Devices Meeting, San Francisco, 10–13 December 2012
38. H. Noguchi, K. Kushida, K. Ikegami et al., A 250-MHz 256b-I/O 1-Mb STT-MRAM with advanced perpendicular MTJ based dual cell for nonvolatile magnetic caches to reduce active power of processors. Paper presented at the IEEE Symposium on VLSI Technology, Kyoto, 12–14 June 2013
39. C.-W. Smullen, V. Mohan, A. Nigam, S. Gurumurthi, M.-R. Stan, Relaxing non-volatility for fast and energy-efficient STT-RAM caches. Paper presented at the 17th IEEE International Symposium on High Performance Computer Architecture, San Antonio, 12–16 February 2011
40. A. Jog, A.-K. Mishra, X. Cong et al., Cache revive: Architecting volatile STT-RAM caches for enhanced performance in CMPs. Paper presented at the 49th ACM/EDAC/IEEE Design Automation Conference, San Francisco, 3–7 June 2012
41. L. Torres, R.-M. Brum, L.-V. Cargnini, G. Sassatelli, Trends on the application of emerging nonvolatile memory to processors and programmable devices. Paper presented at the IEEE Symposium on Circuits and Systems, Beijing, 19–23 May 2013
42. X. Guo, E. Ipek, T. Soyata, Resistive computation: avoiding the power wall with low-leakage, STT-MRAM based computing. Paper presented at the 37th annual International Symposium on Computer Architecture, Saint-Malo, 19–23 June 2010

43. J. Ahn, K. Choi, Lower-bits cache for low power STTRAM caches. Paper presented at the IEEE Symposium on Circuits and Systems, Seoul, 20–23 May 2012
44. J. Jung, Y. Nakata, M. Yoshimoto, H. Kawaguchi, Energy-efficient spin-transfer torque RAM cache exploiting additional all-zero-data flags. Paper presented at the 14th International Symposium on Quality Electronic Design, Santa Clara, 4–6 March 2013
45. S. Yazdanshenas, M. Pribast, M. Fazeli, A. Patooghy, Coding last level STT-RAM cache for high endurance and low power. Computer Architecture Letters, accepted for publication
46. G. Sun, X. Dong, Y. Xie, J. Li, Y. Chen, A novel architecture of the 3D stacked MRAM L2 cache for CMPs. Paper presented at the 15th IEEE International Symposium on High Performance Computer Architecture, Raleigh, 14–18 February 2009
47. P. Zhou, B. Zhao, J. Yang, Y. Zhang, Energy reduction for STT-RAM using early write termination. Paper presented at the IEEE/ACM International Conference on Computer-Aided Design, San Jose, 2–5 November 2009
48. S. Park, S. Gupta, N. Mojumder, A. Raghunathan, K. Roy, Future cache design using STT MRAMs for improved energy efficiency: Devices, circuits and architecture. Paper presented at the 49th ACM/EDAC/IEEE Design Automation Conference, San Francisco, 3–7 June 2012
49. X. Dong, X. Wu, G. Sun, Y. Xie, H. Li, Y. Chen, Circuit and microarchitecture evaluation of 3D stacking magnetic RAM (MRAM) as a universal memory replacement. Paper presented at the 45th ACM/EDAC/IEEE Design Automation Conference, Anaheim, 8–13 June 2008
50. J. Wang, X. Dong, Y. Xie, OAP: an obstruction-aware cache management policy for STT-RAM last-level caches. Paper presented at the 2013 DATE Conference, Grenoble, 18–22 March 2013
51. M. Mao, H. Li, A. Jones, Y. Chen, Coordinating prefetching and STT-RAM based last-level cache management for multicore systems. Paper presented at the 23rd Great Lakes Symposium on VLSI, Paris, 2–3 May 2013
52. R. Bishnoi, M. Ebrahimi, F. Oboril, M.-B. Tahoori, Asynchronous asymmetrical write termination (AAWT) for a low power STT-MRAM. Paper presented at the 2014 DATE Conference, Dresden, 24–28 March 2014
53. K.-W. Kwon, S.-H. Choday, Y. Kim, K. Roy, AWARE (Asymmetric Write Architecture with REdundant blocks): a high write speed STT-MRAM cache architecture. IEEE Trans. VLSI 22(4), 712–720 (2014)
54. K. Ando, S. Fujita, J. Ito et al., Spin-transfer torque magnetoresistive random-access memory technologies for normally off computing. J. Appl. Phys. 115(17), 172607 (2014)
55. P. Chow, S.-O. Seo, J. Rose et al., The design of an SRAM based field-programmable gate array. Part II: circuit design and layout. IEEE Trans. VLSI 7(2), 191–197 (1999)
56. I. Koun, R. Tessier, J. Rose, FPGA architecture: survey and challenges. Foundat. Trends Electron. Design Automat. 2(2), 135–253 (2008)
57. L.-V. Cargnini, Y. Guillemenet, L. Torres, G. Sassatelli, Improving the reliability of a FPGA using fault-tolerance mechanism based on magnetic memory (MRAM). Paper presented at the International Conference on Reconfigurable Computing and FPGAs, Quintana Roo, 13–15 December 2010
58. N. Bruchon, L. Torres, G. Sassatelli, G. Cambon, New non-volatile FPGA concept using magnetic tunneling junction. Paper presented at the IEEE Computer Society Annual Symposium on Emerging VLSI Technologies and Architectures, Karlsruhe, 2–3 March 2006
59. W. Zhao, E. Belhaire, C. Chappert, P. Mazoyer, Spin transfer torque (STT)-MRAM–based runtime reconfiguration FPGA circuit. ACM Trans. Embedded Comput. Syst. 9(2), 14.1–14.16 (2009)
60. P. Somnath, S. Mukhopadhyay, S. Bhunia, A circuit and architecture codesign approach for a Hybrid CMOS–STTRAM nonvolatile FPGA. IEEE Trans. Nanotechnol. 10(3), 385–394 (2011)
61. W. Zhao, E. Belhaire, C. Chappert, B. Dieny, G. Prenat, TAS-MRAM based low power, high speed run-time reconfiguration (RTR) FPGA. ACM Trans. Reconfigurable Technol. Syst. 2(2), 8.1–8.19 (2009)

62. O. Goncalves, G. Prenat, G. Di Pendina, B. Dieny, Non-volatile FPGAs based on spintronic devices. Paper presented at the 50th ACM/EDAC/IEEE Design Automation Conference, Austin, May 29–June 7 2013
63. Y. Guillemenet, L. Torres, G. Sassatelli, N. Bruchon, I. Hassoune, A non-volatile run-time FPGA using thermally assisted switching MRAMS. Paper presented at the International Conference on Field Programmable Logic and Applications, Heidelberg, 8–10 September 2008
64. Y. Guillemenet, L. Torres, G. Sassatelli, Non-volatile run-time field-programmable gate arrays structures using thermally assisted switching magnetic random access memories. IET Comput. Digital Techniques **4**(3), 211–226 (2010)
65. Business Wire, Everspin to demonstrate Spin-Torque MRAM with Altera FPGAs at Flash Memory Summit 2014. http://www.businesswire.com/news/home/20140804005204/en/ Everspin-Demonstrate-Spin-Torque-MRAM-Altera-FPGAs-Flash#.U_r5jahl-UI. Accessed 25 Aug 2014
66. S. Matsunaga, J. Hayakawa, S. Ikeda, K. Miura, H. Hasegawa, T. Endoh, Fabrication of a nonvolatile full adder based on logic-in-memory architecture using magnetic tunnel junctions. Appl. Phys. Exp. **1**(9), 091301 (2008)
67. M. Natsui, D. Suzuki, N. Sakimura et al., Nonvolatile logic-in-memory array processor in 90nm MTJ/MOS achieving 75% leakage reduction using cycle-based power gating. Paper presented at the IEEE International Solid-State Circuits Conference, San Francisco, 17–21 February 2013
68. T. Hanyu, Challenge of MTJ/MOS-hybrid logic-in-memory architecture for nonvolatile VLSI processor. Paper presented at the IEEE International Symposium on Circuits and Systems, Beijing, 19–23 May 2013
69. B.-A. Behtash, D. Datta, S. Salahuddin, S. Datta, Proposal for an all-spin logic device with built-in memory. Nat. Nanotechnol. **5**(4), 266–270 (2010)
70. Y. Zhang, L. Zhang, W. Wen, G. Sun, Y. Chen, Multi-level cell STT-RAM: is it realistic or just a dream? Paper presented at the IEEE/ACM International Conference on Computer-Aided Design, San Jose, 5–8 November 2012
71. T. Ishigaki, T. Kawahara, R. Tekemura et al., A multi-level-cell spin-transfer torque memory with series-stacked magnetotunnel junctions. Paper presented at the Symposium on VLSI Technology, Honolulu, 15–17 June 2010
72. W. Kang, W. Zhao, Z. Wang et al., DFSTT-MRAM: dual functional STT-MRAM cell structure for reliability enhancement and 3-D MLC functionality. IEEE Trans. Magn. **50**(6), 1–7 (2014)
73. H. Cramman, D.-S. Eastwood, J.-A. King, D. Atkinson, Multilevel 3 bit-per-cell magnetic random access memory concepts and their associated control circuit architectures. IEEE Trans. Nanotechnol. **11**(1), 63–70 (2012)
74. X. Bi, M. Mao, D. Wang, H. Li, Unleashing the Potential of MLC STT-RAM Caches. In Paper presented at the IEEE/ACM International Conference on Computer-Aided Design, San Jose, 18–21 November 2013
75. H. Honjo, S. Fukami, K. Ishihara et al., Three-terminal magnetic tunneling junction device with perpendicular anisotropy CoFeB sensing layer. J. Appl. Phys. **115**(17), 17B750 (2014)
76. P.-M. Braganca, J.-A. Katine, N.-C. Emley et al., A three-terminal approach to developing spin-torque written magnetic random access memory cells. IEEE Trans. Nanotechnol. **8**(2), 190–195 (2009)
77. N.-N. Mojumder, S.-K. Gupta, S.-H. Choday, D.-E. Nikonov, K. Roy, A three-terminal dual-pillar STT-MRAM for high-performance robust memory applications. IEEE Trans. Electron Devices **58**(5), 1508–1516 (2011)

Statistical Reliability/Energy Characterization in STT-RAM Cell Designs

Wujie Wen, Yaojun Zhang, and Yiran Chen

1 Introduction

Conventional memory technologies, i.e., SRAM, DRAM, and Flash, have achieved remarkable successes in modern computer industry. Following technology scaling, the shrunk feature size and the increased process variations impose serious power and reliability concerns on these technologies.

In recent years, many emerging nonvolatile memory technologies have emerged above the horizon. As one promising candidate, spin-transfer torque random access memory (STT-RAM) has demonstrated great potentials in embedded memory and on-chip cache designs [1–6] through a good combination of the non-volatility of Flash, the comparable cell density to DRAM, and the nanosecond programming time like SRAM.

In STT-RAM, the data is represented as the resistance state of a magnetic tunneling junction (MTJ) device. The MTJ resistance state can be programmed by applying a switching current with different polarizations. Compared to the charge-based storage mechanism of conventional memories, the magnetic storage mechanism of STT-RAM shows less dependency on the device volume and hence, better scalability. Nonetheless, despite of these advantages, the unreliable write operation and high write energy are to be the major issues in STT-RAM designs. And these design metrics are significantly impacted by the prominent statistical factors of STT-RAM, including CMOS/MTJ device process variations under scaled technology and the probabilistic MTJ switching behaviors [7, 8]. In particular, the randomness of MTJ switching process incurred by the thermal fluctuations may generate the intermittent write failures of STT-RAM cells.

W. Wen • Y. Zhang • Y. Chen (✉)
Department of Electrical & Computer Engineering, University of Pittsburgh,
Pittsburgh, PA 15213, USA
e-mail: YIC52@pitt.edu

© Springer International Publishing Switzerland 2015
W. Zhao, G. Prenat (eds.), *Spintronics-based Computing*,
DOI 10.1007/978-3-319-15180-9_7

Many studies were performed to evaluate the impacts of process variations and thermal fluctuations on STT-RAM reliability [9–11]. The general evaluation method is as follows: First, Monte-Carlo SPICE simulations are run extensively to characterize the distribution of the MTJ switching current I during the STT-RAM write operations, by considering the device variations of both MTJ and MOS transistor; Then I samples are sent into the macro-magnetic model to obtain the MTJ switching time (τ_{th}) distributions under thermal fluctuations; Finally, the τ_{th} distributions of all I samples are merged to generate the overall MTJ switching performance distribution. A write failure happens when the applied write pulse width is smaller than the needed τ_{th}. Nonetheless, the costly Monte-Carlo runs and the dependency on the macro-magnetic and SPICE simulations incur huge computation complexity [12–15], limiting the application of such a simulation method at the early stage of STT-RAM design and optimization. Meanwhile, the modeling of write energy in STT-RAM was also studied extensively [16]. However, many such works only assume that the write energy of STT-RAM is deterministic and cannot successfully take into account its statistical characteristic induced by process variations and thermal fluctuations.

In this chapter, we propose "PS3-RAM"—a fast, portable and scalable statistical STT-RAM reliability/energy analysis method. PS3-RAM includes three integrated steps: (1) characterizing the MTJ switching current distribution under both MTJ and CMOS device variations; (2) recovering MTJ switching current samples from the characterized distributions in MTJ switching performance evaluation; and (3) performing the simulation on the thermal-induced MTJ switching variations based on the recovered MTJ switching current samples. By introducing the sensitivity analysis technique to capture the statistical characteristics of the MTJ switching, and dual-exponential model to efficiently and accurately recover the MTJ switching current samples for statistical STT-RAM thermal analysis, PS3-RAM can achieve multiple orders-of-magnitude ($> 10^5$) run time cost reduction with marginal accuracy degradation under any variation configurations when compared to SPICE-based Monte-Carlo simulations. Finally, we released PS3-RAM from SPICE and macro-magnetic modeling and simulations, and extended its application into the array-level reliability analysis and the design space exploration of STT-RAM.

The structure of this chapter is organized as the follows: Section 2 gives the preliminary of STT-RAM; Section 3 presents the details of PS3-RAM method; Section 4 presents the application of our PS3-RAM on cell and array level reliability analysis and design space exploration; Section 5 shows the deterministic/statistical write energy analysis based on our PS3-RAM; Section 6 discusses the computation complexity; The last section-Appendix gives the detailed theoretical model deduction and its numerical validation for sensitivity analysis.

2 Preliminary

2.1 STT-RAM Basics

Figure 1c shows the popular "one-transistor-one-MTJ (1T1J)" STT-RAM cell structure, which includes a MTJ and a NMOS transistor connected in series. In the MTJ, an oxide barrier layer (e.g., MgO) is sandwiched between two ferromagnetic layers. '0' and '1' are stored as the different resistances of the MTJ, respectively. When the magnetization directions of two ferromagnetic layers are parallel (anti-parallel), the MTJ is in its low (high) resistance state. Figure 1a, b show the low and the high MTJ resistance states, which are denoted by R_L and R_H, respectively. The MTJ switches from '0' to '1' when the switching current drives from reference layer to free layer, or from '1' to '0' when the switching current drives in the opposite.

2.2 Process Variations and Programming Uncertainty of STT-RAM

2.2.1 Process Variations-Persistent Errors

The current through the MTJ is affected by the process variations of both transistor and MTJ. For example, the driving ability of the NMOS transistor is subject to the variations of transistor channel length (L), width (W), and threshold voltage (V_{th}). The MTJ resistance variation also affects the NMOS transistor driving ability by changing its bias condition. The degraded MTJ switching current leads to a longer MTJ switching time and consequently, results in an incomplete MTJ switching before the write pulse ends. This kind of errors is referred to as "persistent" errors, which are mainly incurred by only device parametric variations. Persistent errors can be measured and repeated after the chip is fabricated.

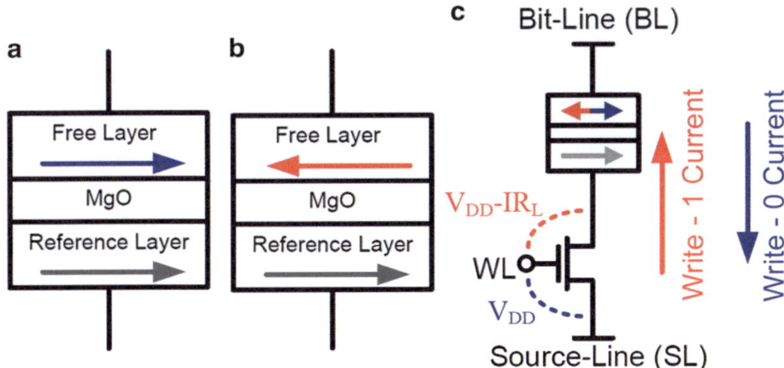

Fig. 1 STT-RAM basics. (**a**) Parallel (low resistance). (**b**) Anti-parallel (high resistance). (**c**) 1T1J cell structure

2.2.2 Thermal Fluctuation-Non-persistent Errors

Another kind of errors is called "non-persistent" errors, which happen intermittently and may not be repeated. The non-persistent errors of STT-RAM are mostly caused by the intrinsic thermal fluctuations during MTJ switching [17]. In general, the impact of thermal fluctuations can be modeled by the thermal induced random field h_{fluc} stochastic Landau-Lifshitz-Gilbert (LLG) equation (1) as [17]:

$$\frac{d\vec{m}}{dt} = -\vec{m} \times \left(\vec{h}_{eff} + \vec{h}_{fluc} \right) + \alpha \vec{m} \times \left(\vec{m} \times \left(\vec{h}_{eff} + \vec{h}_{fluc} \right) \right) + \frac{\vec{T}_{norm}}{M_s} \quad (1)$$

Where $\vec{m}$ is the normalized magnetization vector. Time t is normalized by γM_s; γ is the gyro-magnetic ratio and M_s is the magnetization saturation. $\vec{h}_{eff} = \frac{\vec{H}_{eff}}{M_s}$ is the normalized effective magnetic field. $\vec{h}_{fluc}$ is the normalized thermal agitation fluctuating field at finite temperature which represent the thermal fluctuation. α is the LLG damping parameter. $\vec{T}_{norm} = \frac{\vec{T}}{M_s V}$ is the spin torque term with units of magnetic field. And the net spin torque $\vec{T}$ can be obtained through microscopic quantum electronic spin transport model. Due to thermal fluctuations, the MTJ switching time will not be a constant value but rather a distribution even under a constant switching current.

3 PS3-RAM Method

Figure 2 depicts the overview of our proposed PS3-RAM method, mainly including the sensitivity analysis for MTJ switching current (I) characterization, the I sample recovery, and the statistical thermal analysis of STT-RAM. The first step is to configure the variation-aware cell library by inputting both the nominal design parameters and their corresponding variations, like the channel length/width/threshold voltage of NMOS transistor, as well as the thickness/area of MTJ device. Then a multi-dimension sensitivity analysis will be conducted to characterize the statistical properties of I, followed by an advanced filtering technology—smooth filter, to improve its accuracy. After that, the write current samples can be recovered based on the above characterized statistics and current distribution model. The write pulse distribution will be generated after mapping the switching current samples to the write pulse samples by considering the thermal fluctuations. Finally, the statistical write energy analysis and the STT-RAM cell write error rate can be performed based on the samples of the write current once the write pulse is determined. Array-level analysis and design optimizations can be also conducted by using PS3-RAM.

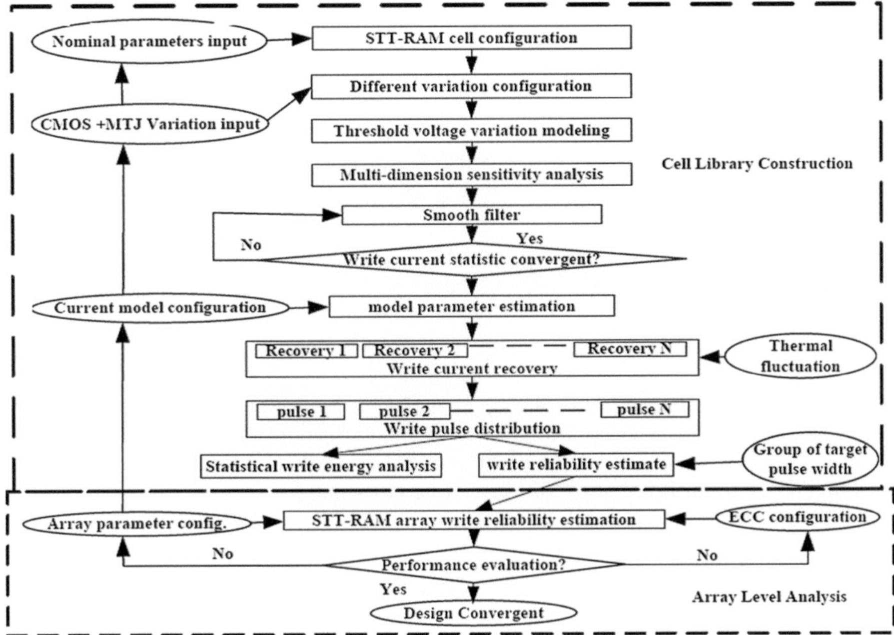

Fig. 2 Overview of PS3-RAM

3.1   Sensitivity Analysis on MTJ Switching

In this section, we present our sensitivity model used for the characterization of the MTJ switching current distribution. We then analyze the contributions of different variation sources to the distribution of the MTJ switching current in details. The definitions of the variables used in our analysis are summarized in Table 1.

3.1.1   Sensitivity Analysis on Variations

1) Threshold voltage variations: The variations of channel length, width and threshold voltage are three major factors causing the variations of transistor driving ability. V_{th} variation mainly comes from random dopant fluctuation (RDF) and line-edge roughness (LER), the latter of which is also the source of some geometry variations (i.e., L and W) [18, 19]. It is known that the V_{th} variation is also correlated with L and W and its variance decreases when the transistor size increases. The deviation of the V_{th} from the nominal value following the change of L (ΔL) can be modeled by [15]:

Table 1 Simulation parameters and environment setting

Parameters	Mean	Standard deviation
Channel length	$\overline{L} = 45$ nm	$\sigma_L = 0.05\overline{L}$
Channel width	$\overline{W} = 90 \sim 1800$ nm	$\sigma_W = 0.05\overline{L}$
Threshold voltage	$\overline{V}_{th} = 0.466$ V	by calculation
MgO thickness	$\overline{T_{thick}} = 2.2$ nm	$\sigma_{T_{thick}} = 0.02\overline{T_{thick}}$
MTJ surface area	$\overline{A} = 45 \times 90$ nm^2	By calculation
Resistance low	$R_L = 1000\,\Omega$	By calculation
Resistance high	$R_H = 2000\,\Omega$	By calculation

$$\Delta V_{th} = \Delta V_{th0} + V_{ds}\, \exp\left(\frac{L}{l'}\right) \cdot \frac{\Delta L}{l'} \tag{2}$$

Then the standard deviation of V_{th} can be calculated as:

$$\sigma_{V_{th}}^2 = \frac{C_1}{WL} + \frac{C_2}{\exp(L/l')} \cdot \frac{W_c}{W} \cdot \sigma_L^2 \tag{3}$$

Here W_c is the correlation length of non-rectangular gate (NRG) effect, which is caused by the randomness in sub-wavelength lithography. C_1, C_2 and l' are technology dependent coefficients. The first term in (3) describes the RDF's contribution to $\sigma_{V_{th}}$. The second term in (3) represents the contribution from NRG, which is heavily dependent on L and W. Following technology scaling, the contribution of this term becomes prominent due to the reduction of L and W.

2) Sensitivity analysis on variations: Although the contributions of MTJ and MOS transistor parametric variabilities to the MTJ switching current distribution cannot be explicitly expressed, it is still possible for us to conduct a sensitivity analysis to obtain the critical characteristics of the distribution. Without loss of generality, the MTJ switching current I can be modeled by a function of W, L, V_{th}, A and T_{thick}. A and T_{thick} are the MTJ surface area and MgO layer thickness, respectively. The 1st-order Taylor expansion of I around the mean values of every parameter is:

$$I(W, L, V_{th}, A, T_{thick}) \approx I\left(\overline{W}, \overline{L}, \overline{V}_{th}, \overline{A}, \overline{T_{thick}}\right) + \frac{\partial I}{\partial W}\left(W - \overline{W}\right) + \frac{\partial I}{\partial L}\left(L - \overline{L}\right)$$
$$+ \frac{\partial I}{\partial V_{th}}\left(V_{th} - \overline{V}_{th}\right) + \frac{\partial I}{\partial A}\left(A - \overline{A}\right) + \frac{\partial I}{\partial T_{thick}}\left(T_{thick} - \overline{T_{thick}}\right) \tag{4}$$

Here W, L and T_{thick} generally follow Gaussian distribution [9], A is the product of two independent Gaussian distributions, V_{th} is correlated with W and L, as shown in (2) and (3).

Because the MTJ resistance $R \propto \frac{e^{T_{thick}}}{A}$ [9], we have:

$$\frac{\partial I}{\partial A}\Delta A + \frac{\partial I}{\partial T_{thick}}\Delta T_{thick} = \frac{\partial I}{\partial R}\left(\frac{\partial R}{\partial A}\Delta A + \frac{\partial R}{\partial T_{thick}}\Delta T_{thick}\right) = \frac{\partial I}{\partial R}\Delta R \quad (5)$$

Equation (5) indicates that the combined contribution of A and T_{thick} is the same as the impact of MTJ resistance. The difference between the actual I and its mathematical expectation μ_I can be calculated by:

$$I(W,L,V_{th},R) - E\big(I(\overline{W},\overline{L},\overline{V}_{th},\overline{R})\big) \approx \frac{\partial I}{\partial W}\Delta W + \frac{\partial I}{\partial L}\Delta L + \frac{\partial I}{\partial V_{th}}\Delta V_{th}$$
$$+ \frac{\partial I}{\partial R}\Delta R \quad (6)$$

Here we assume $\mu_I \approx E\big(I(\overline{W},\overline{L},\overline{V}_{th},\overline{R})\big) = I(\overline{W},\overline{L},\overline{V}_{th},\overline{R})$ and the mean of MTJ resistance $\overline{R} \approx R(\overline{A},\overline{\tau})$. Combining (2), (3), and (6), the standard deviation of I (σ_I) can be calculated as:

$$\sigma_I^2 = \left(\frac{\partial I}{\partial W}\right)^2 \sigma_W^2 + \left(\frac{\partial I}{\partial L}\right)^2 \sigma_L^2 + \left(\frac{\partial I}{\partial R}\right)^2 \sigma_R^2$$
$$+ \left(\frac{\partial I}{\partial V_{th}}\right)^2 \left(\frac{C_1}{WL} + \frac{C_2}{\exp(L/l)}\cdot\frac{W_c}{W}\cdot\sigma_L^2\right) + 2\frac{\partial I}{\partial L}\frac{\partial I}{\partial V_{th}}\rho_1\sqrt{\frac{C_1}{WL}}\sigma_L \quad (7)$$
$$+ 2\frac{\partial I}{\partial W}\frac{\partial I}{\partial V_{th}}\rho_2\sqrt{\frac{C_1}{WL}}\sigma_W + 2\frac{\partial I}{\partial L}\frac{\partial I}{\partial V_{th}}V_{ds}\exp\left(-\frac{L}{l}\right)\frac{\sigma_L^2}{l}$$

Here $\rho_1 = \frac{\text{cov}(V_{th0},L)}{\sqrt{\sigma_{V_{th0}}^2\sigma_L^2}}$ and $\rho_2 = \frac{\text{cov}(V_{th0},W)}{\sqrt{\sigma_{V_{th0}}^2\sigma_W^2}}$ are the correlation coefficients between V_{th0} and L or W, respectively [19]. $\sigma_{V_{th0}}^2 = \frac{C_1}{WL}$. Our further analysis shows that the last three terms at the right side of (7) are significantly smaller than other terms and can be safely ignored in the simulations of STT-RAM normal operations.

The accuracy of the coefficient in front of the variances of every parameter at the right side of (7) can be improved by applying window based smooth filtering. Take W as an example, we have:

$$\left(\frac{\partial I}{\partial W}\right)_i = \frac{I(\overline{W}+i\Delta W,L,V_{th},R) - I(\overline{W}-i\Delta W,L,V_{th},R)}{2i\Delta W} \quad (8)$$

where $i = 1, 2, \ldots K$. Different $\frac{\partial I}{\partial W}$ can be obtained at the different step i. K samples can be filtered out by a windows based smooth filter to balance the accuracy and the computation complexity as:

$$\overline{\frac{\partial I}{\partial W}} = \sum_{i=1}^{K} \omega_i\left(\frac{\partial I}{\partial W}\right)_i \quad (9)$$

Here ω_i is the weight of sample i, which is determined by the window type, i.e., Hamming window or Rectangular window [20].

3) *Variation contribution analysis:* The variations' contributions to I are mainly represented by the first four terms at the right side of (7) as:

$$S_1 = \left(\frac{\partial I}{\partial W}\right)^2 \sigma_W^2, S_2 = \left(\frac{\partial I}{\partial L}\right)^2 \sigma_L^2, S_3 = \left(\frac{\partial I}{\partial R}\right)^2 \sigma_R^2$$

$$S_4 = \left(\frac{\partial I}{\partial V_{th}}\right)^2 \left(\frac{C_1}{WL} + \frac{C_2}{\exp(L/l)} \cdot \frac{W_c}{W} \cdot \sigma_L^2\right)$$

$$(10)$$

As pointed out by many prior-arts [21–24], an asymmetry exists in STT-RAM write operations: the switching time of '0' $\rightarrow$ '1' is longer than that of '1' $\rightarrow$ '0' and suffers from a larger variance. Also, the switching time variance of '0' $\rightarrow$ '1' is more sensitive to the transistor size changes than '1' $\rightarrow$ '0'. As we shall show later, this phenomena can be well explained by using our sensitivity analysis. To the best of our knowledge, this is the first time the asymmetric variations of STT-RAM write performance and their dependencies on the transistor size are explained and quantitatively analyzed.

As shown in Fig. 1, when writing '0', the word-line (WL) and bit-line (BL) are connected to V_{dd} while the source-line (SL) is connected to ground. $V_{gs} = V_{dd}$ and $V_{ds} = V_{dd} - IR$. The NMOS transistor is mainly working in triode region. Based on short-channel BSIM model, the MTJ switching current supplied by a NMOS transistor can be calculated by:

$$I = \frac{\beta\left[(V_{dd} - V_{th})(V_{dd} - IR) - \frac{a}{2}(V_{dd} - IR)^2\right]}{1 + \frac{1}{v_{sat}L}(V_{dd} - IR)}$$

$$(11)$$

Here $\beta = \frac{\mu_0 C_{ox}}{1 + U_0(V_{dd} - V_{th})} \frac{W}{L}$. U_0 is the vertical field mobility reduction coefficient, μ_0 is the electron mobility, C_{ox} is gate oxide capacitance per unit area, a is body-effect coefficient and v_{sat} is carrier velocity saturation. Based on short-channel PTM model [25] and BSIM model [26, 27], we derive $\left(\frac{\partial I}{\partial W}\right)^2$, $\left(\frac{\partial I}{\partial L}\right)^2$, $\left(\frac{\partial I}{\partial R}\right)^2$ and $\left(\frac{\partial I}{\partial V_{th}}\right)^2$ as:

$$\left(\frac{\partial I}{\partial W}\right)_0^2 \approx \frac{1}{(A_1 W + B_1)^4}, \left(\frac{\partial I}{\partial L}\right)_0^2 \approx \frac{1}{\left(\frac{A_2}{W} + B_2 W + C\right)^4}$$

$$\left(\frac{\partial I}{\partial R}\right)_0^2 \approx \frac{1}{\left(\frac{A_3}{W} + B_3\right)^4}, \left(\frac{\partial I}{\partial V_{th}}\right)_0^2 \approx \frac{1}{\left(\frac{A_4}{\sqrt{W}} + B_4\sqrt{W}\right)^4}$$

$$(12)$$

Our analytical deduction shows that the coefficients A_{1-4}, B_{1-4} and C are solely determined by W, L, V_{th} and R. The detailed expressions of coefficients

A_{1-4}, B_{1-4} and C can be found in the appendix. Here R is the high resistance state of the MTJ, or R_H. For a NMOS transistor at '0' $\rightarrow$ '1' switching, the MTJ switching current is:

$$I = \frac{\beta}{2a}\left[(V_{dd} - V_{th} - IR) - \frac{I}{WC_{ox}v_{sat}^2}\right]^2 \tag{13}$$

Here R is the low resistance state of the MTJ, or R_L. We have:

$$\left(\frac{\partial I}{\partial W}\right)_1^2 \approx \frac{1}{(A_5W + B_5)^4}, \left(\frac{\partial I}{\partial L}\right)_1^2 \approx \frac{1}{\left(\frac{A_6}{W} + B_6\right)^2} \\ \left(\frac{\partial I}{\partial R}\right)_1^2 \approx \frac{1}{\left(\frac{A_7}{W} + B_7\right)^4}, \left(\frac{\partial I}{\partial V_{th}}\right)_1^2 \approx \frac{1}{\left(\frac{A_8}{W} + B_8\right)^2} \tag{14}$$

Again, A_{5-8} and B_{5-8} can be expressed as the function of W, L, V_{th} and R and the detailed expressions of those parameters can be found in the appendix in this chapter.

In general, a large S_i corresponds to a large contribution to I variation. When W is approaching infinity, only S_3 is nonzero at '1' $\rightarrow$ '0' switching while both S_2 and S_3 are nonzero at '0' $\rightarrow$ '1' switching. It indicates that the residual values of S_1–S_4 at '0' $\rightarrow$ '1' switching is larger than that at '1' $\rightarrow$ '0' switching when $W \rightarrow \infty$. In other words, '0' $\rightarrow$ '1' switching suffers from a larger MTJ switching current variation than '1' $\rightarrow$ '0' switching when NMOS transistor size is large.

4) **Simulation results of sensitivity analysis:** Sensitivity analysis [28] can be used to obtain the statistical parameters of MTJ switching current, i.e., the mean and the standard deviation, without running the costly SPICE and Monte-Carlo simulations. It can be also used to analyze the contributions of different variation sources to I variation in details. The normalized contributions (P_i) of variation resources, i.e., W, L, V_{th}, and R, are defined as:

$$P_i = \frac{S_i}{\sum_{i=1}^{4} S_i}, \; i = 1, 2, 3, 4 \tag{15}$$

Figures 3 and 4 show the normalized contributions of every variation source at '0' $\rightarrow$ '1' and '1' $\rightarrow$ '0' switching's, respectively, at different transistor sizes. We can see that L and V_{th} are the first two major contributors to I variation at both switching directions when W is small. At '1' $\rightarrow$ '0' switching, the contribution of L raises until reaching its maximum value when W increases, and then quickly decreases when W further increases. At '0' $\rightarrow$ '1' switching, however, the contribution of L monotonically decreases, but keeps being the dominant factor over the simulated W range. At both switching directions, the contributions of R ramps up when W increases. At '1' $\rightarrow$ '0' switching, the normalized contribution of R becomes almost 100 % when W is really large.

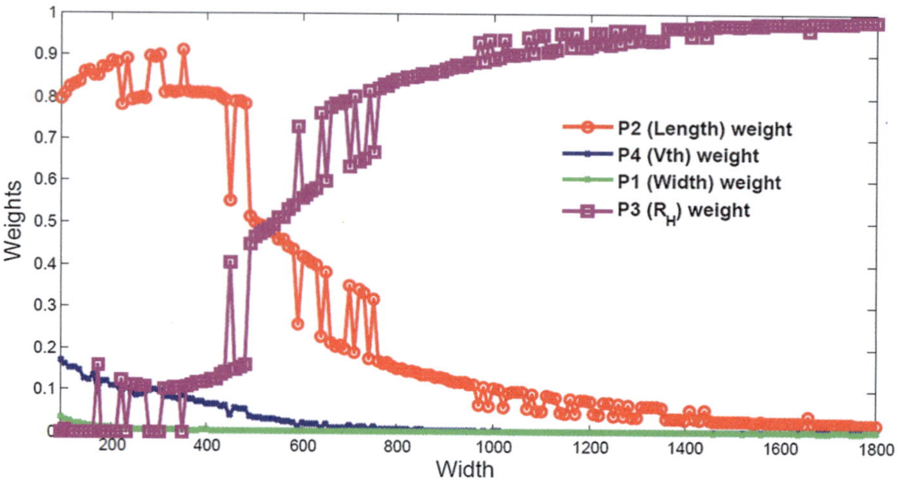

Fig. 3 The normalized contributions under different W at '1' → '0' switching

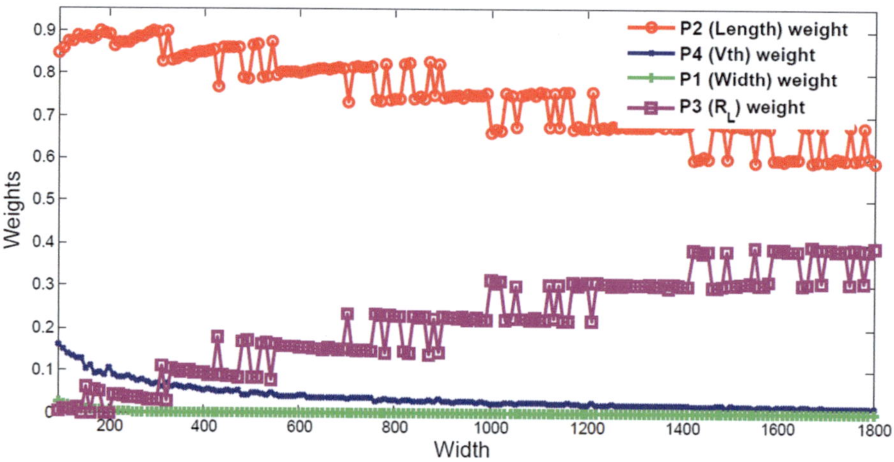

Fig. 4 The normalized contributions under different W at '0' → '1' switching

3.2 *Write Current Distribution Recovery*

After the I distribution is characterized by the sensitivity analysis, the next question becomes how to recover the distribution of I from the characterized information in the statistical analysis of STT-RAM reliability. We investigated the typical distributions of I in various STT-RAM cell designs and found that dual-exponential function can provide the excellent accuracy in modeling and recovering these distributions. The dual-exponential function we used to recover the I distributions can be illustrated as:

$$f(I) = \begin{cases} a_1 e^{b_1(I-\mu)} & I \leq \mu \\ a_2 e^{b_2(\mu-I)} & I > \mu \end{cases} \tag{16}$$

Here a_1, b_1, a_2, b_2 and μ are the fitting parameters, which can be calculated by matching the first and the second order momentums of the actual I distribution and the dual-exponential function as:

$$\int f(I)dI = 1,$$
$$\int If(I)dI = E(I), \tag{17}$$
$$\int I^2 f(I)dI = E(I)^2 + \sigma_I^2$$

Here $E(I)$ and σ_I^2 are obtained from the sensitivity analysis.

The recovered I distribution can be used to generate the MTJ switching current samples, as shown in Fig. 5. At the beginning of the sample generation flow, the confidence interval for STT-RAM design is determined, e.g., $[\mu_I - 6\sigma_I, \mu_I + 6\sigma_I]$ for a six-sigma confidence interval. Assuming we need to generate N samples within the confidence interval, say, at the point of $I = I_i$, a switching current sequence of $[N\,\mathrm{Pr}_i]$ samples must be generated. Here $\mathrm{Pr}_i \approx f(I_i)\Delta$, Δ equals $\frac{12\sigma_I}{N}$, or the step of sampling generation. $f(I_i)$ is the dual-exponential function. Note that N determines both the analysis granularity and the level of the estimated error rate.

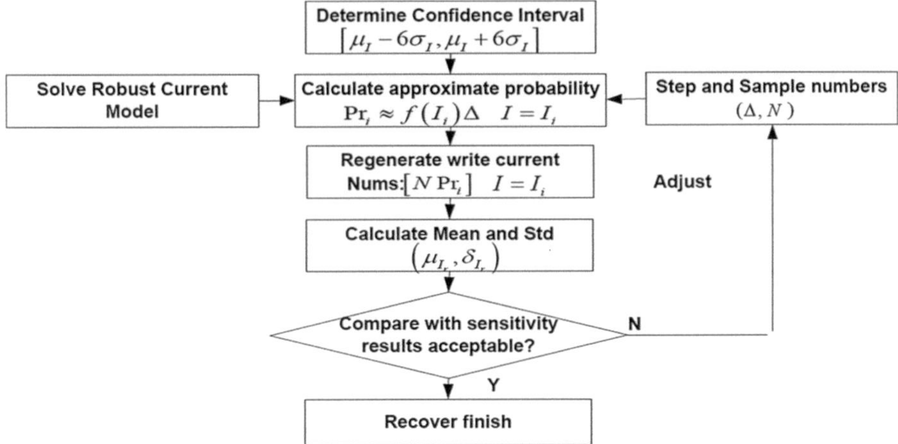

Fig. 5 Basic flow for MTJ switching current recovery

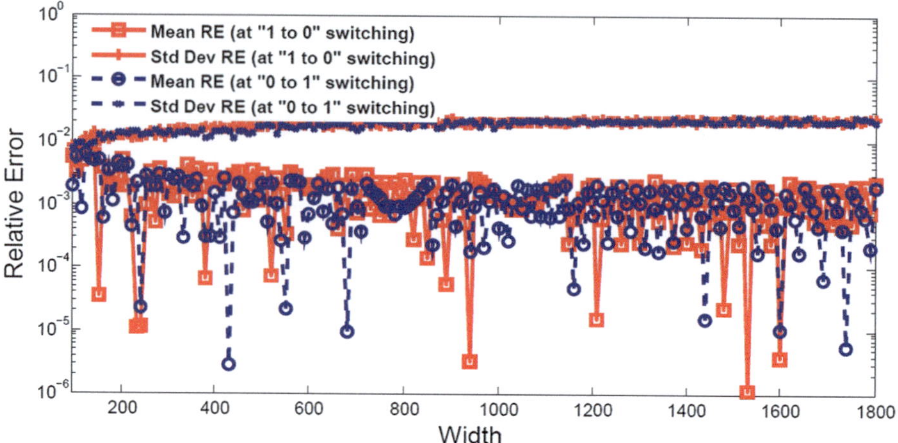

Fig. 6 Relative errors of the recovered I w.r.t. the results from sensitivity analysis

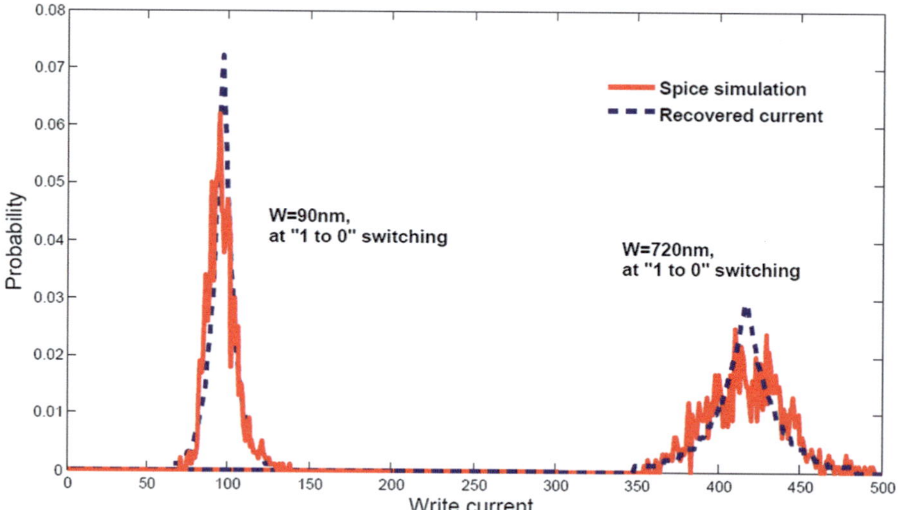

Fig. 7 Recovered I vs. Monte-Carlo result at '1' → '0'

Figure 6 shows the relative errors of the mean and the standard deviation of the recovered I distribution w.r.t. the results directly from the sensitivity analysis (see (6) and (7)). The maximum relative error $<10^{-2}$, which proves the accuracy of our dual-exponential model.

Figures 7 and 8 compare the probability distribution functions (PDF's) of I from the SPICE Monte-Carlo simulations and from the recovery process based on our sensitivity analysis at two switching directions. Our method achieves good accuracy at both representative transistor channel widths ($W = 720$ nm or $W = 720$ nm).

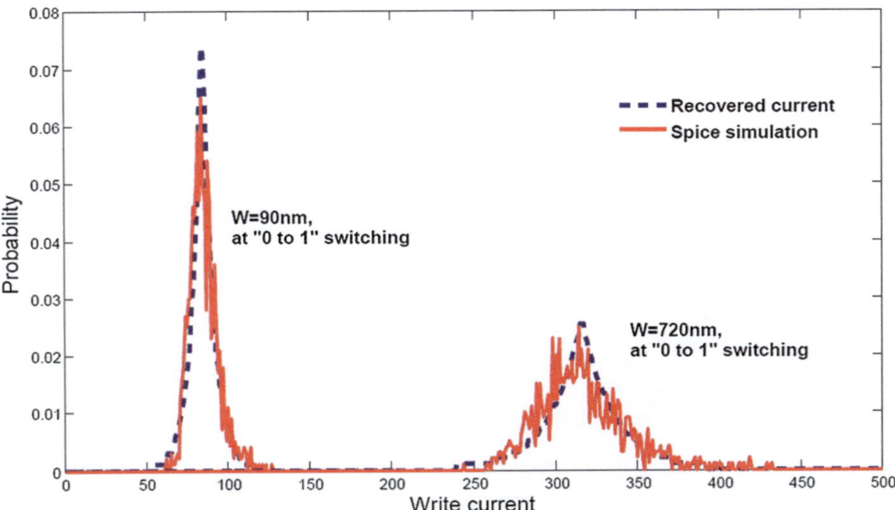

Fig. 8 Recovered I vs. Monte-Carlo result at '0' → '1'

3.3 Statistical Thermal Analysis

The variation of the MTJ switching time (τ_{th}) incurred by the thermal fluctuations follows Gaussian distribution when τ_{th} is below $10 \sim 20$ ns [21]. In this range, the distribution of τ_{th} can be easily constructed after the I is determined. The distribution of MTJ switching performance can be obtained by combining the τ_{th} distributions of all I samples.

4   Application 1: Write Reliability Analysis

In this section, we conduct the statistical analysis on the write reliability of STT-RAM cells by leveraging our PS3-RAM method. Both device variations and thermal fluctuations are considered in the analysis. We also extend our method into array-level evaluation and demonstrate its effectiveness in STT-RAM design optimizations.

4.1 Reliability Analysis of STT-RAM Cells

The write failure rate P_{WF} of a STT-RAM cell can be defined as the probability that the actual MTJ switching time τ_{th} is longer than the write pulse width T_w, or $P_{WF} = P(\tau_{th} > T_w)$. τ_{th} is affected by the MTJ switching current magnitude, the MTJ and MOS device variations, the MTJ switching direction, and the thermal fluctuations. The conventional simulation of P_{WF} requires costly Monte-Carlo runs with hybrid SPICE and macro-magnetic modeling steps. Instead, we can use PS3-RAM to analyze the statistical STT-RAM write performance. The corresponding simulation environment is also summarized in Table 1.

Figures 9 and 10 depict the P_{WF}'s simulated by PS3-RAM for both switching directions at 300 K. For comparison purpose, the Monte-Carlo simulation results

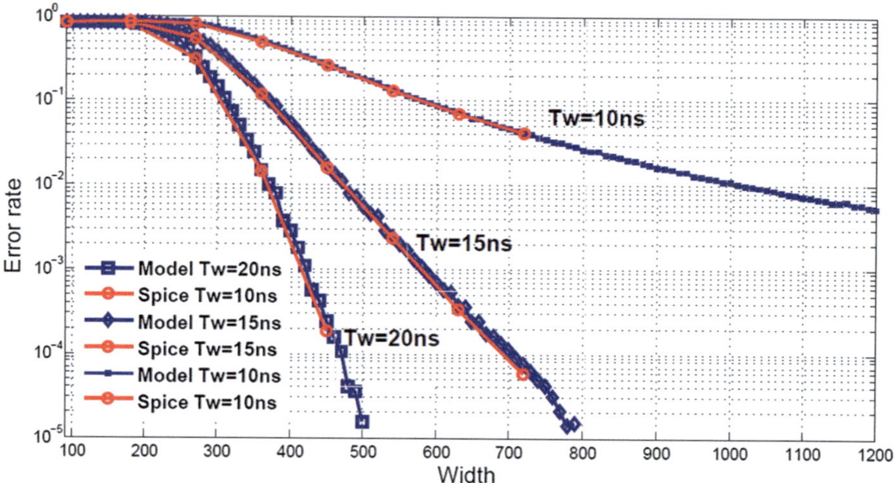

Fig. 9 Write failure rate at '0' → '1' when T = 300 K

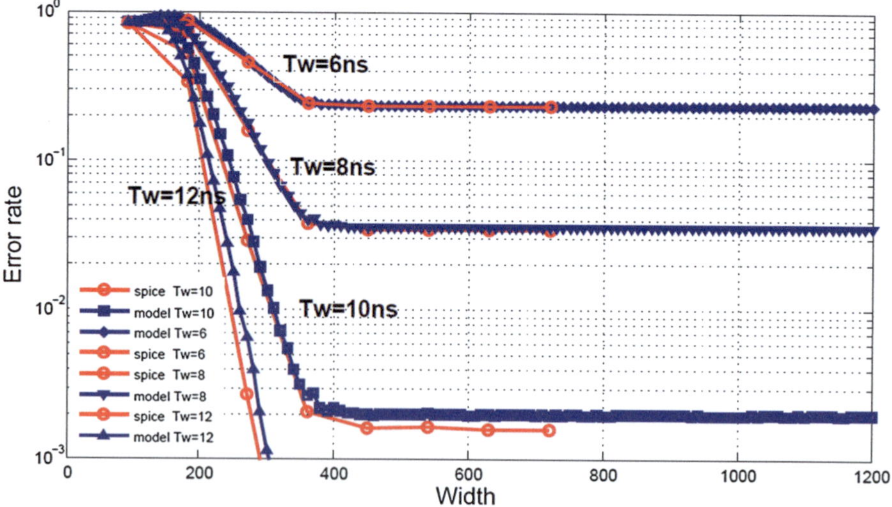

Fig. 10 Write failure rate at '1' → '0' when T = 300 K

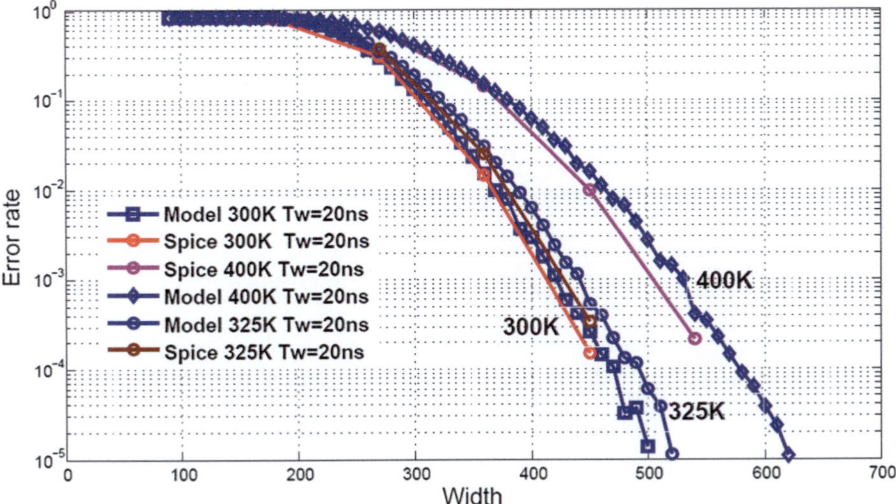

Fig. 11 P_{WF} under different temperatures at '0' → '1'

are also presented. Different T_w's are selected at either switching directions due to the asymmetric MTJ switching performances [21], i.e., $T_w = 10, 15, 20$ ns at '0' → '1' and $T_w = 6, 8, 10, 12$ ns at '1' → '0'. Our PS3-RAM results are in excellent agreement with the ones from Monte-Carlo simulations.

Since '0' → '1' is the limiting switching direction for STT-RAM reliability, we also compare the P_{WF}'s of different STTRAM cell designs under different temperatures at this switching direction in Fig. 11. The results show that PS3-RAM can provide very close but pessimistic results compared to those of the conventional simulations. PS3-RAM is also capable to precisely capture the small error rate change incurred by a moderate temperature shift (from T = 300 to 325 K).

It is known that prolonging the write pulse width and increasing the MTJ switching current (by sizing up the NMOS transistor) can reduce the P_{WF}. In Fig. 12, we demonstrate an example of using PS3-RAM to explore the STT-RAM design space: the tradeoff curves between P_{WF} and T_W are simulated at different W's. For a given P_{WF}, for example, the corresponding tradeoff between W and T_W can be easily identified on Fig. 12.

4.2 Array Level Analysis and Design Optimization

We use a 45 nm 256 Mb STT-RAM design [29] as the example to demonstrate how to extend our PS3-RAM into array-level analysis and design optimizations. The number of bits per memory block $N_{bit} = 256$ and the number of memory blocks $N_{word} = 1M$. To repair the operation errors of memory cells, circuit-level technique-ECC (error correction code) is usually applied [30]. Two types of ECC's with

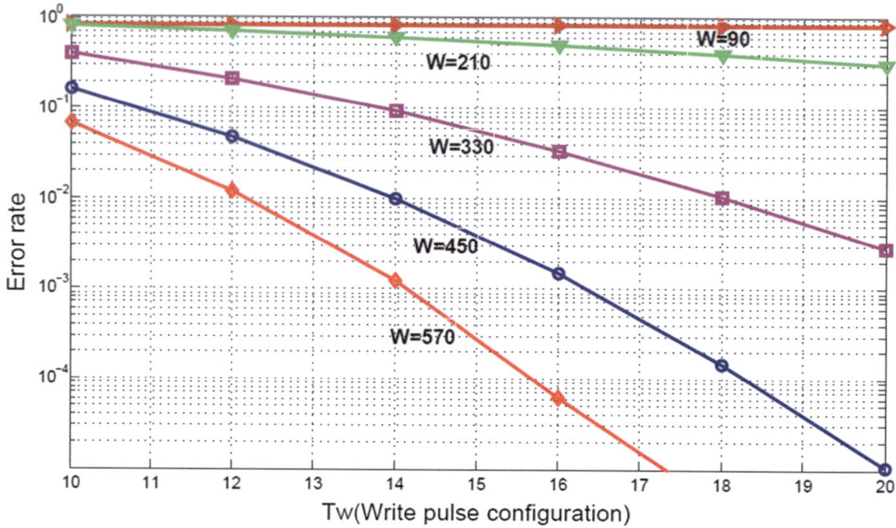

Fig. 12 STT-RAM design space exploration at '0' → '1'

different implementation costs are being considered, i.e., single-bit-correcting Hamming code and a set of multi-bits-correcting BCH codes. We use (n, k, t) to denote an ECC with n codeword length, k bit user bits being protected (256 bit here) and t bits being corrected. The ECC's corresponding to the error correction capability t from 1 to 5 are Hamming code (265; 256; 1) and four BCH codes–BCH1 (274; 256; 2), BCH2 (283; 256; 3), BCH3 (292; 256; 4) and BCH4 (301; 256; 5), respectively. The write yield of the memory array Y_{wr} can be defined as:

$$Y_{wr} = P(n_e \leq t) = \sum_{i=0}^{t} C_n^i P_{WF}^i (1 - P_{WF})^{n-i} \tag{18}$$

Here, n_e denotes the total number of error bits in a write access. Y_{wr} indeed denotes the probability that the number of error bits in a write access is smaller than the error correction capability.

Figure 13 depicts the Y_{wr}'s under different combinations of ECC scheme and W when $T_W = 15$ ns at '0' → '1' switching. The ECC schemes required to satisfy $\sim 100\% \, Y_{wr}$ for different W are: (1) Hamming code for $W = 630$ nm; (2) BCH2 for $W = 540$ nm; and (3) BCH4 for $W = 480$ nm. The total memory array area can be estimated by using the STT-RAM cell size equation $\text{Area}_{cell} = 3(W/L + 1)(F^2)$ [31]. Calculation shows that combination (3) offers us the smallest STT-RAM array area, which is only 88 % and 95 % of the ones of (1) and (2), respectively. We note that PS3-RAM can be seamlessly embedded into the existing deterministic memory macro models [31] for the extended capability on the statistical reliability analysis and the multi-dimensional design optimizations on area, yield, performance and energy.

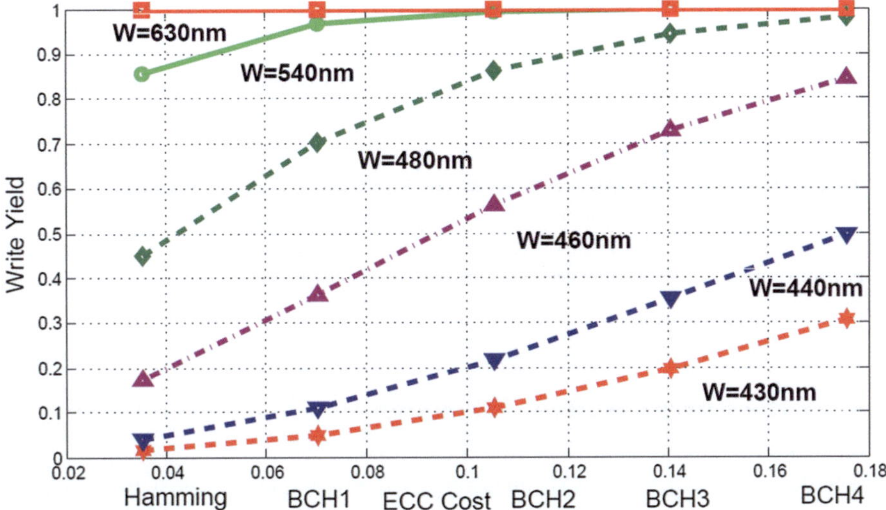

Fig. 13 Write yield with ECC's at '0' → '1', Tw = 15 ns

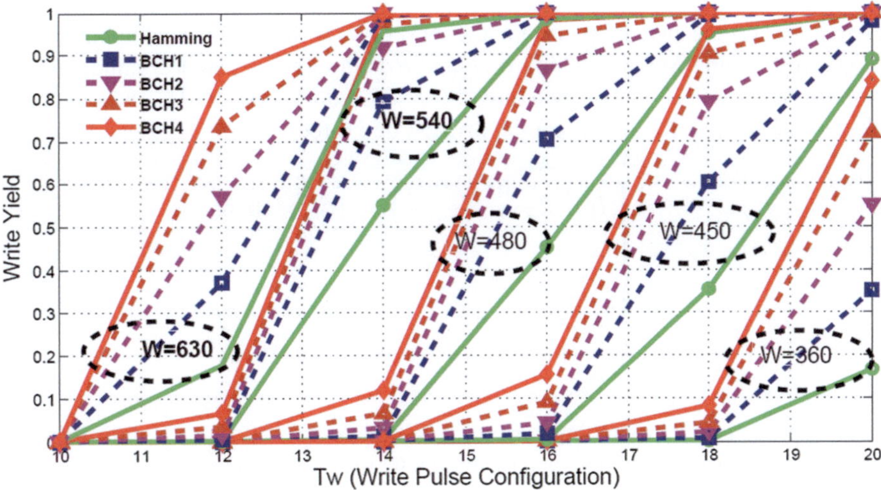

Fig. 14 Design space exploration at '0' → '1'

Figure 14 illustrates the STT-RAM design space in terms of the combinations of Y_{wr}, W, T_w and ECC scheme. After the pair of (Y_{wr}, T_w) is determined, the tradeoff between W and ECC can be found in the corresponding region on the figure. The result shows that PS3-RAM provides a fast and efficient method to perform the device/circuit/architecture co-optimization for STT-RAM designs.

5   Application 2: Write Energy Analysis

In addition to write reliability analysis, our PS3-RAM method can also precisely capture the write energy distributions influenced by the variations of device and working environment. In this section, we first prove that there is a sweet point of write pulse width for the minimum write energy without considering any variations. Then we introduce the concept of statistical write energy of STT-RAM cells considering both process variations and thermal fluctuations, and perform the statistical analysis on write energy using our PS3-RAM method.

5.1   Write Energy Without Variations

The write energy of a STT-RAM cell during each programming cycle without considering process and thermal variations is deterministic and can be modeled by (19) as:

$$E_{av} = I^2 R \tau_{th} \tag{19}$$

Here I denotes the switching current at either '0' $\rightarrow$ '1' or '1' $\rightarrow$ '0' switching, τ_{th} is the corresponding MTJ switching time and R is the MTJ resistance value, i.e., R_L (R_h) for '0' $\rightarrow$ '1'('1' $\rightarrow$ '0') switching. As discussed in prior art [21], the switching process of an STT-RAM cell can be divided into three working regions:

$$I = \begin{cases} I_{C_0}\left(1 - \dfrac{\ln(\tau_{th}/\tau_0)}{\Delta}\right), & \tau_{th} > 10 \text{ ns} \\[2ex] I_{C_0} + C\,\ln\left(\dfrac{\pi}{2\theta}\right)/\tau_{th}, & \tau_{th} < 3 \text{ ns} \\[2ex] \dfrac{P}{\tau_{th}} + Q, & 3 \le \tau_{th} \le 10 \text{ ns} \end{cases} \tag{20}$$

Here I_{C_0} is the critical switching current, Δ is thermal stability, $\tau_0 = 1$ ns is the relax time, θ is the initial angle between the magnetization vector and the easy axis, C, P, Q are fitting parameters.

For a relatively long switching time range ($\tau_{th} \approx 10 \sim 300$ ns), the undistorted write energy P_{av} can be calculated as:

$$E_{av} = I_{C_0}^2\left(1 - \frac{\ln(\tau_{th})}{\Delta}\right)^2 R\tau_{th} = \frac{I_{C_0}^2}{\Delta^2}\left(\Delta - \ln(\tau_{th})\right)^2\tau_{th} \tag{21}$$

In the long switching time range, we have $\ln(\tau_{th}) < 0$. Thus, $(\Delta - \ln(\tau_{th}))^2$ or E_{av} monotonically raises as the write pulse τ_{th} increases and the minimized write energy E_{av} occurs at $\tau_{th} = 10$ ns.

In the ultra-short switching time range ($\tau_{th} < 3$ ns), E_{av} can be obtained as:

$$
\begin{aligned}
E_{av} &= \left(I_{C_0} + C \ln\left(\frac{\pi}{2\theta}\right) \Big/ \tau_{th} \right)^2 R \tau_{th} \\
&= 2 I_{C_0} RC \ln\left(\frac{\pi}{2\theta}\right) + I_{C_0}^2 R \tau_{th} + \frac{C^2 \ln^2(\pi/2\theta) R}{\tau_{th}} \\
&\geq 2 I_{C_0} RC \ln\left(\frac{\pi}{2\theta}\right) + 2\sqrt{I_{C_0}^2 R^2 C^2 \ln^2(\pi/2\theta)} \\
&\geq 4 I_{C_0} RC \ln\left(\frac{\pi}{2\theta}\right)
\end{aligned}
\tag{22}
$$

As (22) shows, the minimum of E_{av} can be achieved when $\tau_{th} = C \ln\left(\frac{\pi}{2\theta}\right) / I_{C_0}$. However, for the ultra-short switching time range (usually $C \ln\left(\frac{\pi}{2\theta}\right) / I_{C_0} > 3\,\text{ns}$), E_{av} monotonically decreases as τ_{th} increases.

Similarly, in the middle switching time range ($3 \leq \tau_{th} \leq 10$ ns), E_{av} can be expressed as:

$$
E_{av} = \left(\frac{P}{\tau_{th}} + Q \right)^2 R \tau_{th} = \left(\frac{P}{\sqrt{\tau_{th}}} + Q\sqrt{\tau_{th}} \right)^2 R \geq 4PQR
\tag{23}
$$

Again, the minimized E_{av} occurs at $\tau_{th} = \frac{P}{Q}$. Here $\frac{P}{Q} \geq 10$ ns based on our device parameters characterization [21]. Thus, the write energy E_{av} in this range monotonically decreases as τ_{th} grows.

According to the monotonicity of E_{av} in the three regions, the most energy-efficient switching point of E_{av} should be at $\tau_{th} = 10$ ns. To validate above theoretical deduction for the sweet point of E_{av}, the SPICE simulations are also conducted. Here the STT-RAM device model without considering process and thermal variations is also adopted from [21].

Figure 15 shows the simulated write energy E_{av} over different write pulse at '0' → '1' switching. As Fig. 15 shows, E_{av} monotonically decreases in the ultra-short switching range and continues decreasing in the middle range, but becomes monotonically increasing after entering the long switching time range. The sweet point of E_{av} occurs around $\tau_{th} = 10$ ns, which validates our theoretical analysis for the write energy without considering any variations.

We also present the simulated $E_{av} - \tau_{th}$ curve under different temperatures in Fig. 16. The trend and sweet point of $E_{av} - \tau_{th}$ curves remain almost the same when the temperature increases from T $= 300$ to 400 K. In fact, the write energy E_{av} decreases a little bit as the temperature increases. The reason is that the driving ability loss of the NMOS transistor (I) dominates E_{av} though the MTJ switching time (τ_{th}) increases when the working temperature raises.

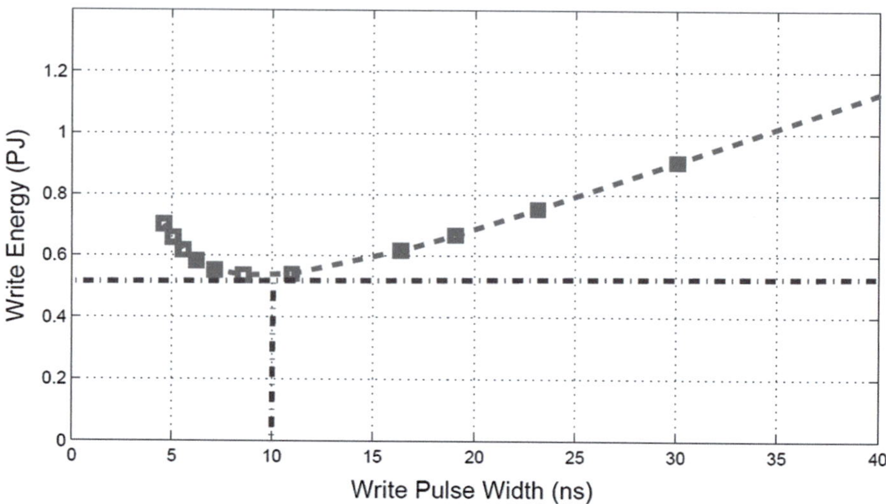

Fig. 15 Average Write Energy under different write pulse width when T = 300 K

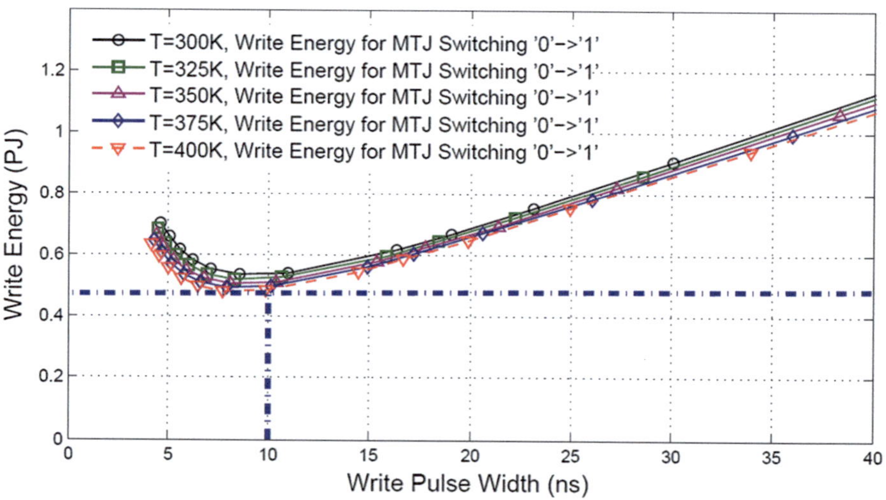

Fig. 16 Average Write Energy vs. write pulse width under different temperature

5.2 PS3-RAM for Statistical Write Energy

As discussed in previous section, the write energy of a STT-RAM cell can be deterministically optimized when all the variations are ignored. However, since the switching current I, the resistance R, and the switching time τ_{th} in (19) may be distorted by CMOS/MTJ process variations and thermal fluctuations, the deterministic value will no longer be able to represent the statistic nature of the write energy of a STT-RAM cell. Accordingly, the optimized write energy at sweet point ($\tau_{th} = 10$ ns) shown in Fig. 15 should be expanded as a distribution.

Similar to the write failure analysis, we conduct the statistical write energy analysis using our PS3-RAM method. We choose the mean of NMOS transistor width $W = 540$ nm. The remained device parameters and variation configurations keep the same as Table 1.

Figures 17 and 18 show the simulated statistical write energy by PS3-RAM for both switching directions at 300 K. For comparison, the SPICE simulation results

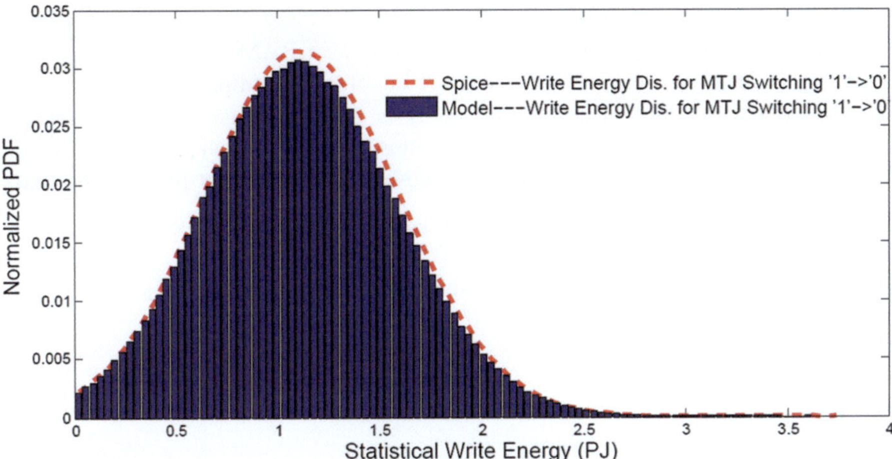

Fig. 17 Statistical Write Energy vs. write pulse width at '1' → '0'

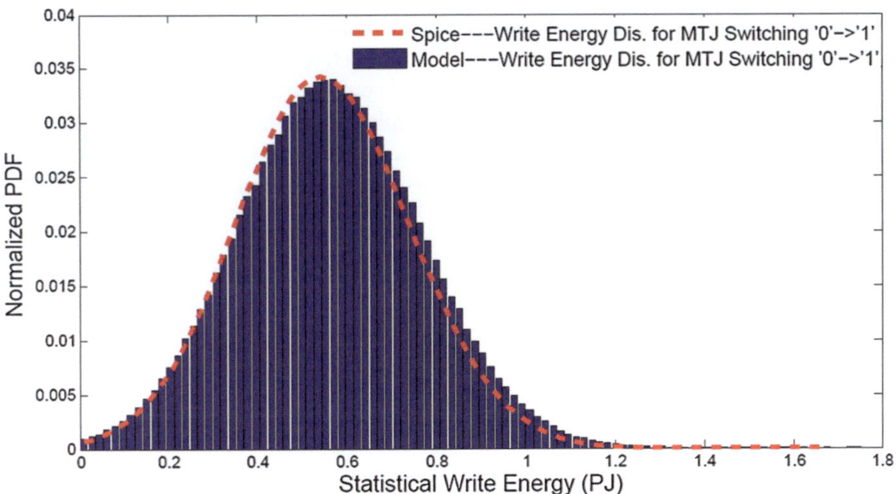

Fig. 18 Statistical Write Energy vs. write pulse width at '0' → '1'

are also presented. As shown in the figures, the distribution of write energy captured by our PS3-RAM method are in excellent agreement with the results from SPICE simulations at both '1' $\rightarrow$ '0' and '0' $\rightarrow$ '1' switching's.

6 Computation Complexity Evaluation

We compared the computation complexity of our proposed PS3-RAM method with the conventional simulation method. Suppose the number of variation sources is M, for a statistical analysis of a STT-RAM cell design, the numbers of SPICE simulations required by conventional flow and PS3-RAM are $N_{std} = N_s^M$ and $N_{\text{PS3-RAM}} = 2KM + 1$, respectively. Here K denotes the sample numbers for window based smooth filter in sensitivity analysis, N_s is average sample number of every variation in the Monte-Carlo simulations in conventional method, $K \ll N_s$. Note that our switching current sample recovery flow does not require any extra Monte-Carlo simulations. The speedup $X_{\text{speedup}} \approx \frac{N_s^M}{2KM}$ can be up to multiple orders of magnitude: for example, if we set $N_s = 100$, $M = 4$, (note: V_{th} is not an independent variable) and $K = 50$, the speed up is around 2.5×10^5.

7 Conclusion

A fast and scalable statistical STT-RAM reliability/energy analysis method called PS3-RAM was developed in this chapter. PS3-RAM can simulate the impact of process variations and thermal fluctuations on the statistical STT-RAM write performance or write energy distributions, without running costly Monte-Carlo simulations on SPICE and macro-magnetic models. Simulation results show that PS3-RAM can achieve very high accuracy compared to the conventional simulation method, while achieving a speedup of multiple orders of magnitude. The great potentials of PS3-RAM in the application of the device/circuit/architecture co-optimization of STT-RAM designs are also demonstrated.

Appendix

In this appendix, the details on the model deduction in sensitivity analysis and the summary of the analytic results involved in the PS3-RAM development are given. Meanwhile, the validation of the analytic results based on Monte-Carlo simulations is also presented. Table 2 [26] summarizes some additional parameters used in this Appendix.

Sensitivity Analysis Model Deduction

The sensitivity analysis model is developed based on the electrical MTJ model and the simplified BSIM model [26, 27]. At '1' $\rightarrow$ '0' switching, the MTJ switching current supplied by an NMOS transistor working in the triode region is:

$$I = \frac{\beta\left[(V_{dd} - V_{th})(V_{dd} - IR) - \frac{a}{2}(V_{dd} - IR)^2\right]}{1 + \frac{1}{v_{sat}L}(V_{dd} - IR)} \tag{24}$$

Here $\beta = \frac{\mu_0 C_{ox}}{1 + U_0(V_{dd} - V_{th})}\frac{W}{L}$. U_0 is the vertical field mobility reduction coefficient, μ_0 is the electron mobility, C_{ox} is gate oxide capacitance per unit area, a is body-effect coefficient and v_{sat} is carrier velocity saturation. The MTJ is in its high resistance state, or $R = R_H$. Based on PTM [25] and BSIM [26], the partial derivatives in (6) can be calculated by ignoring the minor terms in the expansion of (24) as:

$$\left(\frac{\partial I}{\partial W}\right)_0^2 \approx \frac{1}{(A_1 W + B_1)^4}, \quad \left(\frac{\partial I}{\partial L}\right)_0^2 \approx \frac{1}{\left(\frac{A_2}{W} + B_2 W + C\right)^4}$$

$$\left(\frac{\partial I}{\partial R}\right)_0^2 \approx \frac{1}{\left(\frac{A_3}{W} + B_3\right)^4}, \quad \left(\frac{\partial I}{\partial V_{th}}\right)_0^2 \approx \frac{1}{\left(\frac{A_4}{\sqrt{W}} + B_4\sqrt{W}\right)^4} \tag{25}$$

Table 2 Parameter definition

Variable	Definition
U_0	Vertical field mobility reduction coefficient
μ_0	Electron mobility
C_{ox}	Gate oxide capacitance per unit area
a	Body-effect coefficient
v_{sat}	Carrier velocity saturation

Here,

$$A_1 = \sqrt{\frac{\mu_0 C_{ox} V_{dd}(V_{dd} - V_{th})}{L}}R,$$

$$B_1 = \sqrt{\frac{L}{\mu_0 C_{ox} V_{dd}(V_{dd} - V_{th})}},$$

$$A_2 = \frac{L^2}{\mu_0 C_{ox} V_{dd}(V_{dd} - V_{th})}$$

$$B_2 = R^2 \mu_0 C_{ox} \frac{V_{dd} - V_{th}}{V_{dd}},$$

$$A_3 = \frac{L}{\mu_0 C_{ox} \sqrt{V_{dd}}(V_{dd} - V_{th})},$$

$$B_3 = \frac{R}{\sqrt{V_{dd}}}, C = \frac{2LR}{V_{dd}},$$

$$A_4 = \sqrt{\frac{L}{\mu_0 C_{ox} V_{dd}}},$$

$$B_4 = \sqrt{\frac{\mu_0 C_{ox}}{L V_{dd}}}R(V_{dd} - V_{th})$$

At '0' $\rightarrow$ '1' switching, the NMOS transistor is working in the saturation region. The current through the MTJ is:

$$I = \frac{\beta}{2a}\left[(V_{dd} - V_{th} - IR) - \frac{I}{W C_{ox} v_{sat}^2}\right]^2 \tag{26}$$

The MTJ is in its low resistance state, or $R = R_L$. The derivatives can be also calculated as:

$$\left(\frac{\partial I}{\partial W}\right)_1^2 \approx \frac{1}{(A_5 W + B_5)^4}, \left(\frac{\partial I}{\partial L}\right)_1^2 \approx \frac{1}{\left(\frac{A_6}{W} + B_6\right)^2}$$

$$\left(\frac{\partial I}{\partial R}\right)_1^2 \approx \frac{1}{\left(\frac{A_7}{W} + B_7\right)^4}, \left(\frac{\partial I}{\partial V_{th}}\right)_1^2 \approx \frac{1}{\left(\frac{A_8}{W} + B_8\right)^2} \tag{27}$$

by ignoring the minor terms in the expansion of (24). Here,

$$A_5 = \sqrt{\frac{2\mu_0 C_{ox} v_{sat}}{La + \mu_0(V_{dd} - V_{th})}}\, R,$$

$$B_5 = \frac{\mu_0}{2C_{ox} v_{sat}[La + \mu_0(V_{dd} - V_{th})]},$$

$$A_6 = \frac{\mu_0}{2aC_{ox} v_{sat}^2},$$

$$B_6 = \frac{R\mu_0}{a v_{sat}},$$

$$A_7 = \frac{1}{2C_{ox} v_{sat}} \sqrt{\frac{\mu_0}{La v_{sat} + \mu_0(V_{dd} - V_{th})}},$$

$$B_7 = \sqrt{\frac{\mu_0}{La v_{sat} + \mu_0(V_{dd} - V_{th})}}\, R,$$

$$A_8 = \frac{1}{2C_{ox} v_{sat}},\ B_8 = R$$

The contributions of different variation sources to I are represented by:

$$S_1 = \left(\frac{\partial I}{\partial W}\right)^2 \sigma_W^2,\ S_2 = \left(\frac{\partial I}{\partial L}\right)^2 \sigma_L^2,\ S_3 = \left(\frac{\partial I}{\partial R}\right)^2 \sigma_R^2$$
$$S_4 = \left(\frac{\partial I}{\partial V_{th}}\right)^2 \left(\frac{C_1}{WL} + \frac{C_2}{\exp(L/l')} \cdot \frac{W_c}{W} \cdot \sigma_L^2\right) \tag{28}$$

Here S_1, S_2, S_3 and S_4 denote the variations induced by W, L, R (R_H or R_L) and V_{th}, respectively.

Analytic Results Summary

Table 3 shows the monotonicity and the upper or lower bounds of the variation contributions $S_1 - S_4$ as the transistor channel width W increases. Here, "↑", "↓" and "↗↘" denote monotonic increasing, monotonic decreasing and changing as a convex function. $K_1 = \frac{C_1}{L} + \frac{C_2 W_c \sigma_L^2}{\exp(L/l')}$. Table 3 also gives the maximum and minimum values of $S_1 - S_4$ and their corresponding W's.

Validation of Analytic Results

As (27) shows, $\left(\frac{\partial I}{\partial W}\right)^2$, $\left(\frac{\partial I}{\partial L}\right)^2$, and $\left(\frac{\partial I}{\partial R}\right)^2$ solely determine the trends of S_1, S_2, S_3, respectively, when W increases at both switching directions. The corresponding

Table 3 Summary of variation contribution

	Variation	Monoto	Bounds	$W \to \infty$
'0'	S_1	$\uparrow$	$\min S_1 = 0$ $W = \infty$	$S_1 \to 0$
	S_2	$\nearrow\searrow$	$\max S_2 = \left(\dfrac{V_{dd}}{4LR_H}\sigma_L\right)^2$ $W = \dfrac{L}{\mu_0 C_{ox}(V_{dd} - V_{th})R_H}$	$S_2 \to 0$
	S_3	$\uparrow$	$\max S_3 = \left(\dfrac{V_{dd}}{R_H^2}\sigma_{R_H}\right)^2$ $W = \infty$	$\max S_3$
	S_4	$\nearrow\searrow$	$\max S_4 = \dfrac{K_1\mu_0 C_{ox}V_{dd}^2}{16LR_H(V_{dd} - V_{th})}$ $W = \dfrac{L}{\mu_0 C_{ox}R_H(V_{dd} - V_{th})}$	$S_4 \to 0$
'1'	S_1	$\downarrow$	$\min S_1 = 0$ $W = \infty$	$S_1 \to 0$
	S_2	$\uparrow$	$\max S_2 = \left(\dfrac{av_{sat}}{R_L\mu_0}\sigma_L\right)^2$ $W = \infty$	$\max S_2$
	S_3	$\uparrow$	$\max S_3 \approx \left(\dfrac{V_{dd} - V_{th}}{R_L^2}\sigma_{R_L}\right)^2$ $W = \infty$	$\max S_3$
	S_4	$\nearrow\searrow$	$\max S_4 = \dfrac{C_{ox}v_{sat}}{2R_L}K_1$ $W = \dfrac{1}{2C_{ox}v_{sat}R_L}$	$S_4 \to 0$

Monte-Carlo simulation results of S_1, S_2, S_3 are shown in Figs. 19, 20, and 21, respectively.

Figure 19 shows S_1 monotonically decreases to zero as W increases to infinity at both switching directions. Its value at '1' $\to$ '0' switching is always greater than that at '0' $\to$ '1' switching because $A_1 < A_5$.

Figure 20 shows that the variation contribution of L at '0' $\to$ '1' switching is always larger than that at '1' $\to$ '0' switching. The gap between them reaches the maximum when $W \to \infty$.

Figure 21 shows that the contribution from MTJ resistance R becomes dominant in the MTJ switching current distribution when W is approaching infinity. Because $\left(\dfrac{V_{dd} - V_{th}}{R_L^2}\sigma_{R_L}\right)^2 < \left(\dfrac{V_{dd}}{R_H^2}\sigma_{R_H}\right)^2$, the normalized contribution of R is always larger at '1' $\to$ '0' switching than that at '0' $\to$ '1' switching. We note that the additional coefficient $\left(\dfrac{C_1}{WL} + \dfrac{C_2}{\exp(L/l)} \cdot \dfrac{W_c}{W} \cdot \sigma_L^2\right)$ at the right side of (28) after the $\left(\dfrac{\partial I}{\partial V_{th}}\right)^2$ results in the different features of $\left(\dfrac{\partial I}{\partial V_{th}}\right)^2$ from S_4 in our simulations.

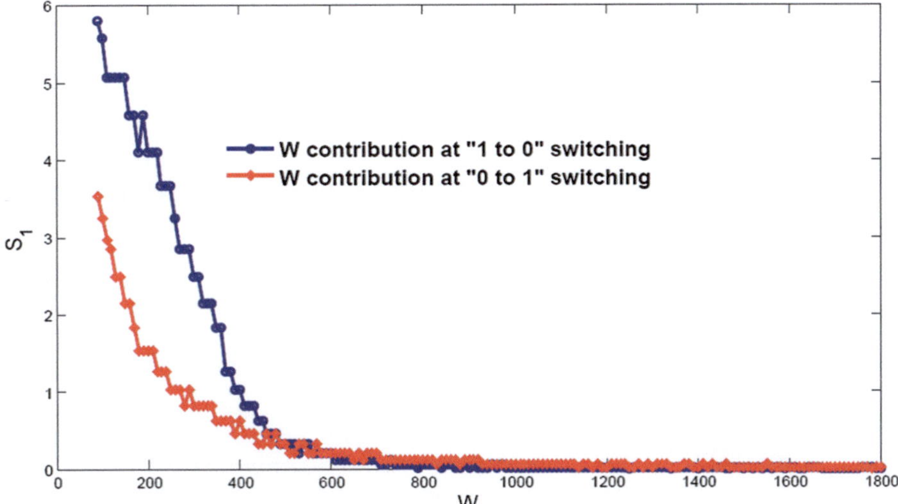

Fig. 19 Contributions from W

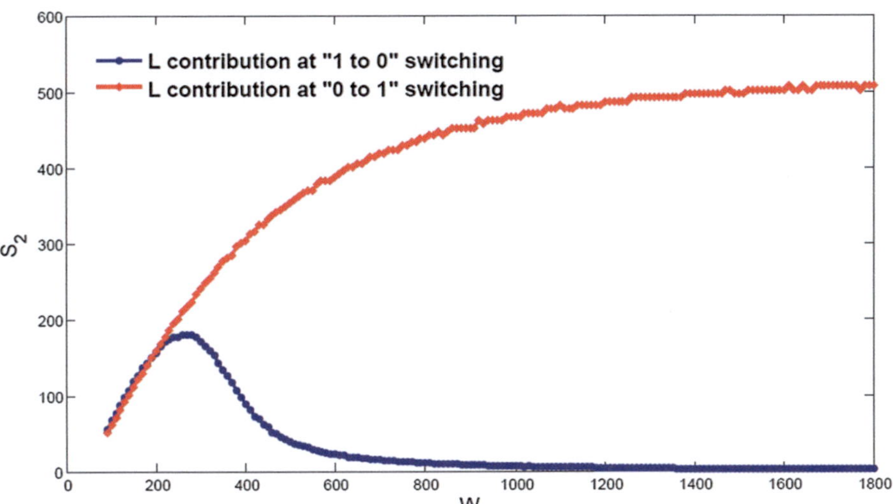

Fig. 20 Contributions from L

Figure 22 shows the values of $\left(\frac{\partial I}{\partial V_{th}}\right)^2$ at both switching directions. At '0' $\rightarrow$ '1' switching, $\left(\frac{\partial I}{\partial V_{th}}\right)^2$ increases monotonically when W grows. At '1' $\rightarrow$ '0' switching, $\left(\frac{\partial I}{\partial V_{th}}\right)^2$ increases first, then quickly decays to zero after reaching its maximum. These trends follow the expressions of $\left(\frac{\partial I}{\partial V_{th}}\right)^2$ at either switching directions very well.

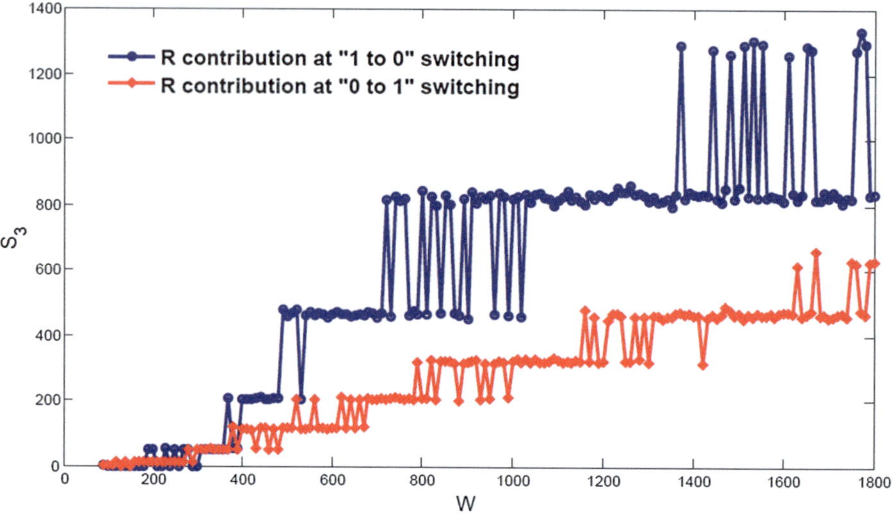

Fig. 21 Contributions from R

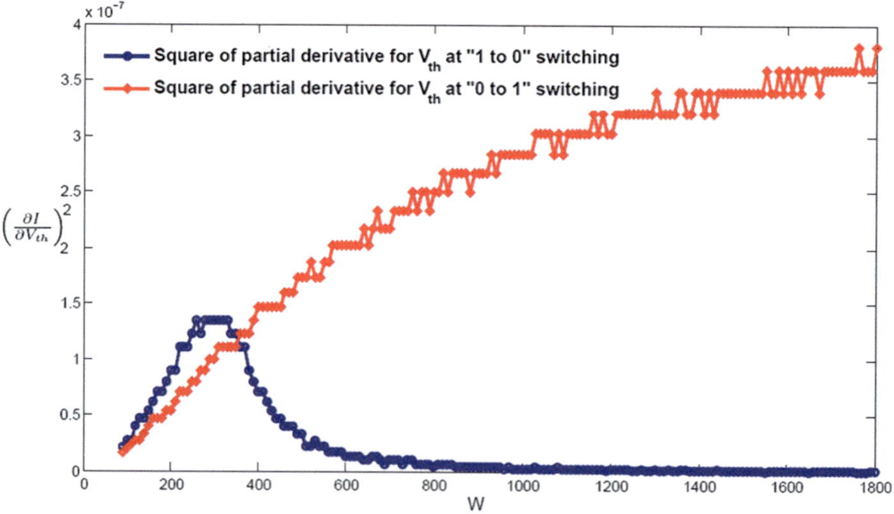

Fig. 22 Square partial derivatives for V_{th}

However, because of the additional coefficient on the top of $\left(\frac{\partial I}{\partial V_{th}}\right)^2$, S_4 does not follow the same trend of $\left(\frac{\partial I}{\partial V_{th}}\right)^2$ at either switching directions. Figure 23 shows that at '0' → '1' switching, S_4 increases first and then slowly decreases when W rises. At

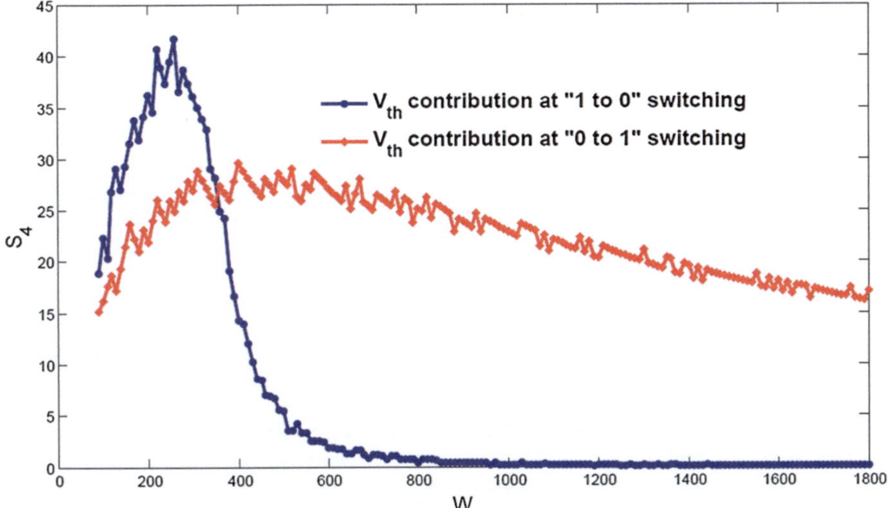

Fig. 23 Contributions from V_{th}

this switching direction, S_4 will become zero when $W \rightarrow \infty$ due to the existence of the additional coefficient $\left(\frac{C_1}{WL} + \frac{C_2}{\exp(L/l')} \cdot \frac{W_c}{W} \cdot \sigma_L^2 \right)$.

All these above results are well consistent with our analytic analysis in Table 3.

References

1. G. Sun, X. Dong, Y. Xie, J. Li, Y. Chen, in *A Novel Architecture of the 3D Stacked MRAM L2 Cache for CMPs*. 15th HPCA. IEEE, 2009, pp. 239–249
2. W. Xu, H. Sun, Y. Chen, T. Zhang, Design of last-level on-chip cache using spin-torque transfer ram (STT-RAM). IEEE Trans. VLSI Syst **19**, 483–493 (2011)
3. P. Zhou, B. Zhao, J. Yang, Y. Zhang, in *Energy Reduction for STTRAM Using Early Write Termination*. ICCAD. ACM, 2009, pp. 264–268.
4. C. Smullen, V. Mohan, A. Nigam, S. Gurumurthi, M. Stan, in *Relaxing Non-volatility for Fast and Energy-Efficient STT-RAM Caches*. High Performance Computer Architecture (HPCA), 2011 IEEE 17th International Symposium, 2011, pp. 50–61.
5. Y. Chen, W.-F. Wong, H. Li, C.-K. Koh, Y. Zhang, W. Wen, On chip caches built on multilevel spin-transfer torque RAM cells and its optimizations. J. Emerg. Technol. Comput. Syst. **9**(2), 16:1–16:22 (2013)
6. Z. Sun, X. Bi, H. Li, W. Wong, Z. Ong, X. Zhu, W. Wu, in *Multi Retention Level STT-RAM Cache Designs with a Dynamic Refresh Scheme*. Proceedings of the 44th Annual IEEE/ACM International Symposium on Microarchitecture, 2011, pp. 329–338
7. Z. Sun, X. Bi, and H. Li, *Process Variation Aware Data Management for STT-RAM Cache Design*, in Proceedings of the 2012 ACM/IEEE International Symposium on Low Power Electronics and Design, 2012, pp. 179–184.
8. W. Wen, Y. Zhang, L. Zhang, Y. Chen, in *Loadsa: A yield-Driven Top-Down Design Method for STT-RAM Array*. Design Automation Conference (ASP-DAC), 2013 18th Asia and South Pacific, 2013, pp. 291–296

9. J. Li, H. Liu, S. Salahuddin, and K. Roy, in *Variation-Tolerant Spin-Torque Transfer (STT) MRAM Array for Yield Enhancement*, in CICC, 2008, pp. 193–196.
10. C. W. Smullen, A. Nigam, S. Gurumurthi, M. R. Stan, in *The STeTSiMS STT-RAM Simulation and Modeling System*. ICCAD, 2011, pp. 318–325
11. Y. Chen, X. Wang, H. Li, H. Xi, Y. Yan, W. Zhu, Design margin exploration of spin-transfer torque ram (stt-ram) in scaled technologies. IEEE Trans. Very Large Scale Integr. VLSI Syst. **18**(12), 1724–1734 (2010)
12. W. Zhao, E. Belhaire, V. Javerliac, Q. Mistral, E. Nicolle, C. Chappert, B. Dieny, in *Macro-Model of Spin-Transfer Torque Based Magnetic Tunnel Junction (MTJ) Device for Hybrid Magnetic/CMOS Design*. Proceeding Of IEEE International Behavioral Modeling and Simulation Conference (IEEE-BMAS), San Jose, 2006, pp. 40–44
13. M. El Baraji, V. Javerliac, W. Guo, G. Prenat, B. Dieny, Dynamic compact model of thermally-assisted switching magnetic tunnel junctions. J. Appl. Phys. **106**, 123906 (2009)
14. W. Guo, G. Prenat, V. Javerliac, M. El Baraji, N. de Mestier, C. Baraduc, B. Dieny, SPICE modelling of magnetic tunnel junctions written by spin transfer torque. J. Phys. D Appl. Phys. **43**, 215001 (2010)
15. G. Prenat, G. di Pendina, K. Torki, B. Dieny, J.-P. Nozières. in *Hybrid CMOS/Magnetic Process Design Kit and Application to the Design of High-Performances Non-Volatile Logic Circuits*. Proceedings of the International Conference on Computer-Aided Design, 2011, pp. 240–245
16. S. Chatterjee, M. Rasquinha, S. Yalamanchili, S. Mukhopadhyay, A scalable design methodology for energy minimization of STTRAM: a circuit and architecture perspective. IEEE Trans. Very Large Scale Integr. VLSI Syst. **19**(5), 809–817 (2011)
17. X. Wang, Y. Zheng, H. Xi, D. Dimitrov, Thermal fluctuation effects on spin torque induced switching: mean and variations. JAP **103**(3), 034507 (2008)
18. R. Singha, A. Balijepalli, A. Subramaniam, F. Liu, S. Nassif. in *Modeling and Analysis of Non-Rectangular Gate for Post-Lithography Circuit Simulation*. 44th DAC, 2007, pp. 823–828
19. Y. Ye, F. Liu, S. Nassif, Y. Cao, in *Statistical Modeling and Simulation of Threshold Variation under Dopant Fluctuations and Line-Edge Roughness*. 45th DAC, 2008, pp. 900–905.
20. F. Harris, On the use of windows for harmonic analysis with the discrete fourier transform. Proc. IEEE **66**(1), 51–83 (1978)
21. Y. Zhang, X. Wang, Y. Chen, in *STT-RAM Cell Design Optimization for Persistent and Non-Persistent Error rate Reduction: A statistical Design View*. ICCAD, 2011, pp. 471–477
22. Y. Zhang, W. Wen, Y. Chen, STT-RAM cell design considering MTJ asymmetric switching. *SPIN* **02**(03), 1240007 (2012)
23. W. Wen, Y. Zhang, Y. Chen, Y. Wang, Y. Xie, in *PS3-RAM: A Fast Portable and Scalable Statistical STT-RAM Reliability Analysis Method*. 49th DAC, 2012, 1187–1192
24. W. Wen, Y. Zhang, L. Zhang, Y. Chen, in *Loadsa: A Yield-Driven Top-Down Design Method for STT-RAM Array*. ASP-DAC, 2013, pp. 291-296
25. P. T. M. (PTM) ASU http://www.eas.asu.edu/ptm/
26. BSIM, UC Berkeley http://www-device.eecs.berkeley.edu/bsim3/
27. B. Sheu, D. Scharfetter, P.-K. Ko, M.-C. Jeng, BSIM: berkeley short-channel IGFET model for MOS transistors. JSSC **22**(4), 558–566 (1987)
28. P. Doubilet, C. Begg, M. Weinstein, P. Braun, B. McNeil, Probabilistic sensitivity analysis using monte carlo simulation. A practical approach. Med. Decis. Making **5**(2), 157–177 (1985)
29. W. Xu, Y. Chen, X. Wang, T. Zhang, in *Improving STT MRAM Storage Density Through Smaller-Than-Worst-Case Transistor Sizing*. 46th DAC, 2009, pp. 87–90
30. W. Kang, W. Zhao, Z. Wang, Y. Zhang, J.-O. Klein, Y. Zhang, C. Chappert, D. Ravelosona, A low-cost built-in error correction circuit design for STT-MRAM reliability improvement. Microelectron. Reliab. **53**, 1224–1229 (2013)
31. C. Xu, D. Niu, X. Zhu, H. S. Kang, M. Nowak, Y. Xie, in *Device Architecture Co-Optimization of STT-RAM Based Memory for Low Power Embedded Systems*. ICCAD, 2011, pp. 463–470.

Synchronized Spin Torque Nano-Oscillators: From Theory to Applications

Mehdi Kabir and Mircea Stan

1 Introduction

The operation of spin torque oscillators takes advantage of two key discoveries in magnetic materials–spin dependent transport and spin transfer torque. These discoveries were made in a structure known as the magnetic tunnel junction (MTJ), which consists of two ferromagnetic metal layers separated by a thin insulating barrier layer (Fig. 1). The insulating layer is so thin (a few nm or less) that electrons can tunnel through the barrier when a bias voltage is applied between the two metal electrodes. An important property of an MTJ is that the tunnelling current depends on the relative orientation of the magnetizations of the two ferromagnetic layers, which can in turn be changed (e.g.) by an applied magnetic field. This phenomenon is called tunneling magnetoresistance (TMR). In this section, we review some of the seminal works which led to current focus on spin torque nano-oscillators.

1.1 *Spin Dependent Transport*

The history of spin-dependent transport in magnetic multi-layers went from obscurity to Nobel Prize winning research. Here we present just a brief outline of that history–full reviews of the research progress are given in the following excellent papers [1–4]. The first work on spin dependent transport began more than 40 years ago with experiments by Meservey et al. [5]. A few years earlier, the BCS theory of

M. Kabir • M. Stan (✉)
Department of Electrical & Computer Engineering, University of Virginia,
Charlottesville, VA 22903, USA
e-mail: mk4kg@virginia.edu; mrs8n@virginia.edu

© Springer International Publishing Switzerland 2015
W. Zhao, G. Prenat (eds.), *Spintronics-based Computing*,
DOI 10.1007/978-3-319-15180-9_8

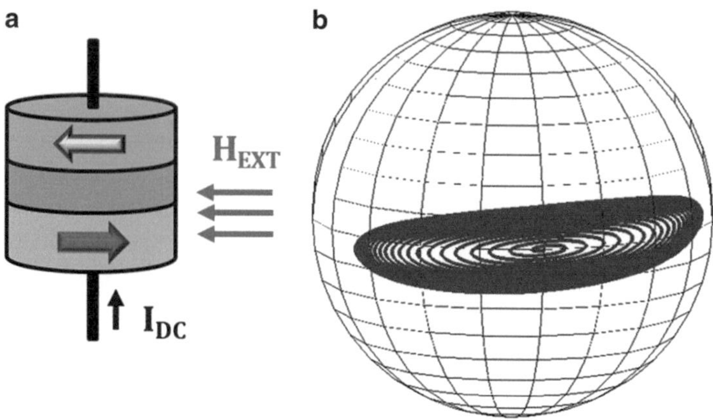

Fig. 1 (a) Structure of a magnetic tunnel junction with a fixed layer, whose magnetization is pinned, and a free layer, whose magnetization can be changed. (b) The stable precession trajectory of the magnetization of an in-plane free layer in response to spin transfer torque

superconductivity had been published [6] and there was great interest in transport properties of electrons through superconducting junctions [7]. Meservey et al. predicted that the energy level of a superconductor in a magnetic field could be split in such a way that it could act as a high precision spin polarization detector. This allowed them to construct a trilayer junction, consisting of ferromagnetic-insulating-superconducting layers, which could detect the conduction changes due to changes in the magnetization orientation of the ferromagnetic layers. This was the first experimental verification that electron transmission could be a function of its spin orientation.

Soon afterwards, Julliere [8] demonstrated tunneling magnetoresistance (TMR) between two ferromagnetic layers separated by an insulating barrier. By using ferromagnets of different coercivities, it was possible to detect the change of resistance between parallel magnetizations and anti-parallel magnetizations. Julliere postulated that the TMR could be calculated based on the spin polarization of the two ferromagnetic layers as given by the equation

$$TMR = \frac{2P_1P_2}{1 - P_1P_2} \tag{1}$$

where $P = \frac{\rho_\uparrow - \rho_\downarrow}{\rho_\uparrow + \rho_\downarrow}$ and $\rho_\uparrow$ and $\rho_\downarrow$ are up-spin and down-spin electron carriers, respectively, in the ferromagnetic layer.

Despite these discoveries, there would be very little research on spin-dependent transport for another decade. Partly this was caused by technologically demanding fabrication process, which makes it difficult to fabricate robust and reliable tunnel junctions. Also, the fact that the reported values of TMR were small (at most a few percent at low temperatures), did not trigger considerable interest in view of sensor/memory applications. The perfecting of molecular beam epitaxy (MBE) and its greater availability during the 1980s [9] allowed for better thin-film fabrication allowing studies of magnetic thin-film structures.

The discovery of giant magnetoresistance (GMR) in spin-valve structures (using metallic spacer layers) ignited interest in spintronic devices. Because of their pioneering work on GMR, Albert Fert [10] and Peter Grünberg [11] were awarded the Nobel Prize in Physics in 2007 [3, 4]. While previous experiments used cryogenic conditions, room temperature operations were shown by Miyazaki and Tezuka [12] that demonstrated the possibility of large values of TMR in MTJs with Al_2O_3 insulating layers, while Moodera et al. [13] developed a fabrication process, which appeared to fulfill the requirements for smooth and pinhole-free Al_2O_3 deposition.

The first accurate theoretical consideration of TMR was made by Slonczewski in 1989 [14]. He considered tunnelling between two identical ferromagnetic electrodes separated by a rectangular potential barrier and by assuming that the ferromagnets can be described by two parabolic bands shifted rigidly with respect to one another to model the exchange splitting of the spin bands. By imposing perfect translational symmetry of the tunnel junction along the layers and matching the wave functions of electrons across the junction, he solved the Schrödinger equation and determined the conductance as a function of the relative magnetization alignment of the two ferromagnetic films. In the limit of thick barrier, he found that the conductance is a linear function of the cosine of angle θ between the magnetic moments of the films:

$$G(\theta) = G_{int}(1 + P^2 \cos\theta) \qquad (2)$$

where G_{int} is the normal conductance through the trilayer interfaces, P is the spin-polarization of the fixed layer and θ is the angular difference in the magnetic orientation between the fixed and free layers. This matches the previous observations in Julliere's equation (Eq. (1)) that parallel magnetizations have higher conductance than anti-parallel orientation.

1.2 Spin Transfer Torque

The other effect, spin transfer torque, was first predicted by Slonczewski [15] and Berger [16] and subsequently verified experimentally in oxide-based MTJs [17, 18]. This behavior is observed in MTJs when a DC current is applied through the junction. One of the ferromagnetic layers (known as the fixed or pinned layer) is used to spin-polarize the DC current to align a majority of electrons in the spin orientation of layer. When these electrons tunnel through the insulating layer and are injected into the second ferromagnetic layer (the free layer), they transfer their angular momentum to electrons in the free layer causing the magnetization to change. Normally, dynamic processes in magnetic layers create damped oscillations which dissipate over time. However, Slonczewski showed that the spin transfer torque effect can be used to cancel out the dampening effect, creating stable oscillations, and in some cases, complete reversal of the magnetization [19] (Fig. 1b). The latter effect can be used to perform the "write" operation of MTJs in magnetic memory devices such as MRAM and STT-RAM [20, 21], while the former can be used to create microwave oscillations in spin torque nano-oscillators (STNO).

Soon after the theoretical development, spin torque nano-oscillators were demonstrated experimentally [19, 22, 23]. The microwave oscillations generated by the STNOs exhibit wide ranging tunability simply by adjusting the DC current and applied magnetic field (Fig. 1a), with frequencies ranging from 0.1 to 100 GHz and linewidths on the order of ~ 100 MHz [24]. More recently, it has been experimentally shown that two STNOs can mutually phase-lock and synchronize [25, 26]. These experiments involved two MTJ structures with a shared free layer which allowed current-injected spinwaves to interact with each other causing them to phase-lock. The ability of spin torque oscillators to synchronize is not only advantageous in increasing the output oscillation power [27] but also provides a way of transferring information and computing with multiple oscillators.

In this paper, we propose an array of STNOs which can perform Non-Boolean computations through synchronization. Specifically, we show how to use the STNO array for pattern recognition, which serves as a first step towards associative memory behaviour [28, 29]. Throughout this paper, we discuss some of the design parameters that have to be considered for effective synchronization of the oscillators. In Sects. 2 and 3, we discuss the dynamics of the STNO and how electrically coupling them can provide an alternative method to synchronization. In Sect. 4 we examine the device geometries which are optimal for coherence of multiple STNOs and introduce the idea of parallel connected DMTJ-STNO arrays. Finally, in Sects. 5 and 6, we implement a hybrid MOSFET/STNO array which can be used to for various applications such as analyzing patterns between two sets of vectors or for RF circuits. We use HSPICE simulations to model the spin torque nano-oscillator behavior. More details about the simulation setup are provided in the appendix.

2 Modeling the Dynamic Behavior of STNOs

The precessional motion of magnetization (**M**) of the free layer of an MTJ, in the presence of an effective magnetic field (**H**$_{\text{eff}}$), can be accurately modeled by the Landau–Lifshitz–Gilbert (LLG) equation. With the introduction of Slonczewski's spin-transfer torque, the LLG equation with the STT term is given by:

$$\frac{\partial \mathbf{M}}{\partial t} = -\gamma_0 \mathbf{M} \times \mathbf{H}_{\text{eff}} + \frac{\gamma_0 \alpha}{M_s} \mathbf{M} \times (\mathbf{M} \times \mathbf{H}_{\text{eff}}) + \eta(\theta) \frac{\mu_B I}{eV} \mathbf{M} \times (\mathbf{M} \times M_{\mathbf{p}}) \quad (3)$$

The first term in Eq. (3) is the precession of the magnetization ($\mathbf{M}$) around an effective magnetic field ($\mathbf{H_{eff}}$) where γ_0 is the gyromagnetic ratio. The second term describes the phenomenological dampening (α) which brings the magnetization in anti-/parallel alignment with $\mathbf{H_{eff}}$ (here M_s is the saturation magnetization of the free layer). Finally, the third Slonczewski term is similar to the dampening term, except the torque pulls the magnetization towards the orientation of the polarizing layer ($\mathbf{M_p}$). The spin torque is modified by the applied current density given by $\left(\frac{I}{V}\right)$, where V is the free layer volume, and also by the spin polarization efficiency factor $\eta(\theta)$ given by [15]:

$$\eta(\theta) = \left[-4 + (1 + P)^3 \ \frac{3 + \cos\theta}{4P^{3/2}}\right]^{-1} \tag{4}$$

where P is the polarization of the fixed layer and the angle θ is the relative orientation between the free and fixed layer.

The effective magnetic field ($\mathbf{H_{eff}}$) is given by:

$$\mathbf{H_{eff}} = \mathbf{H_{ext}} + \mathbf{H_{an}} + \mathbf{H_{dem}} + \mathbf{H_{amp}} \tag{5}$$

where $\mathbf{H_{ext}}$ is the external applied magnetic field. The field $\mathbf{H_{an}}$ represents the magnetocrystalline anisotropies that exist due to the crystal structure. This anisotropy contribution can be further divided into contributions from the easy-axis, perpendicular axis and planar fields:

$$\mathbf{H_{an}} = \mathbf{H_{easy}} + \mathbf{H_{perp}} + \mathbf{H_{planar}} \tag{6}$$

The easy axis field lies in the plane of the free layer while the perpendicular axis is the out-of-plane contribution. Finally, the planar field is the field that tries to pull an out-of-plane magnetization towards the free layer plane. The demagnetization field captures the shape anisotropy of the magnetic material and depends on the geometry. In general, $\mathbf{H_{dem}} = \mathbf{N} \cdot \mathbf{M}$, where $\mathbf{N}$ is a shape dependent term [30]. Finally, $\mathbf{H_{amp}}$ is the amperian field created by the injected current within the magnetic tunnel junction. This circular field will lie in the plane of the free layer and will depend on the magnitude of the current through the wire.

3 Electrically Coupled STNOs

3.1 *Magnetic vs. Electrical Synchronization*

While phase locking via magnetic spinwaves has been shown between two spin torque oscillators, it still remains uncertain if this process can be scaled up for fabrication, or even synchronization, when many STNOs are involved [31].

An alternative to magnetic synchronization is to use the magnetoresistance effect to convert the magnetic oscillations in the free layer into an alternating current signal through the MTJ. Then two or more STNOs can be connected *electrically* to achieve synchronization. Each STNO in an electrically coupled array can operate independently and does not require specialized features (e.g. shared free layer) to achieve synchronization. This allows for a variety of circuit topologies which would be difficult to realize in a purely magnetic system. In spinwave-coupled STNOs, the geometric layout of the oscillators will determine functionality, whereas in electrically-coupled arrays the oscillator connectivity is determined by the more flexible circuit topology. Furthermore, the electrical nature of the signals can be used to integrate the STNOs with MOSFET transistors and other conventional circuit elements. This feature makes a hybrid MOSFET/STNO circuit attractive for computational purposes.

3.2 Synchronization Dynamics

One possible method of increasing the power of spin torque nano-oscillators has been to synchronize them to a common frequency [26]. Electrically synchronizing multiple STNOs has been previously considered for both series and parallel configurations [32, 33]. We have shown previously [34] that in a parallel array of STNOs with out-of-plane fixed and free layers, oscillators with similar frequencies can exchange energies through their AC signal, slowing down or speeding up their oscillations until they achieve synchronization. We now examine the synchronization dynamics of the circuit shown in Fig. 2. Each parallel STNO will have a DC

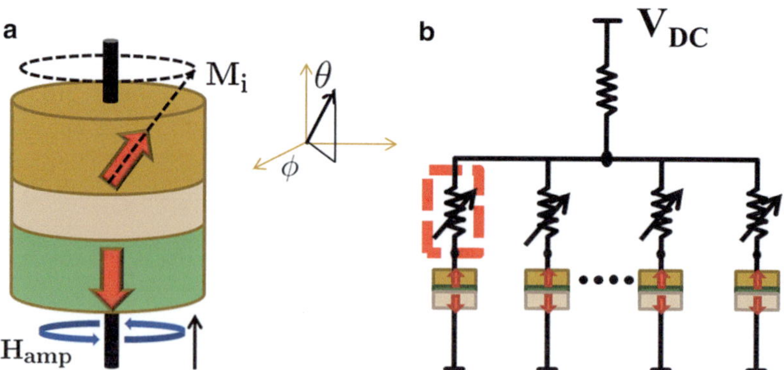

Fig. 2 (**a**) The relative orientation of the magnetization vector and the effective magnetic field which synchronizes the STNOs. (**b**) A general layout of an electrically-coupled parallel-connected STNO array. The variable resistor in the schematic represents the ability to change the current through each parallel branch of the array to set the DC current ($I_{DC,i}$). This produces an oscillating signal ($I_{AC,i}$) which combines on the common node with the signals from the other parallel branches

current, $I_{dc,i}$, which determines the initial frequency of its operation. This in turn will generate an AC current, $I_{ac,i}$, due to the TMR of the magnetic junction. By Kirchoff Current Law, the common node will have a combined AC current of $I_{ac} = \Sigma_i I_{ac,i}$. This AC current will re-distribute the energy throughout the array. To quantify the energy gain/loss in one of the STNO, we look at the $\mathbf{H_{amp}}$ component of the effective magnetic field in the LLG equation (Eq. (3)). In this case, $\mathbf{H_{amp}} = \xi I_{ac}\phi_{\mathbf{ac}}$, where ξ is a material dependent property and $\phi_{\mathbf{ac}}$ is the angular unit vector in-plane to the free layer of the STNO. Therefore, for the i-th STNO, the energy injected/removed by the magnetic field $\mathbf{H_{amp}}$ is given by:

$$E_i = -\mu_0 M_s V \int \mathbf{H_{amp}} \cdot dM_i = -\mu_0 M_s V \xi \int \sum_j I_{ac,j}\phi_{\mathbf{ac,j}} \cdot d\mathbf{M_i}$$

$$= -\sum_j \vartheta_j m_i \cos\left(\phi_i - \phi_j\right) \tag{7}$$

where $\vartheta_j = \pi\mu_0 M_s V K I_{ac,j}$, and $m_i = \sqrt{[m_x^2 + m_y^2 + (m_x^2 - m_y^2)\sin(2\phi_{ac,j})]/2}$ represents the projection of $\mathbf{M_i}$ onto $\phi_{\mathbf{ac},j}$. Finally, ϕ_i and ϕ_j are phase components of the i-th and j-th oscillator respectively.

Now that we have an expression for the energy landscape of STNOs, we can determine how the frequency of oscillation changes to respond to the change in energy. First, we recognize that the phase is related to its frequency as $\phi_i = \phi_{i,0} + 2\pi f_i t$. Therefore, the changes in the phase can be represented as:

$$\frac{d\phi_i}{dt} = 2\pi f_i + \delta f_i = 2\pi f_i + \left(\frac{df_i}{dE_i}\right)E_i = 2\pi f_i - \sum_j \Delta_{ij}\cos(\phi_i - \phi_j) \tag{8}$$

where ϕ_i and f_i are respectively the phase and frequency of the oscillator, and the coupling constant $\Delta_{ij} = -\left(\frac{df_i}{dE_i}\right)\vartheta_j m_i$ is between the two oscillators which depends on the material properties of the device and relative frequency distance. Equation (8) represents the exact dynamics outlined by Kuramoto's model of weakly coupled oscillators [35]. Kuramoto's oscillation model is a general principle which applies to many emergent behaviour phenomena in nature [36]. An important application was first proposed by Hoppensteadt and Izhikevich [37] to model associative memory in neural networks. In the next section, we consider the similarities between the Hoppensteadt model and an array of parallel connected STNOs.

3.3 *STNO Array as an Oscillatory Neurocomputer*

The Hoppensteadt model uses oscillators to implement a neural network. The Non-Boolean nature of the oscillators can realize neural networks with only N-connections as opposed to N^2-connections required by traditional cross-weighted

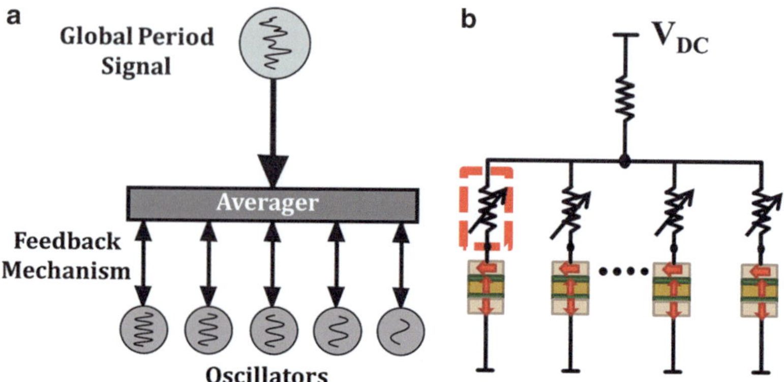

Fig. 3 STNO array as an oscillatory neurocomputer. (**a**) Components of an oscillatory neurocomputer (modified from [37]). (**b**) A general layout of an electrically coupled parallel-connected DMTJ-STNO array. The dual barrier magnetic tunnel junction (DMTJ) produces a harmonic signal with a strong voltage signal

architectures [29]. Because both systems are manifestations of Kuramoto's oscillation model, the STNO array can act as a physical implementation of Hoppensteadt's oscillatory neurocomputer.

Figure 3 shows the similarities between the oscillatory neurocomputer and the STNO array. The STNO array consists of parallel connected devices with the oscillation frequency set by the current running through each branch. The oscillating signals from each parallel branch are summed up in an analogous fashion to the "averager" in the Hoppensteadt scheme and act as a global periodic signal through which the oscillators communicate. The feedback layer is emulated in the STNOs through the magnetization dynamics of the free layer. The overall effect is a change in frequency and phase of the oscillator shown in Eq. (8). When the oscillator frequencies are close they begin to synchronize. In the next section we discuss how such an array can be physically implemented.

4   STNO Geometry and Oscillations

In the previous sections, we have considered the properties of a general parallel connected STNO array. However, to create a physical array there are several device-level considerations that must be taken into account. There are many different geometric configurations of the fixed and free layers of a spin torque oscillator. Each configuration can generate different modes of oscillation depending on the biasing current and applied external magnetic fields. First, we consider two configurations–perpendicular (Fig. 4) and mixed–both of which use perpendicular magnetic anisotropy (PMA) free layers. These geometries were chosen because they can produce oscillations without the need for an external magnetic field [38];

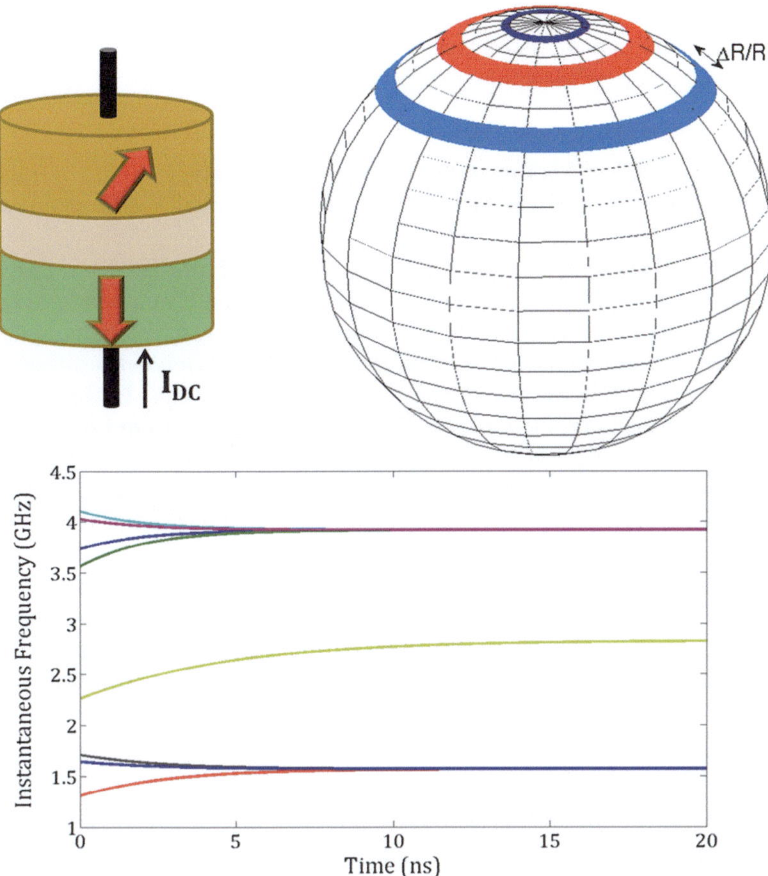

Fig. 4 (*Top*) a perpendicular STNO. This configuration exhibits circular orbits which produce almost harmonic oscillations. However, the amplitude of the electrical signal is weak due to the infinitesimal variation in $\Delta R/R$ when the angle of precession is constant. Here we show multiple orbits corresponding to different DC currents and frequencies. (*Bottom*) transient simulation showing changes in frequency and synchronization of almost harmonic STNOs

this makes it much easier to operate a large-scale array without the need of a localized magnetic field. The mode of oscillation for these two configurations are different which creates a trade-off between synchronization and signal strength.

4.1 Harmonicity and Signal Strength

When considering these two geometries of STNOs, the oscillation mode plays a significant role in the synchronization behaviour and the amplitude of the electrical signal generated by the MTJ. The amplitude of the electrical signal depends on the magnetoresistance of the MTJ. This effect can be estimated as the angular difference between the free and fixed layer given by the equation,

$$\frac{\Delta R}{R} = 1 - P^2 cos(\theta) \tag{9}$$

where R is the average MTJ resistance, ΔR is the deviation from the average, P is the effective polarization by the fixed layer and θ is the angular difference between the free and fixed layer.

In the **perpendicular** geometry, where both the free and fixed layer are PMA materials, the trajectory of the magnetization is approximately circular, producing a simple almost harmonic oscillation. With almost harmonic oscillations, synchronization between multiple oscillators can be achieved with frequency differences as large as 22 % (Fig. 4, bottom) and within ~ 100 periodic cycles. However, because of the circular trajectory, there is very little change in $\Delta R/R$ from Eq. (9), creating very small electrical signals (Fig. 4, top). This, in turn, prevents effective communications between the oscillators in the array.

In a **mixed** configuration the fixed layer is in-plane while the free layer is out-of-plane (Fig. 5). In this geometry, the orbit of oscillation produces a Lissajous curve and thus produces a complex periodic signal. This is advantageous from an electric signal point of view due to the large change in $\Delta R/R$. However, simulation results show that synchronization becomes difficult as the frequency bandwidth of coherence is reduced to 2–3 % with much larger coherence time than for harmonic oscillations (Fig. 5, bottom).

4.2 Dual Barrier MTJ

In order to take advantage of the positive aspects of both configurations, we propose combining the two geometries in a trilayer structure. This dual barrier magnetic tunnel junction (DMTJ) consists of PMA fixed and free layers and an in-plane analyzer layer. During operation, the PMA fixed layer induces harmonic oscillation in the free layer while the in-plane analyzer layer creates large changes in $\Delta R/R$ yielding a large electrical signal.

It should be noted that DMTJs have been realized experimentally [39] though those devices used only in-plane ferromagnets for all three layers. The fabrication of PMA DMTJs is an area of active research (Fig. 7).

Simulations results show that the DMTJ structure maintains near harmonic oscillations and thus can have synchronization frequency bandwidths as large as 18 % with coherence times on the same order as the perpendicular geometry (Fig. 6). At the same time, the DMTJ maintains a large electrical signal comparable to the mixed geometry. With this device geometry, an array of STNOs can be created which is robust enough for computation.

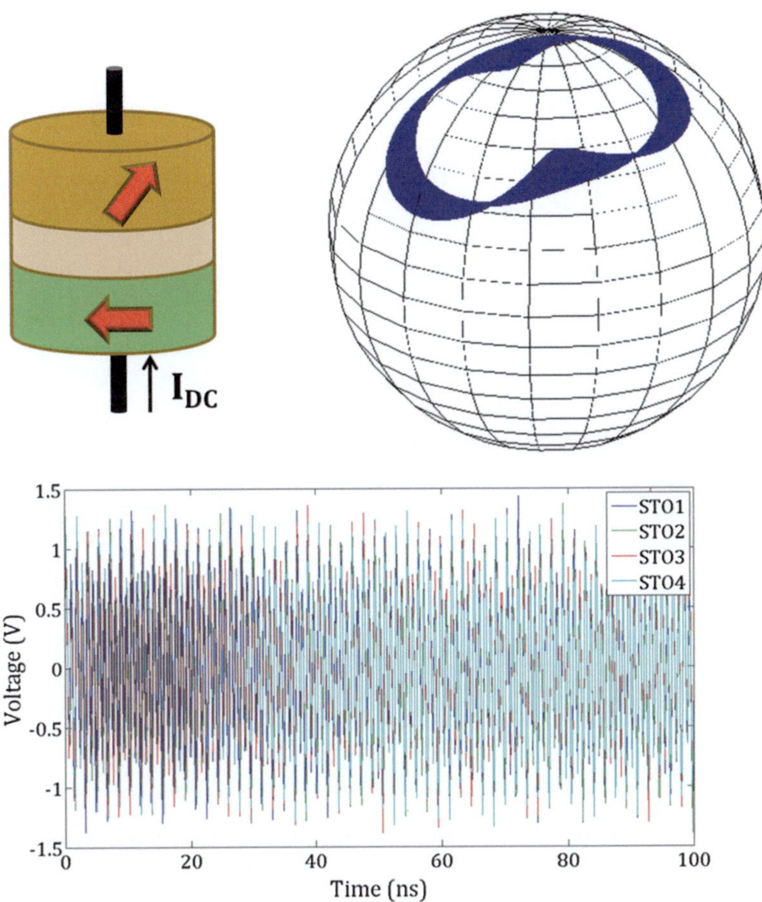

Fig. 5 *Top*: mixed geometry STNO. The mixed geometry STNO exhibits complex orbital precessions which lead to nonharmonic oscillations. *Bottom*: transient analysis shows the complex periodic orbit of four mixed STNOs with similar frequencies. This configuration creates a large electrical signal because of the large variations in magnetoresistance. At the end, all four oscillators are shown to synchronize

5   Application: Pattern Recognition

As a first step to an associative memory circuit, we designed a hybrid MOSFET/ STNO array which can compare a test vector with a reference vector and give an indication of the degree of match (DOM) between the two data sets. As explained above, we use DMTJ spin torque nano-oscillators for the array.

The MOSFET transistors are used to control the frequency of oscillation by modulating the current in each branch depending on the test vectors. Each of the transistors act as source followers thereby maintaining the same voltage across the

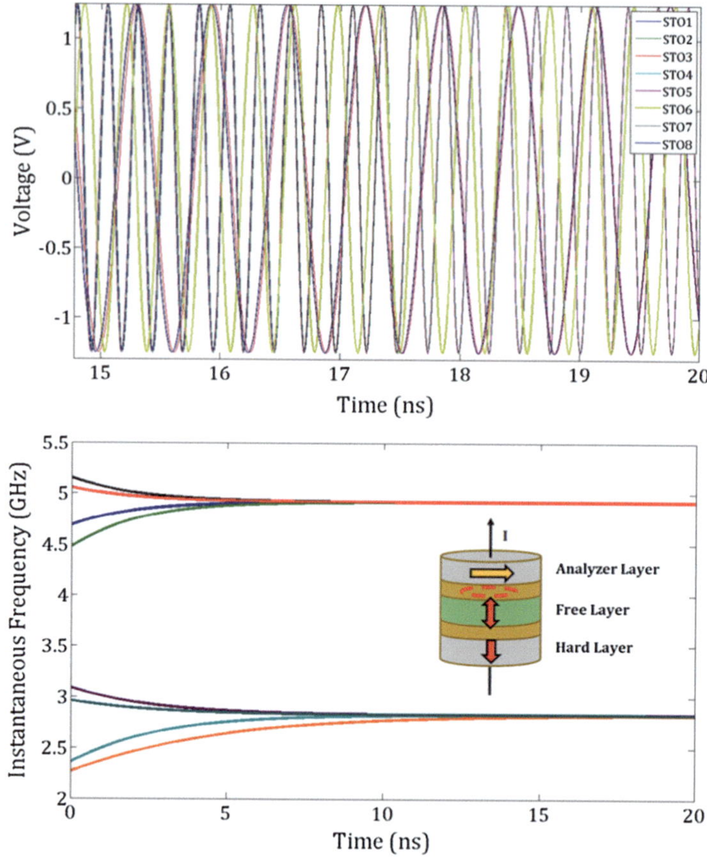

Fig. 6 Dual barrier magnetic tunnel junction. (*Top*) the oscillations in the DMTJ are both harmonic and have large electrical signal. (*Bottom*) the transient frequency behavior of eight DMTJ STNOs as they synchronize

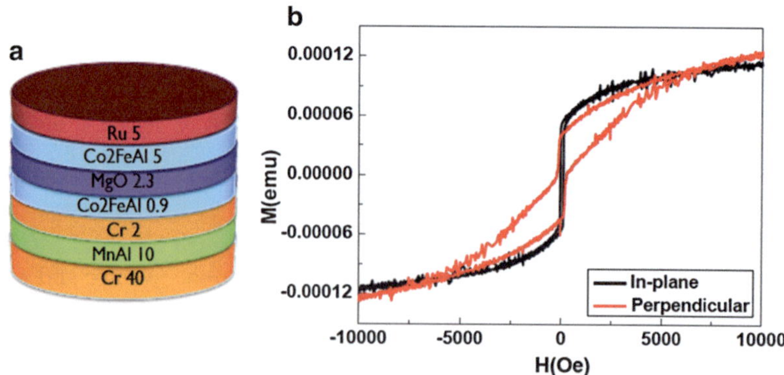

Fig. 7 (**a**) Schematic diagram of a perpendicular spin torque nano-oscillator in design. (**b**) In-plane and out-of-plane hysteresis loops of the multilayer structure

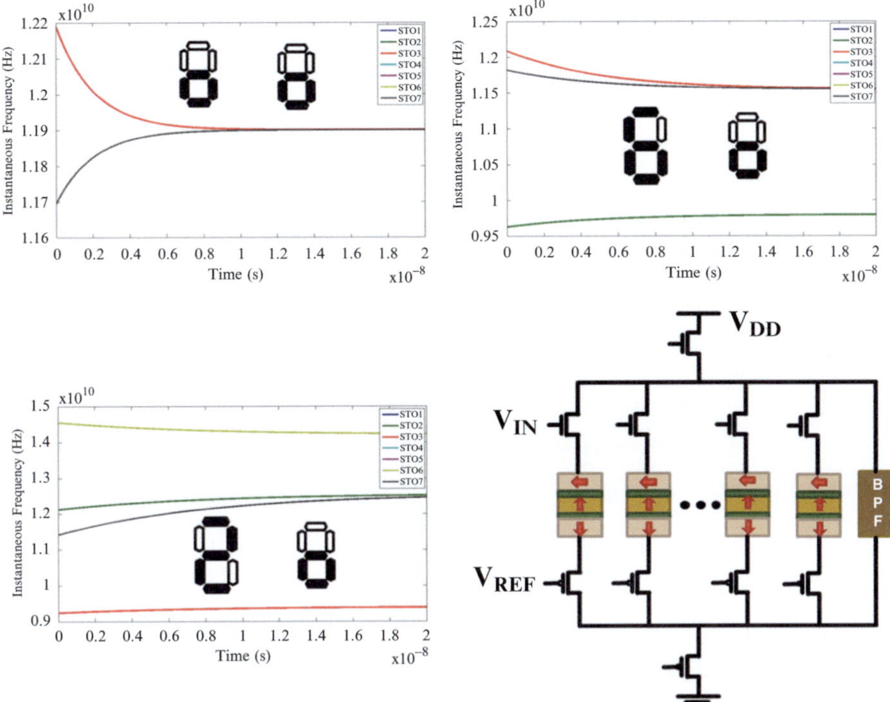

Fig. 8 Hybrid MOSFET/STNO array for pattern recognition. (*Bottom right corner*) the hybrid circuit consists of DMTJ spin torque oscillators sandwiched by NMOS and PMOS source followers. The V_{REF} transistors receive the reference vector, while the V_{IN} handle the input vector. The bandpass filter (BPF) is used to isolate frequencies representing matches. (*Left column*) transient analysis showing the synchronization of STNOs over time

DMTJ if both inputs to the transistors are shifted by an equal amount. This ensures that the frequency of the STNO remains unchanged if the reference and test bits are the same. In the test case shown in Fig. 8, an STNO is assigned to each bit of a 7-segment array. The **Reference** column shows how the reference transistors are assigned-with '0' representing a blank and '1' representing a filled segment. These bits represent patterns which have already been "memorized" by the array. Similarly, the **Input** column represents the bits for the test pattern which will be compared to the memorized patterns.

In the topmost case, we have a test pattern which is an exact match to one of the memorized patterns. The STNO arrays start at two frequencies representing the bits which are both '0's and both '1's. These frequencies are close enough such that they synchronize together until all oscillators operate at the same frequency. The time for all the STNOs to cohere together was approximately 10 ns for oscillators operating at 12 GHz (Fig. 8).

The middle and bottom cases illustrate the scenario where the test pattern is not an exact match to the memorized pattern. As the transient analysis shows, there are

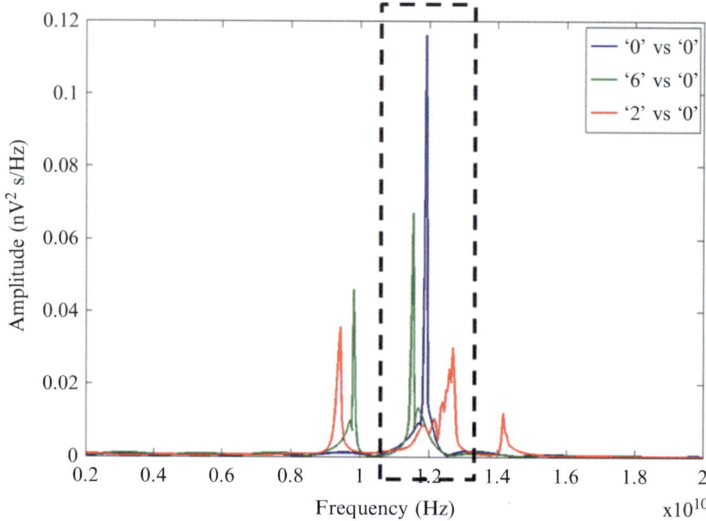

Fig. 9 The Fourier analysis of the parallel node can be translated as the degree of match. Hence '6' is a better match than '2'. The *dashed box* shows the frequencies of the BPF

additional frequencies that appear when there are mismatches. These frequencies are different enough that they do not synchronize completely with the STNOs which represent matches, though their frequencies are affected. In case of mismatches, the oscillators take longer to reach synchronization, sometimes taking more than twice longer than for a perfect match scenario. If we perform a frequency analysis on the parallel node of the STNO array after all the signals have synchronized, we can get an estimate of the DOM between the test and reference vectors. As the number of oscillator synchronize to a single frequency, the output signal power is boosted. Therefore, if we analyze the signal power near frequencies representing matches, we can determine which memorized pattern represents the best match to the test vector. A bandpass filter can be used to eliminate all frequencies except for the ones representing matches (Fig. 9).

This array represents a simple hybrid MOSFET/STNO circuit which can be used to analyze patterns and determine a degree of match. More work needs to be done to implement advanced techniques in pattern recognition such as rotation, scale transformations and edge detection which are part of a more realistic associative memory design. The use of oscillators for these purposes remains an active area of research which suggest that STNOs are a possible device implementation for these applications. Fabrication of these specialized spin torque oscillators also remains a challenge. However, with the recent advancements in PMA spin torque devices, DMTJs could be feasible in the near future. Finally, more simulation and modelling needs to be done on these STNO arrays to determine the robustness of the design to noise and process variations. These analyses will help refine and improve future circuit designs and architectures.

6 Application: Radio Frequency Circuits

One of the most exciting aspects of the STNO devices is the possible application to system-on-chip (SoC) [40, 41]. The nanoscale dimensions, coupled with the possibility of CMOS integration, low power, tunable frequency through magnetic fields and electrical currents, high quality factor Q ($\geq$ 10,000) and the possible control of coupling opens many possibilities beyond the current state-of-the-art RF systems. One such RF application is as filters which take advantage of injection locking which has already been observed in STNOs [42]. Using a current mirror, we can create a circuit which allows the STNO to be tuned to a resonant frequency, and then apply an AC signal to the oscillator. For frequencies close to the center frequency, the STNO acts as either a bandpass or a band-reject filter depending on the circuit topology used. This allows us to do signal processing for a wide range of frequencies (1–100 GHz). This simple circuit acts as notch bandpass/bandstop filter with line widths of 10 MHz.

Although the STNOs can be used to filter narrow bandwidth signals, we are also interested in developing robust techniques to tune the filter strength and bandwidth. One of the ways to increase the filter strength is to improve the current mirror design used to set the STNO resonance oscillation. A stacked cascode structure (Fig. 10) can be used to increase the circuit impedance, this strengthening the coupling of the STNO to the desired frequency. In our simulations, the stacked cascode structure produces a 5 dB improvement over the normal structure (Fig. 10b).

Yet another technique to increase the filter strength is to use STNO synchronization. It has been shown that the oscillation power of the STNOs can be increased through the synchronization of multiple nano-oscillators [27, 32, 43]. In Fig. 11, we use the same principle to boost the power of STNOs. The coherence creates a stronger resonance signal which helps to increase the filter strength. In Fig. 11b, we

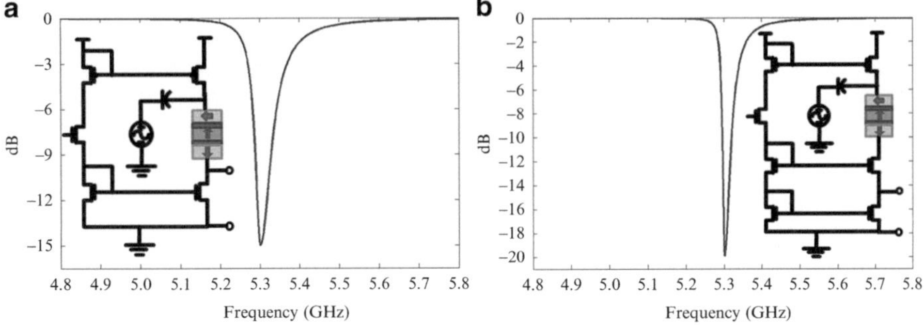

Fig. 10 (**a**) An RF filter using STNO. A current mirror is used to set the DC current, which in turn, adjusts the resonating frequency. An input signal is applied to the STNO, which filters the frequencies near resonance. Depending on circuit topology, both bandpass and bandstop (pictured) can be implemented. (**b**) An RF filter using stacked current mirror. By stacking the footer of the current, we can increase the circuit impedance resulting in a deeper notch filter

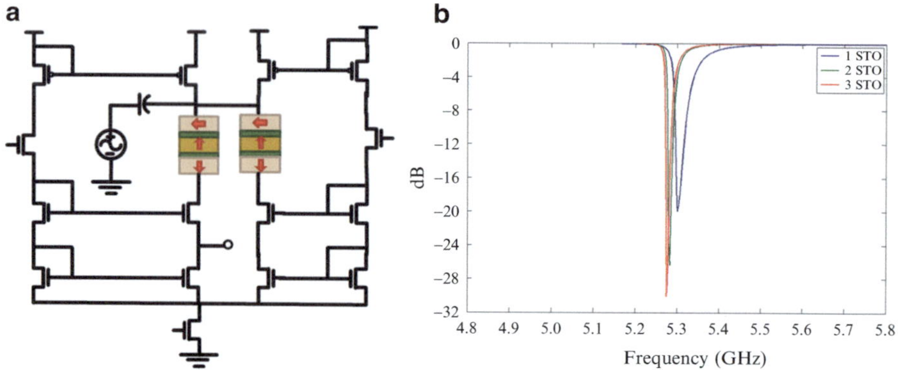

Fig. 11 (**a**) A coupled STNO bandstop filter. The cascode current mirrors set the frequency of oscillation in the STNOs. When the frequencies are close together, the oscillators synchronize to produce a stronger filter. (**b**) The frequency analysis shows a deeper notch for more coupled STNOs. The increase in filter strength is superlinear ($\sim N^2$)

show that a coupled STNO can enhance the filter strength by almost 10 dB over a single STNO filter. However, the increase in strength begins to level out with more STNO since the oscillation power increases quadratically. The bandwidth of the STNO filter can also be adjusted through engineering the thermal stability of the free layer of the magnetic tunnel junction. The thermal stability determines the susceptibility of the free layer to noise. In general, spin torque oscillators are resilient to noise and thus create high Q-factor oscillations. However, the thermal stability (Eq. (1)) can be exploited through device geometry and material properties to make the device more susceptible to noise.

$$\Delta \approx \frac{H_k M_s V}{2k_B T} \tag{10}$$

where H_k and M_s are the free layer magnetic properties and V is the volume of the layer. This effect allows other frequencies to be included in the STNO filter which broadens the filter bandwidth (Fig. 12). However, there is a trade-off between filter bandwidth and the filter strength as the noise affects the integrity of the harmonic oscillations.

In addition to filters, there are also other RF applications which are actively being researched [44]. Because spin-torque devices exhibit an intrinsic instability leading to oscillation, these devices may also be exploited to realize self-oscillating RF mixers. Mixers are the central component of heterodyne receivers, which are the most important sensor architecture used for RF applications. Heterodyne receivers typically make use of a separate local oscillator with which the signal to be detected is mixed. This allows the power of the detected signal to be measured while maintaining phase coherence. STNO devices offer a means of realizing self-oscillating mixers because the magnetic procession in these devices is inherently nonlinear. With the self-mixing characteristic of the device, the oscillation source

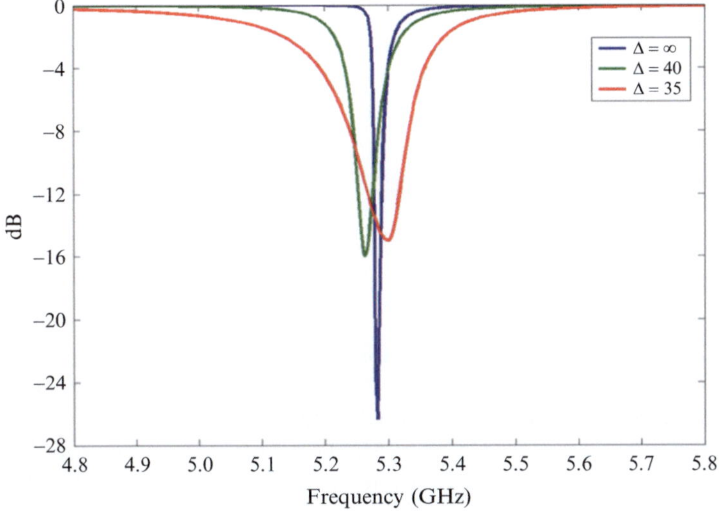

Fig. 12 The effect of noise and thermal stability on STNO filters

and the mixer are combined in one device. In addition, the device is tunable via applied current or magnetic field. All these advantages make STNO device an intriguing choice for realizing RF/millimeter-wave tunable mixers.

These preliminary results provide only a hint of some of the RF applications which are possible with nanoscale STNOs. The SoCs of the future can thus replace large power hungry active and passive CMOS devices with simple nanodevices with better characteristics.

7 Conclusions

In this paper, we have only scratched the surface of the possible applications using spin-torque nano-oscillators. There still exist several engineering challenges, such as fabrication of dual magnetic tunnel junctions and the generated oscillation power of these devices. However, given the theoretical background, rapidly progressing experimental studies and similarities to systems in nature and other fields of study, spin-torque nano-oscillators are very promising for future conventional and non-conventional applications. Furthermore, their nanoscale size and integration with MOSFET technology opens the possibility of "More-than-Moore" type applications.

Acknowledgements The authors would like to thank the Non-Boolean Computation group at Intel and all their collaborators for the useful discussions and feedbacks which helped us refine our ideas. This work has been supported and funded by Intel, DARPA and NSF (under NSF grants EAGER 115-1242802 and INSPIRE ECCS-1344218).

Table 1 Simulation parameters

Material parameter	Value
Saturation mag. (M_s)	1×10^6 A/m
In-plane anisotropy (H_{in})	5×10^5 J/m^3
Out-of-plane anisotropy (H_{out})	2×10^6 J/m^3
Damping parameter (α)	0.01
Spin polarization (P)	0.6

Appendix

All simulations were done using an HSPICE device model of spin torque nano-oscillators.This device model solves the Landau–Lifshitz–Gilbert–Slonczewski (LLGS) equations fora macrospin model. The material properties were extracted from experimental values for Co$_2$FeAl/MgO/Co$_2$FeAl MTJ structures. Some of the simulation properties are listed in Table 1. We also used 45 nm PTM transistor model [45].

References

1. R. Meservey, P.M. Tedrow, Phys. Rep. **238**(4), 173 (1994). doi:10.1016/0370-1573(94)90105-8. http://www.sciencedirect.com/science/article/pii/0370157394901058
2. E.Y. Tsymbal, O.N. Mryasov, P.R. LeClair, J. Phys. Condens. Matter **15**(4), R109 (2003). doi:10.1088/0953-8984/15/4/201. http://iopscience.iop.org/0953-8984/15/4/201
3. A. Fert, Rev. Mod. Phys. **80**(4), 1517 (2008). doi:10.1103/RevModPhys.80.1517. http://link.aps.org/doi/10.1103/RevModPhys.80.1517
4. P.A. Grnberg, Rev. Mod. Phys. **80**(4), 1531 (2008). doi:10.1103/RevModPhys.80.1531. http://link.aps.org/doi/10.1103/RevModPhys.80.1531
5. R. Meservey, P.M. Tedrow, P. Fulde, Phys. Rev. Lett. **25**(18), 1270 (1970). doi:10.1103/PhysRevLett.25.1270. http://link.aps.org/doi/10.1103/PhysRevLett.25.1270
6. J. Bardeen, L.N. Cooper, J.R. Schrieffer, Phys. Rev. **108**(5), 1175 (1957). doi:10.1103/PhysRev.108.1175. http://link.aps.org/doi/10.1103/PhysRev.108.1175
7. B.D. Josephson, Phys. Lett. **1**, 251 (1962). doi:10.1016/0031-9163(62)91369-0. http://adsabs.harvard.edu/abs/1962phl.....1..251j
8. M. Julliere, Phys. Lett. A **54**(3), 225 (1975). doi:10.1016/0375-9601(75)90174-7. http://www.sciencedirect.com/science/article/pii/0375960175901747
9. W.P. McCray, Nat. Nanotechnol. **2**(5), 259 (2007). doi:10.1038/nnano.2007.121. http://www.nature.com/nnano/journal/v2/n5/full/nnano.2007.121.html
10. M.N. Baibich, J.M. Broto, A. Fert, F.N. Van Dau, F. Petroff, P. Etienne, G. Creuzet, A. Friederich, J. Chazelas, Phys. Rev. Lett. **61**(21), 2472 (1988). doi:10.1103/PhysRevLett.61.2472. http://link.aps.org/doi/10.1103/PhysRevLett.61.2472
11. G. Binasch, P. Grnberg, F. Saurenbach, W. Zinn, Phys. Rev. B **39**(7), 4828 (1989). doi:10.1103/PhysRevB.39.4828. http://link.aps.org/doi/10.1103/PhysRevB.39.4828
12. T. Miyazaki, N. Tezuka, J. Magn. Magn. Mater. **139**(3), L231 (1995). http://www.sciencedirect.com/science/article/pii/0304885395900012
13. J.S. Moodera, L.R. Kinder, T.M. Wong, R. Meservey, Phys. Rev. Lett. **74**(16), 3273 (1995). doi:10.1103/PhysRevLett.74.3273. http://link.aps.org/doi/10.1103/PhysRevLett.74.3273

14. J.C. Slonczewski, Phys. Rev. B **39**(10), 6995 (1989). doi:10.1103/PhysRevB.39.6995. http://link.aps.org/doi/10.1103/PhysRevB.39.6995
15. J.C. Slonczewski, J. Magn. Magn. Mater. **159**, L1 (1996). doi:10.1016/0304-8853(96)00062-5. http://adsabs.harvard.edu/abs/1996JMMM..159L...1S
16. L. Berger, Phys. Rev. B **54**(13), 9353 (1996). doi:10.1103/PhysRevB.54.9353. http://link.aps.org/doi/10.1103/PhysRevB.54.9353
17. M. Tsoi, A.G.M. Jansen, J. Bass, W.C. Chiang, M. Seck, V. Tsoi, P. Wyder, Phys. Rev. Lett. **80**(19), 4281 (1998). http://prl.aps.org/abstract/PRL/v80/i19/p4281_1
18. J.Z. Sun, J. Magn. Magn. Mater. **202**(1), 157 (1999). doi:10.1016/S0304-8853(99)00289-9. http://www.sciencedirect.com/science/article/pii/S0304885399002899
19. J.A. Katine, F.J. Albert, R.A. Buhrman, E.B. Myers, D.C. Ralph, Phys. Rev. Lett. **84**(14), 3149 (2000). doi:10.1103/PhysRevLett.84.3149. http://link.aps.org/doi/10.1103/PhysRevLett.84.3149
20. E. Chen, D. Apalkov, Z. Diao, A. Driskill-Smith, D. Druist, D. Lottis, V. Nikitin, X. Tang, S. Watts, S. Wang, S. Wolf, A.W. Ghosh, J. Lu, S.J. Poon, M. Stan, W. Butler, S. Gupta, C.K.A. Mewes, T. Mewes, P. Visscher, IEEE Trans. Magn. **46**(6), 1873 (2010). doi:10.1109/TMAG.2010.2042041
21. C. Smullen, V. Mohan, A. Nigam, S. Gurumurthi, M. Stan, in *2011 IEEE 17th International Symposium on High Performance Computer Architecture (HPCA)* (2011), pp. 50–61. doi:10.1109/HPCA.2011.5749716
22. S.I. Kiselev, J.C. Sankey, I.N. Krivorotov, N.C. Emley, R.J. Schoelkopf, R.A. Buhrman, D.C. Ralph, Nature **425**(6956), 380 (2003). doi:10.1038/nature01967. http://dx.doi.org/10.1038/nature01967
23. A.V. Nazarov, H.M. Olson, H. Cho, K. Nikolaev, Z. Gao, S. Stokes, B.B. Pant, Appl. Phys. Lett. **88**(16), 162504 (2006). doi:10.1063/1.2196232. http://link.aip.org/link/APPLAB/v88/i16/p162504/s1&Agg=doi
24. W. Rippard, M. Pufall, S. Russek, Phys. Rev. B **74**(22) (2006). doi:10.1103/PhysRevB.74.224409. http://link.aps.org/doi/10.1103/PhysRevB.74.224409
25. W. Rippard, M. Pufall, S. Kaka, S. Russek, T. Silva, Phys. Rev. Lett. **92**(2) (2004). doi:10.1103/PhysRevLett.92.027201. http://link.aps.org/doi/10.1103/PhysRevLett.92.027201
26. S. Kaka, M.R. Pufall, W.H. Rippard, T.J. Silva, S.E. Russek, J.A. Katine, Nature **437**(7057), 389 (2005). doi:10.1038/nature04035. http://dx.doi.org/10.1038/nature04035
27. S.E. Russek, W.H. Rippard, T. Cecil, R. Heindl, in *Handbook of Nanophysics: Functional Nanomaterials*, vol. 38 (CRC Press, Boca raton, 2010), pp. 1–22
28. T. Shibata, R. Zhang, S. Levitan, D. Nikonov, G. Bourianoff, in *2012 13th International Workshop on Cellular Nanoscale Networks and Their Applications (CNNA)* (2012), pp. 1–5. doi:10.1109/CNNA.2012.6331464
29. S. Levitan, Y. Fang, D. Dash, T. Shibata, D. Nikonov, G. Bourianoff, in *2012 13th International Workshop on Cellular Nanoscale Networks and Their Applications (CNNA)* (2012), pp. 1–6. doi:10.1109/CNNA.2012.6331473
30. B.D. Cullity, C.D. Graham, *Introduction to Magnetic Materials* (Wiley, New York, 2011)
31. S. Sani, J. Persson, S.M. Mohseni, Y. Pogoryelov, P.K. Muduli, A. Eklund, G. Malm, M. Kll, A. Dmitriev, J. kerman, Nat. Commun. **4** (2013). doi:10.1038/ncomms3731. http://www.nature.com/ncomms/2013/131108/ncomms3731/full/ncomms3731.html
32. J. Grollier, V. Cros, A. Fert, Phys. Rev. B **73**(6) (2006). doi:10.1103/PhysRevB.73.060409. http://link.aps.org/doi/10.1103/PhysRevB.73.060409
33. B. Georges, J. Grollier, V. Cros, A. Fert, Appl. Phys. Lett. **92**(23), 232504 (2008). doi:10.1063/1.2945636. http://scitation.aip.org/content/aip/journal/apl/92/23/10.1063/1.2945636
34. M. Kabir, M. Stan, in *Proceedings of the 51st Annual Design Automation Conference on Design Automation Conference*, DAC'14 (ACM, New York, 2014), pp. 73:1–73:6. doi:10.1145/2593069.2596673. http://doi.acm.org/10.1145/2593069.2596673
35. Y. Kuramoto, *Chemical Oscillations, Waves, and Turbulence* (Courier Dover Publications, New York, 2003)

36. A. Pikovsky, *Synchronization: A Universal Concept in Nonlinear Sciences* (Cambridge University Press, Cambridge, 2003)
37. F.C. Hoppensteadt, E.M. Izhikevich, Phys. Rev. Lett. **82**(14), 2983 (1999). doi:10.1103/PhysRevLett.82.2983. http://link.aps.org/doi/10.1103/PhysRevLett.82.2983
38. Z. Zeng, G. Finocchio, B. Zhang, P.K. Amiri, J.A. Katine, I.N. Krivorotov, Y. Huai, J. Langer, B. Azzerboni, K.L. Wang, H. Jiang, Sci. Rep. **3**, 1426 (2013), pp. 1–3. doi:10.1038/srep01426. http://www.nature.com/srep/2013/130312/srep01426/full/srep01426.html
39. G. Feng, S.v. Dijken, J.F. Feng, J.M.D. Coey, T. Leo, D.J. Smith, J. Appl. Phys. **105**(3), 033916 (2009). doi:10.1063/1.3068186. http://scitation.aip.org/content/aip/journal/jap/105/3/10.1063/1.3068186
40. M. Stan, M. Kabir, J. Lu, S. Wolf, in *2013 IEEE 11th International New Circuits and Systems Conference (NEWCAS)* (2013), pp. 1–4. doi:10.1109/NEWCAS.2013.6573632
41. M. Stan, M. Kabir, S. Wolf, J. Lu, in *2014 IEEE/ACM International Symposium on Nanoscale Architectures (NANOARCH)* (2014), pp. 37–38. doi:10.1109/NANOARCH.2014.6880508
42. W. Rippard, M. Pufall, S. Kaka, T. Silva, S. Russek, J. Katine, Phys. Rev. Lett. **95**(6) (2005). doi:10.1103/PhysRevLett.95.067203. http://link.aps.org/doi/10.1103/PhysRevLett.95.067203
43. B. Razavi, IEEE J. Solid State Circuits **39**(9), 1415 (2004). doi:10.1109/JSSC.2004.831608. http://ieeexplore.ieee.org/lpdocs/epic03/wrapper.htm?arnumber=1327738
44. N. Locatelli, V. Cros, J. Grollier, Nat. Mater. **13**(1), 11 (2013). doi:10.1038/nmat3823. http://www.nature.com/doifinder/10.1038/nmat3823
45. Predictive Technology Model (PTM). http://ptm.asu.edu/. September 2008

Index

© Springer International Publishing Switzerland 2015
W. Zhao, G. Prenat (eds.), *Spintronics-based Computing*,
DOI 10.1007/978-3-319-15180-9

If you have any concerns about our products,
you can contact us on
ProductSafety@springernature.com

In case Publisher is established outside the EU,
the EU authorized representative is:
Springer Nature Customer Service Center GmbH
Europaplatz 3, 69115 Heidelberg, Germany

Printed by Libri Plureos GmbH
in Hamburg, Germany